Natural Science in Archaeology

Series editors

Günther A. Wagner
Christopher E. Miller
Holger Schutkowski

More information about this series at http://www.springer.com/series/3703

Gad El-Qady • Mohamed Metwaly
Editors

Archaeogeophysics

State of the Art and Case Studies

 Springer

Editors
Gad El-Qady
National Research Institute of Astronomy
and Geophysics
Cairo, Egypt

Mohamed Metwaly
National Research Institute of Astronomy
and Geophysics
Helen, Cairo, Egypt

Department of Archaeology
College of Tourism and Archaeology
King Saud University
Riyadh, Saudi Arabia

ISSN 1613-9712
Natural Science in Archaeology
ISBN 978-3-030-07666-5   ISBN 978-3-319-78861-6   (eBook)
https://doi.org/10.1007/978-3-319-78861-6

Printed on acid-free paper

This Springer imprint is published by the registered company Springer International Publishing AG part of Springer Nature.
The registered company address is: Gewerbestrasse 11, 6330 Cham, Switzerland

Preface

Archaeology is a branch of sciences that deals with the study of ancient human activities and remains through the analysis of physical relics for a comprehensive understanding of the ancient cultures. The process of analyzing and interpreting the ancient cultures needs interdisciplinary sciences and methods to achieve the goals of the investigations. The utilized methods and techniques to gather data about the nature of the archaeological remains and relics can vary according to the nature of the site, the degree of physical contrast between the archaeological relics and the surrounding medium, as well as the burial depths of the targets. However, techniques used to explore and gather data can be applied to any time period, starting from the very early relics to the very recent past. The traditional way to explore the burial and hidden archaeological remains was based primarily on utilizing the shovels for digging. However, with the fast improvements in applying different sciences to archaeological work, the thinking for using the trowels and other toolkits can be the second or even the third process in archaeological exploration. As with applying the geophysical techniques in measuring the physical contrast in the archaeological sites, the shape, distribution, and the constructions of the archaeological remains can be obtained in 2D or even in 3D. Such process can be applied in a noninvasive manner and faster than the traditional digging process. Subsurface imaging of the archaeological sites nowadays is a powerful tool for assessment and mapping. Continuous improvements in survey equipment, their performance, and automation have enabled large area surveys with a high data sample density and accuracy. Advances in processing and imaging software have made it possible to detect, display, and interpret subtle patterns of archaeological remains within the geophysical data. Examples of geophysical surveys that do not involve disturbing the soil include magnetometry, microgravity, electrical resistivity, ground-penetrating radar (GPR), electromagnetic, and geothermal.

Another technique that is considered very effective in remotely exploring the archaeological site is aerial archaeology. Almost as soon as the camera was invented, people were ascending the skies in balloons to capture the unique perspective that only the birds see. As most of archaeological remains are land-based features, which were otherwise unnoticed from the ground, they were identified while in the air. Therefore, analyzing such scenes is very important to get more valuable information that cannot be obtained from the ground survey. Recently exploring an archaeological site with aerial photography has become a routine practice for most recognized excavation expeditions across the world. The rapid improvements in all science branches and implementation of new technologies in the archaeological exploration expedition open the door for remotely collecting the data about the archaeological sites. This started with the classical photograph activities to use the satellite imagery and multi- and hyper-spectral data for the identification of the hidden and confused heritage over wide areas.

With increasing interdisciplinarity in sciences and introducing the multidimensional imaging of the archaeological remains using laser and X-ray technologies, it was possible to virtually image different archaeological remains and constructions. Recently, Airborne LiDAR (light detection and ranging) surveys have been applied for archaeological reconnaissance assessment by illuminating the earth surface with pulses of light and measuring the distances that light takes to arrive back to the aircraft. The LiDAR surveys have several advantages all the surface phenomena can be measured; throughout the processing steps, it can separate the digital elevation model for all the surface phenomena including vegetation cover, buildings, and roads; and the earth surface can be traced correctly and accurately. Imagery constructed from high-resolution LiDAR data is useful for detecting the surface topographic patterning that may be associated with archaeological features.

Based on the rapid developments in different sciences, archaeology has reached an exciting and fast progressing period in which many mysteries of our ancient civilizations are rapidly evolving and new knowledge and data sets are being collected and processed. The applications of technologies in archaeology are essential nowadays starting from the first planning of the site, before excavations, during excavation, and post-excavation to dealing with the archaeological remains and samples. Now we have 3D and 4D images for the archaeological sites and materials, virtual images for many of archaeological materials, and complete virtual museums. That makes the data analysis and exchanges more flexible. Currently, many archaeological projects about the neural network are running to identify more critical information about the archaeological materials. In the coming years, we expect to have more new technologies in the archaeological fields. The most promising one is the augmented reality (AR) technology that offers an air touch with the archaeological materials and cultural heritage. It is kind of mixing 3D virtual images with animated patterns that are based on large information and database about the archaeological sites and heritage. Augmented reality and/or virtual restoration mixed with the living monument and available in real time valuing archaeological values for the general public. However, this type of implementation of the technologies in archaeology is just the beginning and we expect in the coming years a great success in resolving many mysterious phenomena using the technologies in archaeology.

In this book, we display some of the recent applications of exploration technologies and restoration for different archaeological environments through applying the archaeogeophysics technologies at different scales. We are aiming to introduce some practical experience to those interested in the application of physical technologies in exploring the archaeological sites.

Cairo, Egypt   Gad El-Qady
                                                                                                                            Mohamed Metwaly

Contents

List of Contributors

Abbas M. Abbas National Research Institute of Astronomy and Geophysics, Cairo, Egypt

T. Abdallatif National Research Institute of Astronomy and Geophysics, Cairo, Egypt

L. Capozzoli CNR-IMAA, C.da s.Loja, Tito, Potenza, Italy

António Correia Department of Physics and Institute of Earth Sciences, University of Évora, Évora, Portugal

Juzhi Deng School of Nuclear Engineering and Geophysics, East China University of Technology, Nanchang, Jiangxi, China

Mahmut Göktuğ Drahor Department of Geophysical Engineering, Engineering Faculty, Dokuz Eylül University, Buca-İzmir, Turkey

Gad El-Qady National Research Institute of Astronomy and Geophysics, Cairo, Egypt

A. E. El Emam National Research Institute of Astronomy and Geophysics, Cairo, Egypt

Alan G. Green Institute of Geophysics, ETH-Hoenggerberg, Zurich, Switzerland

Heinrich Horstmeyer Institute of Geophysics, ETH-Hoenggerberg, Zurich, Switzerland

Ahmed Ismail Boone Pickens School of Geology, Oklahoma State University, Stillwater, OK, USA

I. Lemperger MTA CSFK GGI, Sopron, Hungary

University of West-Hungary, Sopron, Hungary

Man Li School of Nuclear Engineering and Geophysics, East China University of Technology, Nanchang, Jiangxi, China

Hansruedi Maurer Institute of Geophysics, ETH-Hoenggerberg, Zurich, Switzerland

Mohamed Metwaly National Research Institute of Astronomy and Geophysics, Helen, Cairo, Egypt

Department of Archaeology, College of Tourism and Archaeology, King Saud University, Riyadh, Saudi Arabia

A. Mohsen National Research Institute of Astronomy and Geophysics, Cairo, Egypt

David C. Nobes School of Nuclear Engineering and Geophysics, East China University of Technology, Nanchang, Jiangxi, China

Attila Novák Geodetic and Geophysical Research Institute of the Hungarian Academy of Sciences, Sopron, Hungary

University of West-Hungary, Sopron, Hungary

H. H. Odah National Research Institute of Astronomy and Geophysics, Cairo, Egypt

Á. M. Pattantyús Eötvös Loránd Geophysical Institute of Hungary, Budapest, Hungary

José Antonio Peña Instituto Andaluz de Geofísica (IAG), Universidad de Granada, Campus Universitario de Cartuja, Granada, Spain

Dpto. de Prehistoria y Arqueología, Universidad de Granada, Campus Universitario de Cartuja, Granada, Spain

E. Rizzo CNR-IMAA, C.da s.Loja, Tito, Potenza, Italy

Mohsen M. Saleh Conservation Department, Faculty of Archaeology, Cairo University, Giza, Egypt

Sándor Szalai Geodetic and Geophysical Research Institute of the Hungarian Academy of Sciences, Sopron, Hungary

University of West-Hungary, Sopron, Hungary

László Szarka Geodetic and Geophysical Research Institute of the Hungarian Academy of Sciences, Sopron, Hungary

University of West-Hungary, Sopron, Hungary

M. Teresa Teixidó Instituto Andaluz de Geofísica (IAG), Universidad de Granada, Campus Universitario de Cartuja, Granada, Spain

Mihály Varga KBFI-Triász Kft., Budapest, Hungary

University of West-Hungary, Sopron, Hungary

Lynda R. Wallace Akaroa Museum, Akaroa, New Zealand

Jun Yang Jiangxi Administration Bureau of Cultural Relics, Nanchang, Jiangxi, China

Zhiyong Zhang School of Nuclear Engineering and Geophysics, East China University of Technology, Nanchang, Jiangxi, China

Geophysical Techniques Applied in Archaeology

Gad El-Qady, Mohamed Metwaly, and Mahmut Göktuğ Drahor

Abstract

With the increased demand to facilitate the archaeological work either in well-known archaeological sites or the crude sites, geophysical methods plays an important role. The Geophysical methods have been used since 1946 with increasing frequency for archaeological investigations and currently the branch of archaeogeophysics is widely applied. The wide varieties of geophysical methods applied in archaeological work relies principally upon existing reasonable contrast in physical properties between the buried archaeological feature and the surrounding subsoil. Understanding the archaeological properties of the physical contrasts, in terms of density, thermal conductivity, electrical resistance, magnetic or dielectric properties, remains fundamental issues of choosing and applying the discipline geophysical techniques. In this regard, we tried to introduce a brief outline for the common and applicable techniques in archaeological investigations. The physical principles and field instrumentation involved for the acquisition of data with each method are considered, as well as some common results from the worldwide case studies. Generally, the archeogeophysical survey results can be used to guide excavation and to give archaeologists insight into the patterning of non-excavated parts of the site as well as it is often used where preservation of the sensitive sites is the aim rather than excavation.

Keywords

Archaeogeophysics · Non-invasive techniques · Physical contrast · Excavation · Archaeological remains

G. El-Qady
National Research Institute of Astronomy and Geophysics, Helen, Cairo, Egypt
e-mail: gadosan@nriag.sci.eg

M. Metwaly (✉)
National Research Institute of Astronomy and Geophysics, Helen, Cairo, Egypt

Department of Archaeology, College of Tourism and Archaeology, King Saud University, Riyadh, Saudi Arabia

M. G. Drahor
Department of Geophysical Engineering, Engineering Faculty, Dokuz Eylül University, Buca-İzmir, Turkey

1.1 General

In this chapter, the general principles of shallow geophysical surveying particularly in archaeological fields will be briefly introduced. Geophysics is a science that involves the application of physical theories and measurements to investigations of the earth. The earliest geophysical measurements were conducted in the late 1900s, initially with magnetism and gravity. Tremendous improvements in instrumentation in the early years of the twentieth century led to rapid

Table 1.1 Geophysical techniques employed in environmental and archaeological investigations

Technique	Sensitive property	Vertical resolution	Horizontal resolution	Application in environmental and archaeological studies
Thermal	Thermal conductivity	Poor	Moderate	Yes
Magnetic	Magnetic susceptibility	Moderate	Good	Yes
Gravity	Density	Moderate	Moderate	Rarely
Self potential	Electrokinetic effect	Moderate	Moderate	No
Electrical resistivity	Electrical resistivity	Moderate	Good	Yes
Electromagnetic	Electrical resistivity	Moderate	Good	Yes
Induced polarization	Electrical chargeability	Moderate	Moderate	No
Georadar	Dielectric permittivity	Good	Good	Yes
Seismic	Seismic velocity/ density	Good	Good	Yes

progress in geophysical applications with target sizes ranging from a few centimetres to several kilometres. Engineering and environmental geophysics concentrates on investigations of near-surface lithologies and structures. This geophysical branch is concerned with engineering applications, exploration for groundwater and minerals, and locating mine shafts and buried cavities and archaeological relics (Sheriff 1991; Reynolds 1997). In addition, shallow geophysical methods approved that they are powerful tools for archaeological prospection. In general, the depth of investigation for archaeological purposes is less than 5 m.

Two broad classes of geophysical prospecting techniques are available. The first is passive and the other is active. The passive techniques do not introduce any changes to the environment. Instead, variations in natural physical phenomena reflect the presence of various subsurface structures. Classical passive techniques include aerial photography, thermal, magnetic, gravity and self-potential prospecting methods. The active geophysical techniques include the electrical, electromagnetic, induced polarization, ground-penetrating radar (georadar or GPR) and seismic methods. For all of these methods, external fields are locally created and applied to the surface of the earth. The shallow subsurface can

be investigated for environmental and archaeological purposes using a variety of geophysical techniques with different degrees of resolution (Table 1.1).

According to the nature of the survey site and the specific objectives of the survey, the geophysical techniques are chosen to investigate the most appropriate physical contrasts of the subsurface. Moreover, the spatial and temporal sampling intervals should be selected to give the required depth penetration and resolution. Choosing these intervals may require some idea about the nature, depth and size of the target relative to the surrounding medium.

1.2 Thermal Methods

The temperature at the surface of the soil is one physical parameter upon which an archaeological prospecting method may be based (Wynn 1997). Variations of surface temperature may reveal the presence of subsurface structures (Fig. 1.1). Buried stone blocks, for example, lose thermal energy at different rates than the surrounding topsoil, resulting in subsurface temperatures variations. Thermal surveys may be carried out on the ground or from aircraft utilizing the simple equipped plan and/or using the drones. Surface

Fig. 1.1 Example for the visible spectrum image on the *left* and the equivalent thermal image on the *right* (after http:// www.armadale.org.uk/phototech03.htm, Accessed 2018)

Fig. 1.2 Thermal anomalies at the eastern side of the Great Pyramid of Egypt (Khufu or Cheops Pyramid), after (http://www.ancient-origins.net/news-history-archaeology/thermal-scanegyptian-pyramids-reveals-mysterious-anomaly-great-pyramid-020616, Accessed 2018)

temperatures may be determined from probes pushed into the ground (Benner and Brodkey 1984) or from thermal infrared imagery (Fig. 1.2). Airborne surveys have the advantage that very large areas may be covered quickly. For this reason, thermal prospecting in archaeology requires further studies (Wynn 1997). Most drones are now fitted with a camera of some kind for providing strong HD images and 3D plotting for the surface and subsurface thermal changes. This has so far proved useful in land surveying for mining and in monitoring large construction sites, but the drones need another level of capability to go deeply into subsurface.

1.3 Magnetic Methods

The study of the earth magnetism is the oldest branch of geophysics. It is a fast and dependable technique for mapping the distribution of archaeological remains in the shallow subsurface (Scollar et al. 1990; Clark 1986, 1990; Gaffeny et al. 1991; Reynolds 1997). During and after World War II, modern non-mechanical magnetometers (e.g., fluxgate, proton-precession, optically-pumped and cryogenic magnetometer) were developed. They measure the three individual components (vertical and two horizontal components) and/or the total magnetic field strength. Magnetics is very useful method to detect the burial archaeological objects containing magnetic properties in archaeological prospection studies. Magnetic fields caused by thermoremanent magnetisation or magnetic susceptibility differences within the subsoil are measured by magnetometers. The measurable magnetic field variations in the earth surface are observed over the pottery, brick, ceramic and tile deposits, burning pits, metallic objects, structural bases consist of rocks with magnetic properties and some kind of bacteria reduce the ferrous materials found in storage pits and organic environment in archaeological sites. Burned features such as kilns, bricks and other materials that contain clay can produce strong magnetic anomalies after the burning process, allowing these features to be easily mapped by magnetic surveys. In this process, weakly ferromagnetic iron oxide haematite is changed to the strongly ferromagnetic iron oxide maghaemite, and the magnetisation intensity of the materials is increased by this conversion (Drahor 2011). The comprehensive description about this process is revealed by Le Borgne (1955, 1960), Tite (1972), Tite and Mullins (1971) and Mullins (1977). In addition, the variation of magnetic susceptibility of the soil is also important factor to determine the burial archaeological context. Thus, many archaeological structures such as burial walls, roads, foundations, kilns, furnaces, and other burned structures; garbage areas and tombs can be easily determined using magnetic prospection technique, which is the very beneficial in archaeological site investigation. However, this technique is rarely used in the surveys carried out in urban sites due to the background magnetic noise resulting from the geological materials and the presence of ferromagnetic materials and/or disturbed soils. In addition, the entity of electrical currents, the affluence of neighbouring high frequency electromagnetic signals and the plenty of man-made objects affect directly the magnetic data in urban sites (Drahor 2011).

The use of two magnetometers mounted at different heights allows the recording of vertical magnetic gradients in nano-Tesla per meter (nT/m). Such measurements usually reflect the distribution of archaeological remains at shallow depths (Fig. 1.3; Sheriff 1984; Telford et al. 1990). The vertical-gradient magnetic field is given by:

$$\partial F/\partial Z = (F_2 - F_1)/\Delta Z, \qquad (1.1)$$

where F_1 and F_2 are the readings at the upper and lower sensors, and ΔZ is the separation between the sensors.

Magnetic data are mostly collected using gradient measuring technique by different gradiometers (fluxgate or caesium) for fixing shallow subsurface features in archaeological sites. Modern gradiometers are highly sensitive to magnetic susceptibility changes in the subsurface environment. The sensitivity characteristic of gradiometers is very important factor to detect the some burial archaeological objects because of having very small anomalies (several nT). Also, the usage of gradient technique importantly removes the influences occurred from the presence of magnetic disturbances and temporal variations in magnetic field. The survey area should be divided into regular grids (10 by 10 m, 20 by 20 m or 50 by 50 m) previous to magnetic data collection. The data must be acquired with small measuring and line intervals using parallel or zigzag measuring techniques within the grids (Clark 1990; Drahor et al. 2008b). Therefore, large areas (1 ha or more) can measure in per day using modern gradiometers. However, the measuring and line spacing should be very small during magnetic data acquisition in urban investigations, even though it is time-consuming. In archaeological applications, the line and measuring intervals should be determined considering the dimensions, depth and extensions of buried features. In general, the line spacing is selected between 0.5 and 1 m, while the measuring spacing is designated within a range of 10 and 50 cm. Then the data collected from the grids creating in accordance with the survey plan should be combined to obtain a large-scale magnetic image of investigation site. After the data combination process, overall data were imaged as the grey-scale colouring. The dark tones demonstrate high magnetic anomalies, while the light

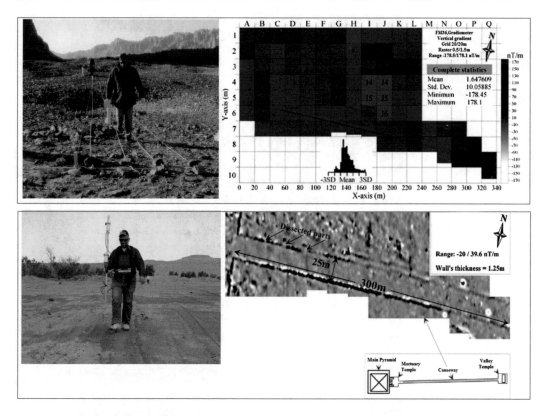

Fig. 1.3 *Upper*, shade-plot of the raw magnetic data acquired at the eastern region of the pyramid of Amenemhat II in Dahshour, *lower*, the processed magnetic data showing the causeway and mortuary temple (modified after Abdallatif et al. 2010)

tones indicate low magnetic anomalies generally. The gradiometric data were primarily manipulated to zero mean traverse and grid functions to remove the striping effects in traverses and the edge discontinuities between the grids, respectively. Afterwards the despiking was used to remove "iron spikes" and similar disturbances from the gradiometer data. Then data were enhanced by clip, low- and high-pass filtering techniques to increase the data quality (Drahor et al. 2008b). Example of the utilizing of magnetic data obtained from the Šapinuva archaeological site is presented in Fig. 1.4. The image shows that the magnetic contrast is considerably good to detect the possible archaeological relics in the area. The various archaeological structures and features, which have high and moderate magnetic values, are clearly seen.

Another example of application of the vertical gradient magnetic survey to explore the hidden structures associated with white pyramids of Amenemhat II in Dahshour area, Egypt led to discover important archaeological features of different shapes and sizes, mostly made of mud bricks (Fig. 1.5). This includes the causeway that connects the mortuary temple with the valley temple in the Middle Kingdom of the 12th Dynasty, the mortuary temple and its associated rooms (Abdallatif et al. 2010). Also, the combined application of the analytic signal and Euler method for automatic interpretation of magnetic data provides additional information about the source geometry, location and the amount of soil debris to be removed upon excavations.

1.4 Gravity Methods

Gravity methods have been tried in archaeology at various times since about 1955 (Linington

Fig. 1.4 Large-scale magnetic gradiometer image obtained from Šapinuva archaeological site, Central Anatolia, Turkey (from Drahor et al. 2015)

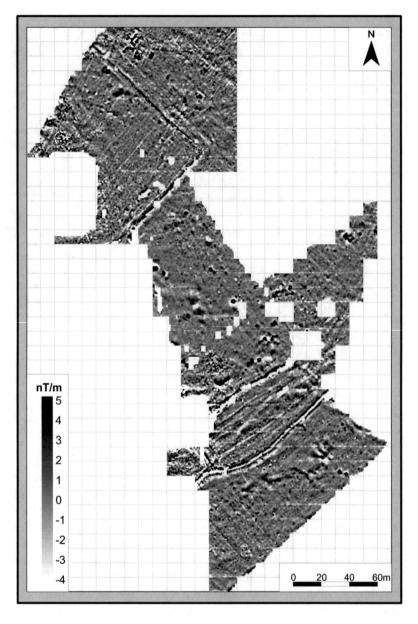

1966; Kolendo et al. 1973; Fajklewicz et al. 1982). These methods depend on density differences between the targets and the surrounding soil (Fig. 1.6). Archaeological gravity signatures are typically very small (in the milligal range), close to the instrumental noise threshold. Centimetre-precision accuracy in elevation is absolutely necessary when applying gravity methods to archaeology. This and similar accuracy requirements in calculating terrain corrections have significantly dampened the interest of geophysicists who might consider applying this method in archaeology.

In fact, the gravity anomalies obtained from archaeological sites are generally small and it usually close to instrumental noises. This method is enabled to determine some architectural remains, filled caverns, caves and voids, crypts, buried

Fig. 1.5 Processed vertical magnetic gradient data, (**a**) the causeway and the mortuary temple of the pyramid of Amenemhat II in Dahshour, Egypt. D1, D2, D3 and D4 delineate the main discoveries, (**b**) application of the analytical signal method to the gradiometer data and (**c**) depth estimation after the application of 3D Euler deconvolution (after Abdallatif et al. 2010)

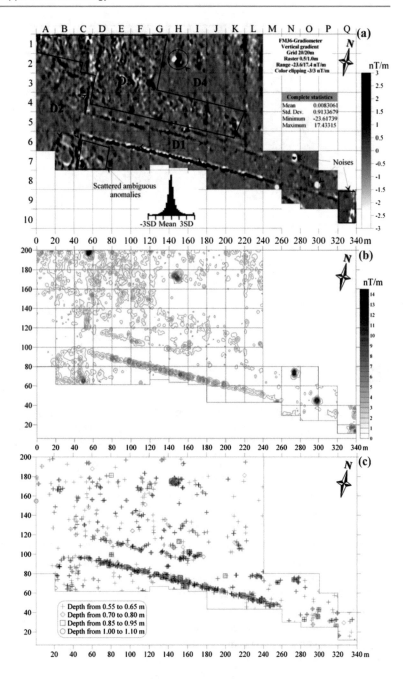

tombs. We measure the variations in the gravitational field at the ground surface corresponded to density changes in the subsurface relics in archaeological sites. Some successful microgravity application examples can be found on some related papers (Blizkovsky 1979; Fais et al. 2015; Fajklewicz 1976; Lakshmanan and Montlucon 1987; Linford 1998; Linington 1966;

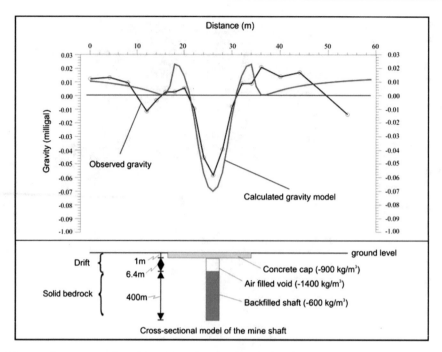

Fig. 1.6 Microgravity survey confirms the position of an abandoned mine shaft and provides information on the degree of backfill (after TerraDat 2003)

Orfanos and Apostopoulos 2011; Owen 1983; Panisova and Pasteka 2009; Panisova et al. 2013; Pašteka and Zahorec 2000). This non-invasive method has very time consuming application in archaeology, and therefore this case is also caused the limited usage of this technique. In spite of this negativity of the technique, it may to be useful as a non-destructive tool for the detection of man-made cavities and buried tombs and similar archaeological targets.

In this section, a microgravity investigation example carried out in St. Catherine's church is given in Fig. 1.7 (Panisova et al. 2013). The aim of this study was to determine and characterise of the archaeological remains inside the church floor. The microgravity survey was given some useful information about the subsurface features such as the known aristocratic crypt. Thus, Panisova et al. (2013) determined the position and the size of this crypt using the microgravity data. During the data processing, a novel modelling approach to microgravity data was employed to define the subsurface characteristics inside the church floor. In Fig. 1.7, a residual

Bouguer anomaly map obtained by planar trend subtraction from the Bouguer anomaly map is given. The residual Bouguer anomaly values are changed between 0.03 and –0.04 mGal. The negative residual gravity anomaly is showed the location of the known aristocratic crypt. This study indicated that the microgravity investigation may be useful to determine the some burials such as crypt, burial chamber and similar archaeological structures.

1.5   Self Potential Methods

Self-potential methods are the least expensive geophysical techniques used in archaeological exploration. The self-potential (SP) method is to measure the changes in natural potentials, which can be measured between any two points on the Earth's surface (i.e., natural potentials from less than a millivolt to greater than one volt), and it is one of the oldest methods in geophysics. Today, this method is still applied to reveal various problems in applied geophysics. The SP

Fig. 1.7 Residual Bouguer anomaly map obtained by planar trend subtraction from the Bouguer anomaly map at the St Catherine's Monastery, Slovakia (from Panisova et al. 2013)

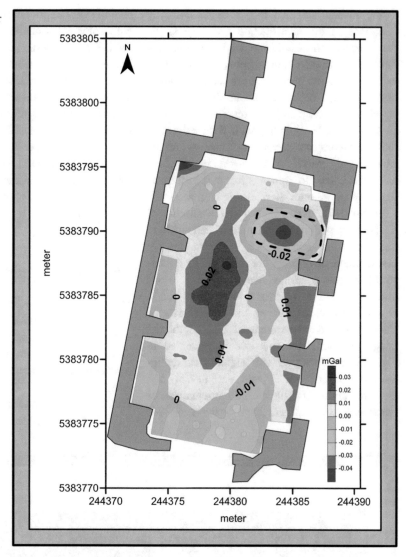

anomalies are caused by temperature, pressure gradient, porosity, fluid migration, resistivity variation and the moisture content of soil. These physical changes are the main causes of ion accumulations along the discontinuities. There are three important SP phenomena (electrokinetic, electrochemical and other SP effects), which are produced a significant anomaly that shows the buried archaeological features at archaeological sites. According to some investigations performed in archaeological sites,

this method is given useful information about the buried structures, altered soils, burnt materials and oxidised metallic objects (Drahor 2004; Drahor et al. 1996; Wynn and Sherwood 1984). The equipment consists of a simple digital voltmeter, cables and low-noise non-polarizing electrodes. SP equipment used in archaeology is much cheaper than other geophysical instrument and equipment for field practitioners and archaeologists. However, data collection is considerably troublesome in archaeological sites due to the covered soil that may show importantly variations in the investigation area. Thus, the data acquisition should be highly systematic throughout the survey. In addition, the processing and interpretation of SP data are rather curtail to obtain useful information about the subsurface structures in archaeological sites.

The principle difficulty with SP methods is that archaeological responses may be much smaller (i.e. micro to milli-volt range) than anomalies due to other natural phenomena (e.g. flowing groundwater, telluric currents, etc.).

Acemhöyük archaeological site is an artificial hill from the period of the Assyrian Trade Colony and one of the largest hills in Anatolia. The SP investigation was carried out using gradient measuring technique as an area with (1 × 1 m) gridding and 2 m electrode intervals in 1994. During the data acquisition, the non-polarizable Cu–CuSO$_4$ electrodes were used. The SP gradient image obtained from this site is presented in Fig. 1.8. The positive and negative anomalies are changed between 39 and –36 mV/m, and they are very compatible with the orientations of burial archaeological structures in the area. This result is very important in determination of burial archaeological structures using SP method. After the SP study, an archaeological excavation was implemented on a selected test site, which was a 5 × 5 m dimension. During this study, two kilns and bases of several structures (wall, pit, oven, oxidized copper fragments, etc.) were discovered. In conclusion, the SP studies performed at Acemhöyük archaeological site were revealed the presence of burnt walls and pits, kilns and soils in the excavated site where the SP anomalies have high positive values (Drahor 2004; Drahor et al. 1996).

1.6   Electrical Resistivity Methods

The electrical resistivity method, which is non-invasive, attractive and sensitive for shallow surveys, has an important role in the application of near surface geophysics. Resistivity is the first investigation technique using in archaeological prospection studies. The method can easily be detected the burial archaeological objects in the subsoil according to its electrical resistivity

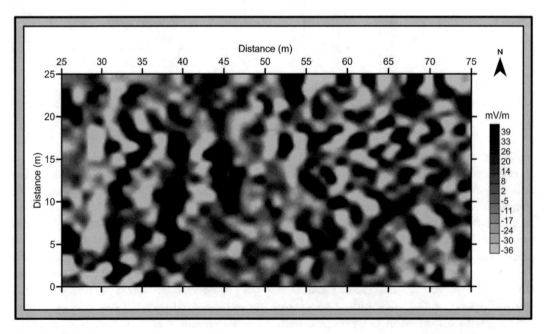

Fig. 1.8 The gradient SP image of AE area in Acemhöyük archaeological site, Central Anatolia, Turkey (from Drahor 2004)

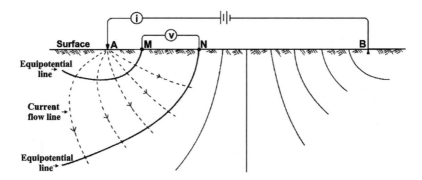

Fig. 1.9 Current lines and equipotential for a pair of current electrodes A and B on a homogeneous half-space. M and N are the potential electrodes

values. Soil resistivity measurements are performed by four electrodes, in both current and potential (Fig. 1.9). The measurable value of resistivity measuring is the apparent resistivity. This value depends on the type of electrode configuration, the burial depth of target, the geometry of the feature and climatic variations inside the soil. Many configuration types use in resistivity surveys, but their apparent resistivity values change according to their geometric positions. The most popular configuration is the twin probe in archaeological investigations. However, Wenner, dipole-dipole, pole-pole, Schlumberger and square arrays also use in site investigations. However, Wenner, dipole–dipole, pole–pole, Schlumberger and square arrays also use in site investigations. Thus, many archaeological structures such as burial walls, roads, foundations, ditches and other similar structures, garbage areas and tombs can easily be determined using resistivity method.

Depending on layer resistivity and thickness, current will penetrate to greater depths as the current electrode separation increases (Fig. 1.10). By making measurements with increasing current electrode separations, resistivity versus depth profiles can be obtained. Using different electrode configurations, resistivity methods may also be used for mapping defined horizons, or for 2- and 3-D imaging of the subsurface (Figs. 1.10 and 1.11; Loke and Barker 1996; Stummer et al. 2002, 2003; Drahor and Öztürk 2011).

The electrical resistivity tomography (ERT) technique is one of the most popular investigations in archaeological investigations in recent years. The acquisition of tomographic resistivity data is performed along a measuring line by sequence of a selected array for create a pseudo-section (2D) or obtained from a selected area by different arrangements of electrode configurations for 3D surveys. To obtain the true resistivities and depths of buried targets, the apparent resistivity data can be processed using 2D and 3D inversion algorithms based on the finite difference or finite element methods for forward model calculations (Drahor 2011; Drahor et al. 2008b; Papadopoulos et al. 2007). In archaeological site investigation, the well-known inversion technique of resistivity data is the robust (L1 norm) approximation, which is used in the presence of sharp shallow structures. This technique provides more realistic images for sharp-edged structures, such as the relics of ancient buildings in archaeological applications (Drahor 2011; Drahor et al. 2008a). Thus it allows for the producing of more accurate resistivity models of the archaeological features and its environs for different electrode array layouts.

To produce a compatible model, the data must be collected with high quality, and the interval of data collection for measuring points and lines should be small according to the possible target dimensions in archaeological applications. These intervals should be selected between 0.5–2 m and 0.25–1 m for lines and measuring points,

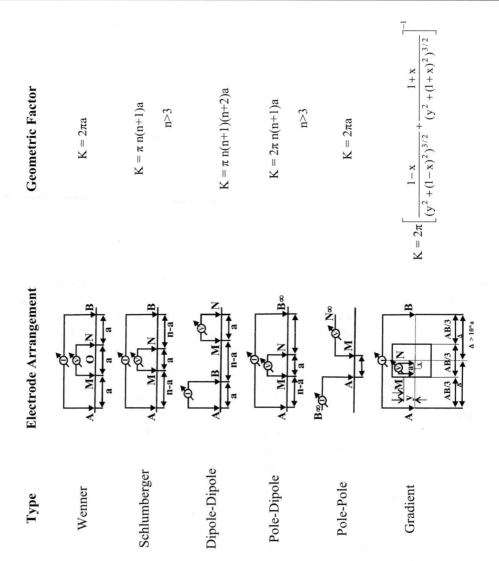

Fig. 1.10 Common electrode arrangements used in DC resistivity surveying and their geometric factors (modified after Knödel et al. 1997)

respectively. Today, the computer-controlled multichannel resistivity meters, that allow implementing the usage of various electrode configurations, are mostly used. These instruments also enable the improvements of data acquisition procedures in electrical imaging. As a result, the measurement time has decreased remarkably by using of fully automated data acquisition systems that allow us the application of various configurations (Binley et al. 1996; Drahor 2011). Furthermore, the "Pulled Array Continuous Electrical Profiling" (PACEP) and Multidepth Continuous Electrical Profiling (MUCEP) systems have been developed for spatially dense measurements over large areas in archaeological and soil investigations (Christensen and Sørensen 1998; Panissod et al. 1997). The application of a mobile electrical resistivity enables to make the continuous measurements that have high spatial resolution over a large range of scales. This data acquisition technique presents significant advantages for

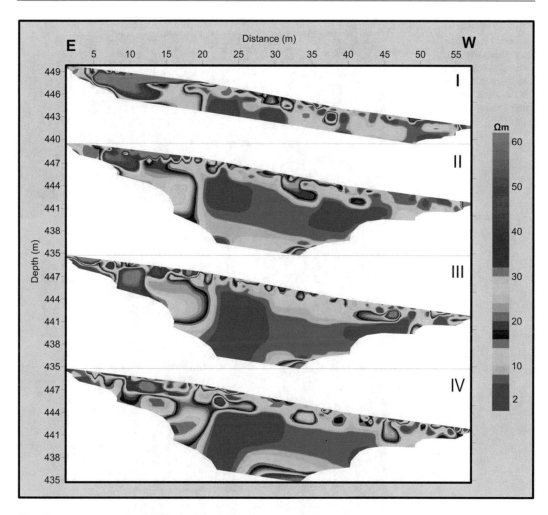

Fig. 1.11 ERT model sections obtained from parallel lines to each other in Sultantepe archaeological site, Southeastern part of Turkey (Drahor and Öztürk 2011)

researchers who investigate shallow targets and covered soil in archaeology, agronomy, waste management, and civil and environmental investigations. In addition, this data acquisition technique allows to non-destructive mapping, temporal monitoring, various scale applications, data acquisition facilities, highly sensitive measurements and numerical modelling advancements (Drahor 2011). Therefore, we can obtain more detailed 2D and/or 3D tomographic visualisations of subsurface archaeological structures. It concludes that this technique has an obvious advantage according to conventional resistivity mapping techniques, and thus we can create more realistic images connected with the

true depth and resistivity information for embedded archaeological remains.

However, this technique also has limitations, such as contact between the soil and the electrodes, calibration, the duration of measurements, the adequacy between heterogeneity and configuration and the non-uniqueness of the solution in the inversion process. Recent technological improvements in resistivity have enabled rapid resolution both in space and time (Fig. 1.12). The duration of data acquisition has been shortened with newly available data loggers and multiplexers. Such equipment is necessary in future prospects to perform 3D measurements with a high spatial resolution or to apply electrical

Fig. 1.12 *Upper*,
processed image for
magnetic gradient data,
lower processed image for
resistance data collected at
tell El-Dabaa area, Egypt
(after Taha et al. 2011)

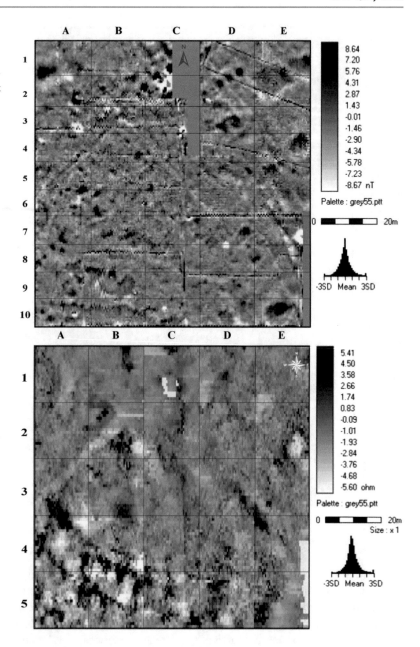

resistivity measurements to transfer processes (Drahor 2011; Samouelian et al. 2005).

Example of using the electrical resistance scanning technique to explore the archaeological remains under the cultivated land of tell El-Dabaa area (is believed to be the capital of the Hyksos,) shows many significant results especially when the resistance results compared with another data sets like magnetic gradient (Fig. 1.12). The site has many different types of archaeological remains such as tombs, palaces, houses and temples. These archaeological remains constructed from different types of materials such as fire-brick, mud-brick and/or stones.

1.7   Electromagnetic Methods

Electromagnetic methods have been extensively developed and adapted over the past three

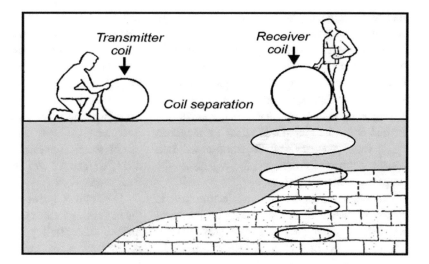

Fig. 1.13 Layout of frequency-domain coils along measured profile

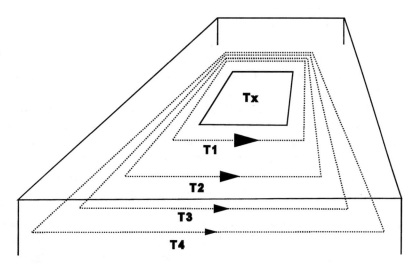

Fig. 1.14 Central-loop sounding configuration with transient current flow in the ground

decades for the lateral and vertical mapping of resistivity variations (Kaufmann and Keller 1983; Nabighian and Macanae 1991). They can be broadly divided into two major groups: frequency domain and time domain. In frequency-domain (FDEM) surveying (Fig. 1.13), the transmitter current varies sinusoidally with time at a fixed frequency, selected on the basis of the desired depth of exploration. By comparison, time-domain (TDEM) surveying (Fig. 1.14) involves the transmission of symmetric periodic square waves. After every second quarter-period, the transmitter current is abruptly reduced to zero

for a one-quarter period, whereupon it flows in the opposite direction. When a current is passed through a rectangular loop, commonly laid on the ground, a primary magnetic field spreads into the ground. By rapidly reducing the transmitter current to zero, the changing primary magnetic field will induce eddy currents in the subsurface that are depended on the distribution of the subsurface resistivity distributions. The eddy currents will generate a changing magnetic field that can be detected by a receiver coil at the surface. The voltage generated in the receiver coil, which is proportional to the time-rate-of-change of the secondary magnetic field created by the eddy currents, is measured.

Secondary magnetic fields decay quickly in poor conductors and slowly in good conductors. By measuring the decay of the magnetic field, an estimate of subsurface resistivities can be made. Voltages measured by TDEM equipment are depended on the amount of decaying magnetic field. These voltages can be transformed into apparent resistivity values to represent the properties of the subsurface.

Although the output of an electromagnetic survey is similar to that produced by electrical resistivity techniques, there are several advantages to the electromagnetic methods. They do not need direct coupling to the ground, they may provide higher resolution information and they may be cost effective. A typical TDEM resistivity sounding configuration includes a square loop of wire laid on the ground (Fig. 1.12). The side length of the loop depends on the desired depth of exploration. For shallow depths (less than 40 m) in relatively resistive ground the length may be as small as 5–10 m.

Utilizing electromagnetic techniques in archaeological works have many types, the slingram measuring technique is commonly used by low-frequency transmitter–receiver pair to obtain the information about the apparent resistivity, its apparent susceptibility and the characteristic signatures of buried metallic objects and subsurface (Tabbagh 1986). Therefore, the conductivity (quadrature phase) or magnetic susceptibility (in-phase) or both parameters can be measured. Thus, we estimate the magnetic and conductivity parameters of the subsurface. Some field tests have performed to obtain the archaeological interpretation using this data (Benech and Marmet 1999; Cole et al. 1995; Linford 1998). In recent years, there has been an increase in the number of published EM surveys (Johnson 2006; Perssona and Olofsson 2004; Simon et al. 2014; Venter et al. 2006; Witten et al. 2000).

The very low frequency (VLF) electromagnetic method has been managed to image embedded conductive near surface characteristics. This method uses the signals transmitted by powerful radio stations that have frequency bands varying between 15 and 30 kHz. The VLF is generally fast and cost-effective investigation method in near surface applications, and it has used to determine the different near surface geophysical problems, such as ground water, archaeology, the determination of soil content and contamination studies (Benson et al. 1997; Drahor 2006; Monteiro Santos et al. 2006). To improve the interpretation, VLF data have been inverted in recent years (Baranwal 2007; Baranwal et al. 2011; Kaikkonen and Sharma 1998; Sharma and Kaikkonen 1998).

The multi-frequency EM instruments allow us the measuring of the apparent resistivity and magnetic susceptibility for different depth investigations. A multi-frequency result from Greece is presented in Fig. 1.15. The presented example shows the raw data obtained using various frequencies (5010, 13770 and 22530 Hz) in this site. The in-phase component of the raw data is given on the top, and the changes in the area are clearly displayed. As can be seen from the figure, in-phase values are changed together with the increasing of the frequency. The quadrature component of this technique is presented in the bottom. The shape of quadrature component values is not changed together with the increasing of frequency.

In this part, a VLF-R study performed in Sardis archaeological site is presented to investigate the detectability of VLF method in archaeology. The data were collected with 2 m measuring and 4 m profiling intervals using 5 m dipole interval in 20 × 20 m grids. In this study, two different frequencies, 19.6 and 23.4 kHz transmitted from Oxford, England and Rhauderfehn, Germany respectively were used. The data were imaged by grey-scale colouring after correction process, and these images are given in Fig. 1.16. As can be seen from the Fig. 1.16, there are three main anomaly groups in the VLF-R images. First anomaly is found between S180–S200 and E760–E776 coordinates approximately. Second anomaly group appears between S200–S220 and E740–E760, and the third is observed between S160–S180 and E740–E760 coordinates. General extends of anomalies are in N–S and E–W directions. As a result, we can determine that the apparent VLF-R anomalies are very similar in both images, and the anomalies are generally

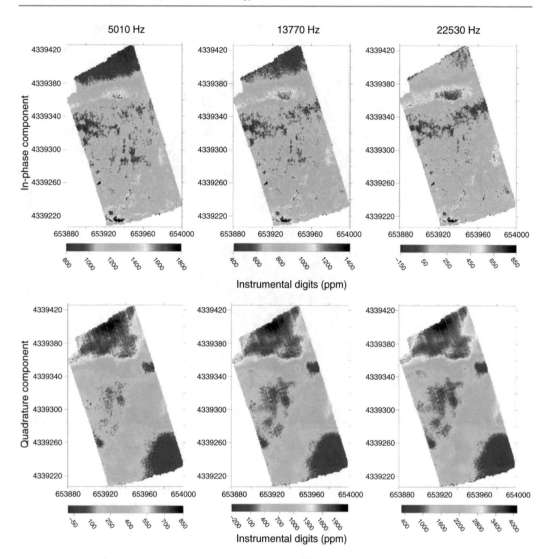

Fig. 1.15 Image of GEM-2 raw data for 5010, 13770 and 31290 Hz for in-phase (*top*) and quadrature out-of-phase component of the signal (*bottom*) (Simon et al. 2014)

found in same places. The resulted image from 19.6 kHz is shown as more evident to interpret the buried archaeological structures than the result of 23.4 kHz (Fig. 1.16). These results revealed that the VLF-R study might be effective to determine the buried archaeological targets in archaeological prospection.

1.8   Georadar Methods

Ground Penetrating Radar (GPR) method is widely used in the localization of different buried objects in near-surface investigations. GPR is one of the important near-surface geophysical methods using in archaeological prospection applications. This method with high resolution for shallow investigations is highly sensitive to determine all metallic and/or non-metallic materials. Thus, the method is commonly used in many applications (engineering, environmental, archaeology, geology and geotechnics) of near-surface geophysical investigation. It has been implemented in archaeological prospection to define the effective depth information and localization of buried archaeological structures

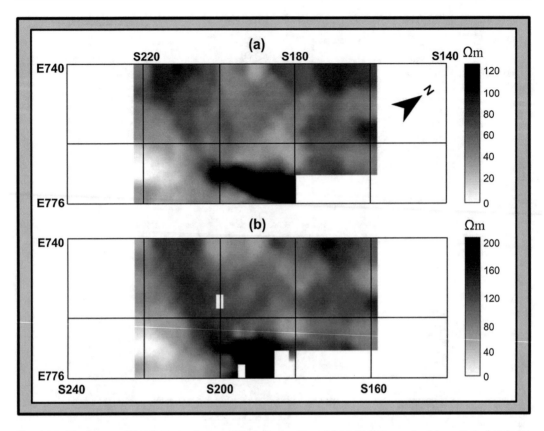

Fig. 1.16 The apparent VLF-R images for two different transmitters. (**a**) 19.6 kHz transmitted from Oxford, UK, (**b**) 23.4 kHz transmitted from Rhauderfehn, Germany (Drahor 2006)

and to constitute 3D image reconstructions of archaeological features in subsurface since 30 years. The method is both fast and cost-effective for the survey of extensive archaeological sites. In addition, it is less sensitive to urban noises than other geophysical methods, and since therefore it has been extensively used due to its flexibility and ease of use in urbanized terrain. Therefore, many archaeological structures such as walls, trenches, ditches, voids, foundations, roads and etc. can be determined by GPR profiling studies. Unfortunately, sedimentary formations and embedded archaeological materials inside these layers involve the different physical and chemical variations such as humidity, ion and water content of porous medium, dielectric properties, electrical conductivity and magnetic permeability changes. Thus, the velocity of electromagnetic wave propagation is highly affected from these changes. The target bodies can easily

be determined, if there is an adequate dielectric contrast between buried objects and covering soil (Davis and Annan 1989; Peters et al. 1994). Otherwise, the archaeological structures that have similar dielectric values cannot be determined, and we could not obtain a sufficient reflection from the interface of target body. In addition, archaeological and urban sites have complex subsurface features and context that may constitute complicated radargrams, which have important difficulties during the interpretation stage (Basile et al. 2000; Drahor 2011).

The high frequency radar pulses produced from a surface antenna is transmitted into the subsurface, and signals reflected from buried reflective materials are plotted as a radargram form. Thus, many thousands of radar reflections are recorded in these radargrams obtained from profiles within a grid, and a 3-D visualization of buried archaeological structures can be constituted. GPR gives us constructive and valuable information about the subsurface characteristic by the help of 2D and 3D representations in archaeological site investigation (Fig. 1.17). In recent years, sophisticated instruments and equipments have been developed, and the data acquisition and processing techniques have importantly been improved (Booth et al. 2008; Conyers 2015; Grasmueck et al. 2006; Valle et al. 2001; Wang and Oristaglio 2000). One of the instrumental and data acquisition improvements is the invention of multi-frequency GPR technique, which is often applied to define and characterise the archaeological context and the buried structures. The choice of antenna frequency in multi-frequency GPR systems is a fundamental parameter to obtain powerful survey results (Booth et al. 2008; Marsiglio et al. 2003). The multi-frequency GPR systems may give reliable and realistic subsurface images together with the suitable penetration depth and data resolution. Multi-frequency GPR has several important advantages including high performance in terms of pulse repetition, frequency and scan rate speed investigation, high detection capability and productivity. In addition, multi-frequency GPR is sensitive to the soil type classification in means of rapid ground and high resolution coverages to detect the small objects allowing effective 3D subsoil reconstruction in archaeological sites. This method also enables the simultaneous use of antennae at different frequencies, and therefore increased range depth for the low frequencies of targets without any loss in their resolution characteristics for high frequencies (Drahor 2011).

Today, radar instruments and equipments are rapidly developed for profiling and tomographic investigations. In addition, GPR studies may provide more effective results using 3-D survey techniques in archaeological applications. The resolution of GPR is much better than the other geophysical techniques in the description of near surface features. GPR also provides sensitive depth information in terms of the determination of buried objects. The measuring and line spacing of GPR data should be small in archaeological investigations. In archaeological surveys, the measuring spacing should be selected from several cm to several tens of cm, while the line spacing should be within a range of 10 cm and 1 m. However, the acquisition of dense radar data sets can be time-consuming (Drahor 2011) (Fig. 1.17).

1.9 Seismic Methods

Seismic methods depend on differences in the velocity and density of elastic waves in soils and rocks. Seismic velocities, in turn, depend on the bulk modulus, shear modulus and density. When a seismic wave encounters a change in physical properties, refractions (bending of the wave paths) and reflections (echos) are generated. The seismic refraction method has been widely used to resolve diverse problems (Dobrin 1976; Parasnis 1997). Because of the long wavelengths (tens of meters) that are involved in refraction surveying, the resolution is generally low with respect to typical shallow archaeological targets. The method consists of producing elastic waves by striking the ground with a large hammer or firing small shotgun shells or explosive charges, and measuring arrival times of waves at various points along the profile.

Seismic reflections are caused by elastic waves that are partially reflected at discontinuities in the mechanical properties of the earth. Such discontinuities may be due to the presence of voids or solid walls in less dense fill. The time of return of the echo provides information on the distances to the discontinuities. In contrast to the refraction method, wavelengths of the order of 1–10 m provide higher resolution information.

Seismic methods can provide a constructive alternative to determine some specific kinds of archaeological targets such as burial graves, walls, etc. It is also possible to understand the archaeological context using high-resolution seismic data by the help of reflection surveys and/or refraction tomography in archaeological sites.

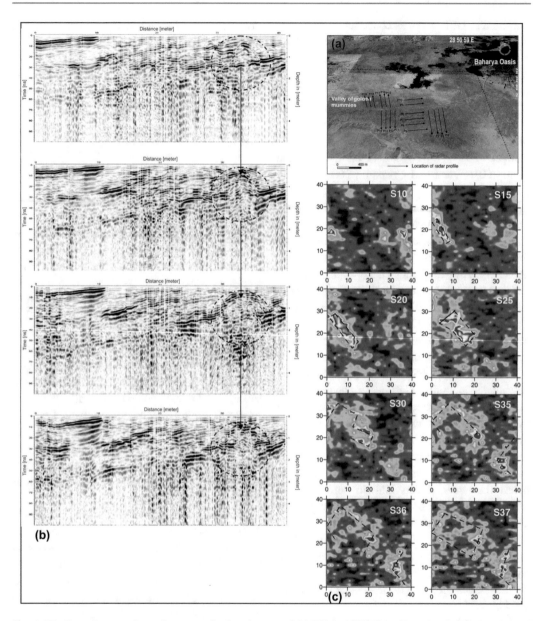

Fig. 1.17 Ground penetrating radar survey for locating the buried tombs Bahariya Oasis, Egypt, (**a**) the 2d GPR profiles, (**b**) location of the profiles along the study area, and (**c**) different GPR time slices showing the locations of underground tombs (after Shaaban et al. 2009)

1.9.1 Shallow 2-D Seismic Reflection Acquisition

Shallow seismic reflection acquisition methods have evolved significantly over the past two decades to overcome the shortcomings of the optimum window technique introduced by Hunter et al. (1984). These common midpoint (CMP) or multi-fold techniques (Fig. 1.18) were developed in Canada (Mair and Green 1981; Green and Mair 1983), the Netherlands (Doornenbal and Helbig 1983; Jongerius and Helbig 1988) and the United States (Steeples and Knapp 1982; Steeples 1984; Miller and Steeples 1986; Steeples and Miller 1990). Since multifold data generally provide better

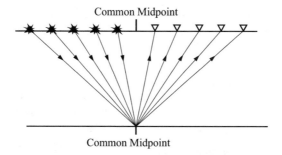

Fig. 1.18 Common midpoint (CMP) method. Explosion *symbols* and *triangles* show the source and receiver positions, respectively. For a flat lying interface, reflections on CMP traces with diverse source-receiver offsets originate from a single common midpoint (CMP)

and more reliable results than single-fold data (Green et al. 1995), CMP surveying is now applied in the vast majority of shallow seismic reflection investigations.

Several factors need to be considered when defining the acquisition geometry for shallow subsurface exploration (Knapp and Steeples 1986a, b; Büker et al. 1998a, b):

- To avoid spatial aliasing, source or receiver offsets need to be set below a maximum value Δx (Yilmaz 2001):

$$\Delta x = v/2f_{max} \sin\theta, \qquad (1.2)$$

where

fmax = max threshold frequency,
v = minimum velocity, and
θ = dip reflector angle

- Imaging structures at traveltimes <50 ms requires numerous closely spaced sources and receivers. A relatively large number of near-offset traces are needed to distinguish reflections from source-generated noise, such that appropriate top mutes may be defined.
- Maximum source-receiver offsets are controlled primarily by the necessity to obtain reliable velocity-depth estimates for the deepest features of interest.

Mapping the shallow subsurface requires higher resolution than needed in classical seismic surveying. Moreover, the application of high resolution seismic techniques in archaeological exploration is another kind of challenge. First, we have to be sure that there are a considerable physical contrast between the individual archaeological targets and the surrounding burial medium. Also, most of archaeological targets lie in the uppermost few metres of the earth. Exceptions, like deep tunnels or tombs, can be located at greater depths. Therefore the most interesting archaeological features are located in the soil/rubble zone above the first compacted bedrock layer or even located as erected distortions of the bedrock surface itself, like burial chambers. The researcher should be sure that his archaeological target can be seismically detectable and differs significantly in either velocity or density from its surroundings medium (Fokin et al. 2012). The selected of seismic methods for detecting specific archaeological targets should be carefully considered and then the field acquisition and processing strategy should be adapted.

The use of sources that generate a broad range of frequencies can enhance resolution. Frequencies > 100 Hz are particularly important in shallow seismic surveying. Depending on the conditions at the survey site, hammers and shotguns may produce wide frequency band signals. To reduce the effect of air-blast energy, sources can be buried or silenced (Miller 1992). Lateral resolution can also be increased to limits defined by the first Fresnal zone by decreasing source and receiver spacing.

1.9.2 Processing of Shallow Seismic Data

Most processing operations employed routinely in deep seismic surveying can be applied to shallow seismic data sets. To avoid generating artefacts and to achieve accurate images of near-surface structures, some special aspects of processing shallow seismic data need to be considered.

Source-generated noise may be difficult to separate from shallow reflections using single- or multi-channel filtering because of the similarity of their frequencies and velocities or because of spatial aliasing of ground-coupled air and surface waves. In addition to standard filtering, shallow reflections may be enhanced by application of spectral balancing or deconvolution. In cases where these processing operations do not work, first arrivals and tailing reverberations may have to be muted from shot or CMP gathers in order to avoid mis-processing and mis-interpretation of these signals as shallow reflections (Robertsson et al. 1996a, b).

Stretch muting can be applied to shallow seismic reflection data in order to reduce the effect of specific frequency bands. Improper choice of stretch mutes can reduce markedly the dominant frequency and bandwidth of stacked signals, thus affecting vertical resolution. If not treated carefully, excessively stretched reflection wavelets at early travel-times may be mis-interpreted as stacked refractions or subtle stratigraphic changes.

Static corrections is an important processing step in seismic reflection surveying. Since shallow reflections are generally less continuous and frequencies much higher than for deeper investigations, greater accuracy is required to derive appropriate static corrections for near-surface surveys.

Migration may produce only very minor changes to very shallow seismic reflection data. Black et al. (1994) have suggested that migration may be unnecessary when dealing with low near-surface velocities (<1000 m/s) and gently dipping reflections at very shallow depths (<50 m). Under such conditions, the migration operator will alter the position of seismic signals very little. Since application of multiple filters may result in processing artefacts, such as high-frequency sideband-lobe effects, careful analyses of the resulting seismic sections and recognition of near-surface reflections after each intermediate processing step are essential for establishing the reliability of high-resolution seismic images of shallow subsurface (Steeples et al. 1997).

References

Abdallatif TF, El Emam AE, Suh M, El Hemaly IA, Odah HH, Ghazala HH, Deebes HA (2010) Discovery of the causeway and the mortuary temple of the Pyramid of Amenemhat II using near-surface magnetic investigation, Dahshour, Giza, Egypt. Geophys Prospect 58:307–320

Ancient origin (2016) HIP Institute, Faculty of Engineering, Cairo/Ministry of Antiquities. http://www.ancient-origins.net/news-history-archaeology/thermal-scan-egyptian-pyramids-reveals-mysterious-anomaly-great-pyramid-020616

ARMADALE (2016) West Lothian Archaeological Trust Scottish Charity No. SC043118. http://www.armadale.org.uk/phototech03.htm

Baranwal VC (2007) Integrated interpretation of VLF data with other geophysical data and study of two-dimensional VLF modeling and inversion. Ph.D. Thesis, Department of Geology and Geophysics, IIT Kharagpur, India

Baranwal VC, Franke A, Börner RU, Spitzer K (2011) Unstructured grid based 2-D inversion of VLF data for models including topography. J Appl Geophys 75:363–372

Basile V, Carrazzo MT, Negri S, Nuzzo L, Quarta T, Villani AV (2000) A ground penetrating radar survey for archaeological investigations in an urban area (Lecce, Italy). J Appl Geophys 44:15–32

Benech C, Marmet E (1999) Optimum depth of investigation and conductivity response rejection of the different electromagnetic devices measuring apparent magnetic susceptibility. Archaeol Prospect 6:31–45

Benner SM, Brodkey RS (1984) Underground detection using differential heat analysis. Archaeometry 26:21–26

Benson AK, Payne KL, Stubbenz MA (1997) Mapping groundwater contamination using dc resistivity and VLF geophysical methods—a case study. Geophysics 62:80–86

Binley A, Shaw B, Henry-Poulter S (1996) Flow pathways in porous media: electrical resistance tomography and dye staining image verification. Meas Sci Technol 7:384–390

Black R, Steeples D, Miller R (1994) Migration of shallow seismic reflection data. Geophysics. 59:402–410

Blizkovsky M (1979) Processing and applications in microgravity surveys. Geophys Prospect 27:848–861

Booth AD, Linford NT, Clark RA, Murray T (2008) Three-dimensional, multioffset ground-penetrating radar imaging of archaeological targets. Archaeol Prospect 15:93–112

Büker F, Green AG, Horstmeyer H (1998a) Shallow seismic reflection study of a glaciated valley. Geophysics 63:1395–1407

Büker F, Green AG, Horstmeyer H (1998b) Shallow 3-D seismic reflection surveying: data acquisition and preliminary processing. Geophysics 63:1434–1450

Christensen NB, Sørensen KI (1998) Surface and borehole electric and electromagnetic methods for hydrogeophysical investigations. Eur J Environ Eng Geophys 3:75–90

Clark AJ (1986) Archaeological geophysics in Britain. Geophysics 51:1404–1413

Clark AJ (1990) Seeing beneath the soil. Batsford, London

Cole MA, Linford NT, Payne AW, Linford PK (1995) Soil magnetic susceptibility measurements and their application to archaeological site investigation. In: Beavis J, Barker K (eds) Science and site: archaeological sciences conference 1993, Bournemouth University, Bournemouth, pp 144–162

Conyers LB (2015) Analysis and interpretation of GPR datasets for integrated archaeological mapping. Near Surf Geophys 13:645–651

Davis JL, Annan AP (1989) Ground-penetrating radar for high-resolution mapping of soil and rock stratigraphy. Geophys Prospect 37:531–551

Dobrin MB (1976) Introduction to geophysical prospecting, 3rd edn. McGraw-Hill, New York

Doornenbal JC, Helbig K (1983) High-resolution reflection seismics on a tidal flat in the Dutch delta—acquisition, processing and interpretation. First Break 1:9–20

Drahor MG (2004) Application of the self-potential method to archaeological prospection: some case studies. Archaeol Prospect 11:77–105

Drahor MG (2006) Integrated geophysical studies in the upper part of Sardis archaeological site, Turkey. J Appl Geophys 59:205–223

Drahor MG (2011) A review of integrated geophysical investigations from archaeological and cultural sites under encroaching urbanisation in İzmir, Turkey. Phys Chem Earth Parts A/B/C 36:1294–1309

Drahor MG, Öztürk C (2011) A report on magnetic gradiometry, electrical resistivity tomography (ERT) and induced polarization tomography (IPT) studies in the Sultantepe archaeological site in south-eastern region of Turkey. GEOIM LTD, 2011ARKEO1-01, 40p (internal report, in Turkish)

Drahor MG, Akyol AL, Dilaver N (1996) An application of the self-potential (SP) method in archaeogeophysical prospection. Archaeol Prospect 3:141–158

Drahor MG, Berge MA, Kurtulmuş TÖ, Hartmann M, Speidel MA (2008a) Magnetic and electrical resistivity tomography investigations in a Roman Legionary camp site (Legio IV Scythica) in Zeugma, Southeastern Anatolia, Turkey. Archaeol Prospect 15:159–186

Drahor MG, Kurtulmuş TO, Berge MA, Hartmann M, Speidel MA (2008b) Magnetic imaging and electrical resistivity tomography studies in a Roman Military installation found in Satala archaeological site from northeastern of Anatolia, Turkey. J Archaeol Sci 35:259–271

Drahor MG, Öztürk C, Ortan B, Berge MA, Ongar A (2015) A report on integrated geophysical investigation in the Šapinuva archaeological site in Central Anatolia of Turkey. GEOIM LTD. internal report no. 2015ARKEO1-05, 90 p (in Turkish)

Fais S, Radogna PV, Romoli E, Klingele EE (2015) Microgravity for detecting cavities an archaeological site in Sardinia (Italy). Near Surf Geophys 13:495–502

Fajklewicz ZJ (1976) Gravity vertical gradient measurements for the detection of small geologic and anthropogenic forms. Geophysics 41:1016–1030

Fajklewicz A, Glinski A, Sliz J (1982) Some applications of the underground tower gravity vertical gradient. Geophysics 47:1688–1692

Fokin IV, Basakina IM, Kapustyan NK, Tikhotskii SA, Schur D Yu (2012) Application of travel time seismic tomography for archaeological studies of building foundations and basements. Seismic Instrum 48 (2):185–195

Gaffeny C, Gater J, Ovenden S (1991) The use of geophysical techniques in archaeological evaluations. Technical paper number 9, Institute of field Archaeologists, Birmingham

Grasmueck M, Weger R, Horstmeyer H (2006) Full-resolution 3D GPR imaging. Geophysics 70(1):K12–K19

Green AG, Mair JA (1983) Subhorizontal fractures in a granitic pluton: their detection and implications for radioactive waste disposal. Geophysics 48:1428–1449

Green AG, Pugin A, Beres M, Lanz E, Büker F, Huggenberger P, Horstmeyer H, Grasmück M, De Iaco R, Holliger K, Maurer H (1995) 3-D high-resolution seismic and georadar reflection mapping of glacial, glaciolacustrinel and glaciofluvial sediments in Switzerland. In: Ann Symp Environ Eng Geophys Soc (SAGEEP), extended abstracts, pp 419–434

Hunter JA, Pullan SE, Burns RA, Gagne RM, Good RL (1984) Shallow seismic reflection mapping of the overburden-bedrock interface with the engineering seismograph: some simple techniques. Geophysics 49:1381–1385

Johnson JK (ed) (2006) Remote sensing in archaeology: an explicitly North American perspective. University of Alabama Press, Tuscaloosa, AL

Jongerius P, Helbig K (1988) Onshore high-resolution seismic profiling applied to sedimentology. Geophysics 53:1276–1283

Kaikkonen P, Sharma SP (1998) 2-D nonlinear joint inversion of VLF and VLF-R data using simulated annealing. J Appl Geophys 39:155–176

Kaufmann AA, Keller GV (1983) Frequency and transient soundings, methods in geochemistry and geophysics, vol 16. Elsevier, Amsterdam. 685p

Knapp RW, Steeples DW (1986a) High-resolution common-depth-point seismic reflection profiling: instrumentation. Geophysics 51:276–282

Knapp RW, Steeples DW (1986b) High-resolution common-depth-point seismic reflection profiling: field acquisition and parameter design. Geophysics 51:283–294

Knödel K, Krummel H, Lange G (1997) Geophysik. Springer, Berlin

Kolendo J, Przenioslo J, Lciek A, Jagodzinski A, Taluc S, Porzezynski S (1973) Geophysical prospecting for the historic remains of Carthage, Tunisia (abs.). In: Proceedings of the Society of Exploration Geophysicists 43rd annual international meeting, Mexico City, October 1973, 30 p

Lakshmanan J, Montlucon J (1987) Microgravity probes the Great Pyramid. Leading Edge 6:10–17

Le Borgne E (1955) Susceptibilité magnétique anormale du sol superficial. Annales de Géophysique 11:399–419

Le Borgne E (1960) Influence du feu sur les propriétés magnétique du sol et sur celles du schiste et du granit. Annales de Géophysique 16:159–195

Linford NT (1998) Geophysical survey at Boden Vean, Cornwall, including an assessment of the microgravity technique for the location of suspected archaeological void features. Archaeometry 40:187–216

Linington RE (1966) Test use of a gravimeter on Etruscan chamber tombs at Cerveteri. Prospezioni Archeologiche 1:37–41

Loke MH, Barker RD (1996) Rapid least-squares inversion of apparent resistivity pseudosections by a quasi-Newton methods. Geophys Prospect 44:131–152

Mair JA, Green AG (1981) High-resolution seismic reflection profiles reveal fracture zones within a "homogeneous" granite batholith. Nature 294:439–442

Marsiglio L, Pipan M, Forte E, Dal Moro G, Finetti I (2003) Multi-frequency and multi-azimuth polarimetric GPR for buried utilities detection. In: EAGE 65th conference & exhibition, Stavanger, Norway, 2–5 June 2003

Miller RD (1992) Normal moveout stretch mute on shallow-reflection data. Geophysics 57:1502–1507

Miller RD, Steeples DW (1986) Shallow structure from a seismic-reflection profile across Borah Peak, Idaho, fault scarp. Geophys Res Lett 13:953–956

Monteiro Santos FA, Mateus A, Figueiras J, Gonçalves MA (2006) Mapping groundwater contamination around a landfill facility using the VLF-EM method—a case study. J Appl Geophys 60:115–125

Mullins CE (1977) Magnetic susceptibility of the soil and its significance in soil science: a review. J Soil Sci 28:223–246

Nabighian MN, Macanae JC (1991) Time domain electromagnetic prospecting methods. In: Nabighian MN (ed) Electromagnetic methods in applied geophysics, vol 2. Society of Exploration Geophysicists, Tulsa, OK, pp 427–514

Orfanos C, Apostopoulos G (2011) 2D–3D resistivity and microgravity measurements for the detection of an ancient tunnel in the Lavrion area, Greece. Near Surf Geophys 9:449–457

Owen TE (1983) Detection and mapping of tunnels and caves. In: Fitch AA (ed) Development in geophysical exploration methods, vol 5, 161258, Wiley, 209–221

Panisova J, Pasteka R (2009) The use of microgravity technique in archaeology: a case study from the St. Nicolas Church in Pukanec, Slovakia. Contrib Geophys Geodes 39(3):237–254

Panisova J, Frastia M, Wunderlich T, Pasteka R, Kusnirak D (2013) Microgravity and ground-penetrating radar investigations of subsurface features at the St Catherine's Monastery, Slovakia. Archaeol Prospect 20:163–174

Panissod C, Dabas M, Jolivet A, Tabbagh A (1997) A novel mobile multipole system (MUCEP) for shallow (0–3 m) geoelectrical investigation: the 'Vol-de-Canards' array. Geophys Prospect 45:983–1002

Papadopoulos NG, Tsourlos P, Tsokas GN, Sarris A (2007) Efficient ERT measuring and inversion strategies for 3D imaging of buried antiquities. Near Surf Geophys 5:349–362

Parasnis DS (1997) Principles of applied geophysics, Pure and applied geophysics, vol 152, 5th edn. Chapman and Hall, London, pp 184–186

Pašteka R, Zahorec P (2000) Interpretation of microgravimetrical anomalies in the region of the former church of St. Catherine, Dechtice. Contrib Geophys Geodes 30:373–387

Perssona K, Olofsson B (2004) Inside a mound: applied geophysics in archaeological prospecting at the Kings' Mounds, Gamla Uppsala, Sweden. J Archaeol Sci 31:551–562

Peters LP Jr, Daniels JJ, Young JD (1994) Ground penetrating radar as a subsurface environmental sensing tool. Proc IEEE 82:1802–1822

Reynolds J (1997) An introduction to applied and environmental geophysics. Wiley, Chichester

Robertsson JOA, Holliger K, Green AG (1996a) Source-generated noise in shallow seismic data. Eur J Environ Eng Geophys 1:107–124

Robertsson JOA, Holliger K, Green AG, Pugin A, De Iaco R (1996b) Effects of near-surface waveguides on shallow high-resolution seismic refraction and reflection data. Geophys Res Lett 23:495–498

Samouelian A, Cousin I, Tabbagh A, Bruand A, Richard G (2005) Electrical resistivity survey in soil science: a review. Soil Tillage Res 83:173–193

Scollar I, Tabbagh T, Hesse A, Herzog I (1990) Archaeological prospecting and remote sensing. Cambridge University Press, Cambridge

Shaaban FA, Abbas MA, Atya MA, Hafez MA (2009) Ground-penetrating radar exploration for ancient monuments at the Valley of Mummies—Kilo 6, Bahariya Oasis, Egypt. J Appl Geophys 68:194–202

Sharma SP, Kaikkonen P (1998) Two-dimensional non-linear inversion of VLF-R data using simulated annealing. Geophys J Int 133:649–668

Sheriff RE (1984) Encyclopedic dictionary of exploration geophysics. Society of Exploration Geophysicists, Tusla, OK. 323 p

Sheriff RE (1991) Encyclopedic dictionary of exploration geophysics, 3rd edn. SEG Geophysical References Series 1, Tusla, OK. 384 p

Simon FX, Tabbagh A, Thiesson J, Donati JC, Sarris A (2014) Complex susceptibility measurement using multi-frequency Slingram EMI instrument. In: Near surface geoscience 2014, 20th European meeting of environmental and engineering geophysics, Athens, Greece, 14–18 Sept 2014

Steeples DW (1984) High-resolution seismic reflections at 200 Hz. Oil Gas J 82:86–92

Steeples DW, Knapp RW (1982) Reflection from 25 feet or less. In: 52nd annual international meeting, Society of Exploration Geophysicists, expanded abstracts, pp 469–471

Steeples DW, Miller RD (1990) Seismic reflection methods applied to engineering, environmental and groundwater problems. In: Ward S (ed) Geotechnical and environmental geophysics, vol I: Review and tutorial. Society of Exploration Geophysicists, Tulsa, OK, pp 1–30

Steeples DW, Green AG, McEvilly TV, Miller RD, Doll WE, Rector JW (1997) A workshop examination of shallow seismic reflection surveying. Leading Edge 16:1641–1647

Stummer P, Maurer H, Horstmeyer H, Green AG (2002) Optimization of DC resistivity data acquisition: real-time experimental design and a new multielectrode system. IEEE Trans Geosci Rem Sens 40:2727–2735

Stummer P, Maurer H, Green AG (2003) Experimental design: electrical resistivity data sets that provide optimum subsurface information. Geophysics 69(1):120–139. https://doi.org/10.1190/1.1649381

Tabbagh A (1986) Applications and advantages of the Slingram EM method for archaeological prospecting. Geophysics 51:576–584

Taha AI, El-Qady G, Metwaly MA, Massoud U (2011) Geophysical investigation at Tell El-Dabaa "Avaris" archaeological site. Mediterr Archaeol Archaeometry 11(1):51–58

Telford WM, Geldart LP, Sheriff RE (1990) Applied geophysics, 2nd edn. Cambridge University Press, Cambridge

TerraDat (UK) (2003) www.terradat.co.uk

Tite MS (1972) Methods of physical examination in archaeology, Studies in archaeological science, Seminar Press, London. 319 pp, 124 figs

Tite MS, Mullins C (1971) Enhancement of the magnetic susceptibility of soils on archaeological sites. Archaeometry 13:209–219

Valle S, Zanzi L, Sgheiz M, Lenzi G, Friborg J (2001) Ground penetrating radar antennas: theoretical and experimental directivity functions. IEEE Trans Geosci Rem Sens 39(4):749–758

Venter ML, Thompson VD, Reynolds MD, Waggoner JC Jr (2006) Integrating shallow geophysical survey: archaeological investigations at Totogal in the Sierra de los Tuxtlas, Veracruz, Mexico. J Archaeol Sci 33:767–777

Wang T, Oristaglio M (2000) 3D simulation of GPR survey over pipes in dispersive soils. Geophysics 65:1560–1568

Witten AJ, Thomas E, Levy TE, Adams RB, Won IJ (2000) Geophysical surveys in the Jebel Hamrat Fidan, Jordan. Geoarchaeology 15:135–150

Wynn CJ (1997) http://www.terraplus.com/papers/wynn.html

Wynn JC, Sherwood SI (1984) The self-potential (SP) method: an inexpensive reconnaissance and archaeological mapping tool. J Field Archaeol 11:195–204

Yilmaz Ö (2001) Seismic data analysis: processing, inversion, and interpretation of seismic data. Society of Exploration Geophysicists, Tulsa, OK. 2027 pp

Integrated Geophysical Investigations in Archaeological Sites: Case Studies from Turkey

2

Mahmut Göktuğ Drahor

Abstract

With the increased demand to facilitate the archaeological work either in well-known archaeological sites or the crude sites, geophysical methods plays an important role. The Geophysical methods have been used since 1946 with increasing frequency for archaeological investigations and currently the branch of archaeogeophysics is widely applied. The wide varieties of geophysical methods applied in archaeological work relies principally upon existing reasonable contrast in physical properties between the buried archaeological feature and the surrounding subsoil. Understanding the archaeological properties of the physical contrasts, in terms of density, thermal conductivity, electrical resistance, magnetic or dielectric properties, remains fundamental issues of choosing and applying the discipline geophysical techniques. In this regard, we tried to introduce a brief outline for the common and applicable techniques in archaeological investigations. The physical principles and field instrumentation involved for the acquisition of data with each method are considered, as well as some common results from the worldwide case studies. Generally, the archeogeophysical survey results can be used to guide excavation and to give archaeologists insight into the patterning of non-excavated parts of the site as well as it is often used where preservation of the sensitive sites is the aim rather than excavation.

Keywords

Archaeogeophysics · Non-invasive techniques · Physical contrast · Excavation · Archaeological remains

2.1 Introduction

Geophysical investigation is one of the non-destructive remote sensing techniques in archaeology. It is sensitive the variations occurred from physical and chemical diversities in archaeological targets and the subsurface. Also it has a powerful effect in the identification of historical texture of cities and urban sites, and it gives us fundamental information about objectives in examined sites. This implementation could be constructive for the archaeologists to execute their managing and planning strategies during the archaeological excavation. The integration of non-destructive near-surface geophysical prospection methods, which harbours a great potential to maximize the information gained about structures buried in the shallow subsurface, has seen a rapid increases over the past two decades, especially in the field of archaeology

M. G. Drahor (✉)
Department of Geophysical Engineering, Engineering Faculty, Dokuz Eylül University, Buca-İzmir, Turkey
e-mail: goktug.drahor@deu.edu.tr

© Springer International Publishing AG, part of Springer Nature 2019
G. El-Qady, M. Metwaly (eds.), *Archaeogeophysics*, Natural Science in Archaeology,
https://doi.org/10.1007/978-3-319-78861-6_2

(Cardarelli and Di Filippo 2004, 2009; Diamanti et al. 2005; Drahor 2006, 2011; Drahor et al. 2011, 2015a; Kvamme 2006; Papadopoulos et al. 2012; Piro et al. 2000; Sarris et al. 2013; Vafidis et al. 2005). The aim of the integrative application is to acquire more reliable interpretative results concerning the characteristics of subterranean distribution with complex soil features in archaeological sites, which can in general cause difficulties regarding geophysical prospection due to the often complex archaeological contexts including a diverse range of physical inhomogeneity in the subsurface. The integrative application of geophysical investigation techniques provides us a very useful approach for the location, detection and imaging about the subsurface archaeological features. Also, it enables us to enhance the interpretation and visualisation quality using different data sets all together. The success of each geophysical method depends on the respective physical contrast between the subsoil and the buried archaeological remains, which is site dependent, as well as on the dimensions and depths of the archaeological structures of interest. The heterogeneity of the local subsurface, the presence of bedrock and the distribution of anthropogenic remains all affect the detectability of the target structures.

The large-scale geophysical examples in archaeological sites is also seen in last decade (e.g. Dabas et al. 2000; Drahor et al. 2007; Gaffney et al. 2000, 2004; Tsokas et al. 1994). The main purpose of these investigations is to reveal the probable layout of buried archaeological remains for effective and extensive excavation strategies. Magnetic, GPR and resistivity are most commonly implemented as integrative geophysical prospection methods in archaeological sites because they are fast and they provide data that are suitable for constructing an image of subsurface archaeological relics (Drahor 2006; Drahor et al. 2008a, b; Gaffney et al. 2004; Neubauer and Eder-Hinterleitner 1997; Seren et al. 2004). In addition, the integrative techniques have been used in the fields of restoration and conservation to get fundamental information about reliable diagnosis of constructional damages, such as cracks and fissures, and to determine of general characteristic of burial structures and objects inside parts of cultural heritage buildings (Drahor et al. 2011; Leucci et al. 2007; Leucci and Negri 2006; Papadopoulos et al. 2009; Tsokas et al. 2008).

The geophysical equipment, instruments and software have quickly been improving in terms of data resolution, storage, sensitivity and fast data acquisition. The usage of integrated geophysical techniques regarding to these developments is considerably increased to get the detailed archaeological interpretation. In the future, the negative influences that affect geophysical data could be reduced with the help of the new generation of instruments and equipments that provide higher resolution and precision of the data. Also, the success of the determination of archaeological relics and their interpretation will be increased according to development of novel signal and image processing techniques together with the innovative computing and modelling researches (Drahor 2011).

2.2 Integrated Geophysical Methods

The goal of the integrated prospection approach is to obtain more reliable results and data interpretations concerning subsurface characteristics conducted at archaeological sites with complex soil features. We know that archaeological sites can in general cause difficulties regarding geophysical prospection due to the often complex archaeological contexts including a diverse range of physical inhomogeneity in the subsurface. The success of each geophysical method depends on the respective physical contrast between the subsoil and the buried archaeological structures, which is site dependent, as well as on the dimensions and depths of the archaeological structures of interest. The heterogeneity of the local subsurface, the presence of bedrock and the distribution of anthropogenic remains all affect the detectability of the target structures. Thus, the integrated applications of geophysical techniques have been highly increased the interpretation due to their advantages and the powerful results that provide greater information about the depth, dimension, extension and characteristics of subsurface targets and infrastructural environments. In addition, methodological developments in data acquisition and processing developing by the help of instrumental improvements and novel software possibilities give us important contribution to obtain the sophisticated interpretation. Therefore, integrated geophysical applications have become the important prospection approaches in archaeology throughout the world (Drahor 2011), and their usage have been increased day after day (Drahor 2006, 2011; Drahor et al. 2008a, b, 2015a; Gaffney et al. 2004, 2015; Neubauer and Eder-Hinterleitner 1997; Seren et al. 2004; Simon et al. 2015). Magnetic gradiometry, GPR and ERT are the more applied techniques in integrated geophysical investigations of archaeological sites (Arato et al. 2015; Drahor et al. 2015a; Linford et al. 2015; Matera et al. 2015; Simyrdanis et al. 2015; Tsokas et al. 2015; Wilken et al. 2015). However, seismic refraction and reflection, micro-gravity, self-potential (SP), induced-polarisation (IP), multiple analysis of surface waves (MASW), very low frequency electromagnetic (VLF) are occasionally used in an integrated manner (De Domenico et al. 2006; Drahor 2004, 2006, 2011; Drahor et al. 1996, 2009, 2015a; Gaffney et al. 2015; Papadopoulos et al. 2012; Simon et al. 2015).

Integrated geophysical surveys are used in two manners in archaeology. The first usage is the "*conventional approach*" of geophysical methods. In this approach, magnetic gradiometry, ground penetrating radar and electrical resistivity/resistance are the main investigation techniques. Second application is also the "*tomographic approach*" to geophysical data. Tomographic approach has been importantly increased in the application of archaeological geophysics since 10 years. Therefore we can obtain most useful and valuable information gathered by using of integrated geophysical techniques, which could be important for the management and planning of archaeological sites.

2.3   Integrated Geophysical Investigation Examples

Integrated geophysical investigations give us more interpretative results for locating, detecting and imaging of buried archaeological context and subsurface changes in archaeological sites. The general characteristics of buried archaeological relics and surrounded soils are importantly affected the success of the geophysical investigations. Also the existence of bedrock and the distribution of anthropogenic materials are a considerably amount influenced the results. In this chapter, the integrated results obtained from wide range of geophysical investigations performed in some archaeological sites in Turkey are given.

2.3.1   Roman Legionary Camp Sites in Turkey; Satala and Zeugma

There were four important Roman Legionary camp sites found in the eastern frontier, which was a Euphrates river between Roman and Parthia, in Roman Empires. These are XV Apollinaris, XII Fulminata, XVI Flavia Firma and IV Scythica from north to south respectively (Fig. 2.1). The integrated geophysical investigations were carried out in two important Roman Legions; Legio XV Apollinaris, which was settled in vicinity of Satala city, and Legio IV Scythica, which was found in the surroundings of Zeugma city.

2.3.1.1   Geophysical Surveys in Legio XV Apollinaris in Satala

The archaeological site of Satala was localized on the crossing of two important routes in Roman Empire. One was extended in E-W direction from Ankyra via Nicopolis and Satala into Northern Armenia and northern Persia. The second route was connected in N-S direction from Trapezus to Melitene, Samosata and Zeugma, which were another major military centers into Northern Syria (Fig. 2.1). For this reason, two of the

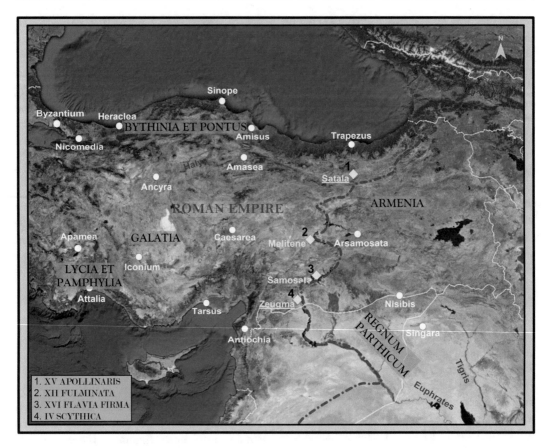

Fig. 2.1 Roma legionary camp sites along the Euphrates. (1) XV Apollinaris in Satala, (2) XII Fulminata in Melitene, (3) XVI Flavia Firma in Samosata and (4) IV Scythica in Zeugma

important Roman Military Legions (Legio XV Apollinaris and Legio XVI Flavia firma) in the Roman East were settled in Satala, which is situated about 25 km from Kelkit, in Anatolias Northeast. Only very limited archaeological surveys in and around Satala were managed. One of them is the study carried out by Lightfoot (1991). Therefore, we have very limited information about site and its surroundings, and particularly to investigate the areas of Legio XV Apollinaris and Legio XVI Flavia firma are important to determine the role of these legions within the Roman military presence in the East. Because these legions garrisoned for different periods at Satala were monitored to the major routes in northeastern Anatolia. Therefore, an integrated geophysical survey including the magnetic scanning and electrical resistivity tomography survey was managed in the northeastern part

of possible legion area defined by Lightfoot (1991). Today some fortress walls are seen on the surface of the investigation site, and apart from that there are no any archaeological remains in this area (Fig. 2.2).

The geophysical surveys in area showed as Fig. 2.2 were performed in two steps. In the first step, magnetic gradiometer survey was managed into 52 grids (approximately, 2.1 ha) of 20 by 20 musing 0.5×1 m grid intervals. The size of the investigation area was 120×180 m, and gradiometric data were gathered from zigzag traverses in the area. Overall data were combined to get detailed and large-scale image of the investigation site. The gray-scale magnetic image of the investigation site is presented in Fig. 2.3. In the image, the dark tones display high magnetic anomalies, while the light tones demonstrate low magnetic anomalies. In the magnetic image, the

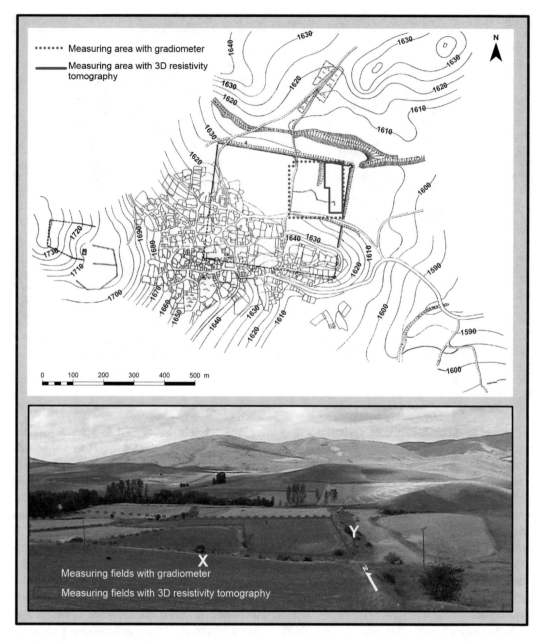

Fig. 2.2 The site and geophysical surveys plans and photo of Roman Legion site in Satala archaeological site (modified from after Lightfoot 1991 and Drahor et al. 2008b)

fundamental anomalies are mostly oriented in the direction of NNW-SSE and E-W. Particularly, the regular anomalies are observed on the eastern side of the magnetic image. A significant struc- ture showed by (1) clearly reveals the presence of a building with the regular ground plan. Another attracted structure (2) is seen in the southwestern part of the regular structure (1). The two high magnetic anomalies (3) appear in the west and middle of the survey site. Additionally, two sepa- rate high magnetic anomalies (4 and 5) between the dummy areas are clearly seen in the image, and these anomalies should be consisted of some features occurred by a fire or resulted from a large

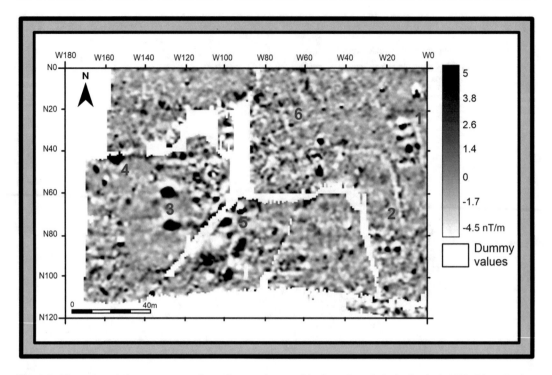

Fig. 2.3 The gray-scale low-pass magnetic gradiometer image of the investigated site in Satala (modified from Drahor et al. 2008b)

amount of magnetized materials, such as bricks or metal deposits, perhaps even dense Roman armours, weapons or other metallic objects. The anomalies of other buried structures and/or archaeological features (6) are appeared in the northern part of the magnetic image (Fig. 2.3). Magnetic gradiometer study indicated that the archaeological relics are mostly localised in shallow depths. In addition, the rectangular structures emergent in eastern part of the area may have been directly connected to the Roman military installation.

In the second step, the electrical resistivity tomography (ERT) technique was applied in the eastern part of the area. The 2D ERT data were collected using Wenner electrode configuration (spacing 1 m) in this part appeared very regular magnetic anomalies. The 3D data were also implemented by the combining of overall 2D ERT lines, and thereafter the combined data for 3D processing were evaluated by using the 3D robust inversion technique. ERT results revealed that the archaeological structures with high

resistivity might be located in shallow depths, except for some features situated in the middle of the investigation site. They are clearly seen at overall depth slices presented in Fig. 2.4. The first slice at 0.125 m clearly demonstrates the walls oriented N-S and E-W can also be showed an important Roman military structure, which is about 18 × 25 m dimension. In addition, the effect of ploughing is observed with N-S elongated traces (possibly plough lines) (Fig. 2.4b). The archaeological relics with N-S, E-W and NNW-SSE direction are appeared in the second slice at depth 0.51 m. The rectangular structure divided into three different parts in the northeastern section of the area (1) is very convenient to the anomalies of the archaeological remain observed in the magnetic gradiometer image that shows high magnetic values in some parts of the area (Fig. 2.4a, c). This increase in the magnetic anomaly could have been involved some burned materials and/or metallic objects (Fig. 2.4a). The archaeological structures in the southern part of the investigation site are oriented in W-E and N-S

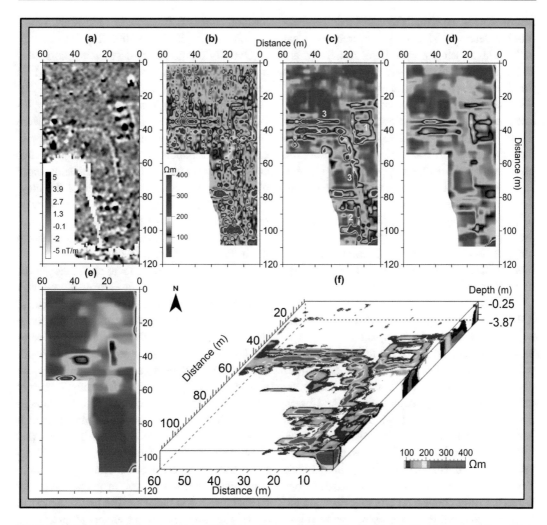

Fig. 2.4 (a) Magnetic gradiometer image of ERT investigation site. Electrical resistivity tomography slices obtained from (b) 0.125 m, (c) 0.51 m, (d) 1.09 m and (e) 1.77 m depths. The 3D volumetric representation was clipped 0.79 and 3.9 m (f). The threshold value ranged from 90 to 400 Ωm. Low resistivity values were also rendered as transparent (modified from Drahor et al. 2008b)

direction. Also one rectangular structure (2) is clearly visible. Additionally, the fortification wall (3) obtained from ERT result can be easily displayed. Figure 2.4d shows the third slice at 1.09 m. In this ERT slice, an important variation in the extension of archaeological structures is appeared. Some archaeological walls and structures at a deeper part of the area are appeared in this slice, while the traces occurred from shallow structures become blurred in the image. The fourth slice at 1.77 m is presented in Fig. 2.4e. As can be seen from this figure, the very limited

traces of the archaeological features are also remained. Overall ERT values obtained from different depths were combined to get the 3D volumetric visualization of the investigated site, and thus the possible archaeological context was presumably constituted according to ERT volumetric representation. In Fig. 2.4f, a possible archaeological context constituted from 3D volumetric threshold representation is presented, and it gives us clearly more detailed picture of the relevant archaeological structures. The threshold calibration process provides useful information in

3D, improving the visualization of subsoil archaeological features. Therefore, the archaeological walls and structures with high resistivity are clearly separated from the soil and its components for this area. As a result, magnetic imaging and ERT surveys performed in Satala present that the integrated application of geophysics enables better understanding the archaeological context in the subsurface. Thus we can conclude that, the used methods in this study were considerably successful in outlining the subsurface features of the investigated site. Therefore, the combination of magnetic imaging and ERT studies could be the important key in the interpretation of archaeological surveys, and this integration provides a more elaborate picture of the investigated archaeological site (Drahor et al. 2008b).

2.3.1.2 Geophysical Surveys in Legio IV Scythica in Zeugma

Zeugma, which had located in the main crossing point on the Euphrates River in southeastern frontier of Roman Empire, was one of the most critical frontier cities in Mesopotamia. In the Roman period a legionary camp, Legio IV Scythica, was garrisoned near the city. This legionary site was incorporated 6000 soldiers, approximately (Hartmann and Speidel 2003; Speidel 1998). Zeugma played a crucial role on the eastern frontier as the most convenient crossing place on the Euphrates River of Roman Empire, and therefore it had an important Roman military presence for many centuries (Fig. 2.5).

The large-scale magnetic survey at Zeugma is included 320 × 300 m sizes. Therefore, a field approximately 8.52 ha in area was measured using magnetic gradiometer technique with a number of grids measuring 20 × 20 m. During the data acquisition, the gradiometer data were obtained by zigzag traverses at a 0.5 m sampling interval along parallel north–south orientated traverses separated by 1 m. Electrical resistivity tomography (ERT) surveys were carried out in two different phases. First phase was the 2D ERT studies implemented along 2D parallel consecutive measuring lines having 2 m line intervals in 2002 survey period. Second phase was the 3D ERT investigations performed in four separate

locations (areas I, II, III and IV) in Roman military site during 2005 and 2006 (Fig. 2.6).

Overall gradiometer data were combined obtaining a large-scale image represented in grey-scale format of the investigation area. The area of investigation is in one of the most important pistachio-growing regions of south eastern Anatolia; this on-going agricultural activity caused some important measurement problems. One of them was the scattered metallic waste observed on the surface at the time of the magnetic survey. Visible metallic objects were removed from the investigation during the measuring process, but some will be buried beneath the surface (Drahor et al. 2008a).

The magnetic gradiometer applications give us good results about the Roman military installations at Zeugma. The magnetic contrast is very good in the large-scale image of the investigation site (Fig. 2.7). The general orientation of the buried archaeological structures is NW–SE and NE–SW directions in the military site according to magnetic gradient studies. Therefore we can determine that the subsurface archaeological structures are given a regular subsurface plan in the legion site. Five different locations that show the importance about the various architectural and archaeological structure plans were determined according to magnetic anomaly characterisation in this area. The first area marked as (I) is become evident in southern section of the area. In this section, we can observe the existence of two distinctive magnetic anomaly orientations, and these anomalies could be showed the two different subsurface archaeological structures, which may relate to various archaeological phases of the military constructions. The three high magnetic anomaly zones separated by almost equal intervals from each other are also appeared with NE–SW orientation near the centre of area I. These anomalies might be derived from the wide range of strongly magnetic material, such as brick or metal materials inside the army structures. The archaeological trenching fieldworks revealed that a large assemblage of Roman helmets and armour having strong magnetic properties was discovered. Another anomaly group (II) is located to the south east of the

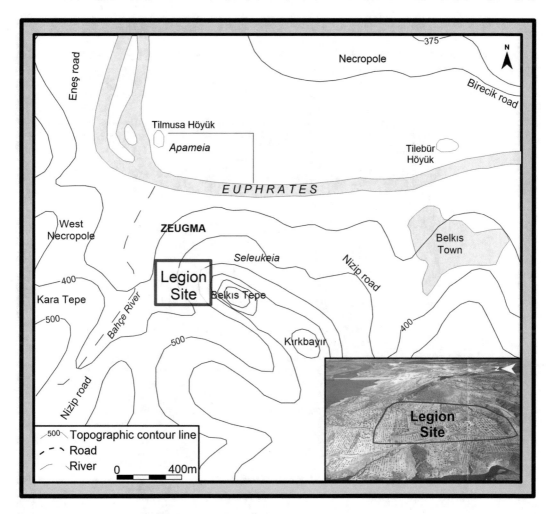

Fig. 2.5 The topographical map of Zeugma archaeological site and its environs before the Birecik dam construction (simplified from Wagner 1976) and a *photo* that shows the legion site and its surroundings (modified from Drahor et al. 2008a)

area I. The positive high magnetic anomalies generally seem on the northwestern and southeastern edges of this encircled group. This anomaly group has a different character to that cited above, and it could be a Roman civil structure (villa or temple). In area III, although the magnetic anomalies are very complicated, it has commonly a rectangle plan. Also two main Roman roads (indicated by thin black arrow) intersected each other are clearly visible inside this area. Gradiometer image show that some blocks containing houses and public areas might be settled surrounding these Roman roads. Therefore

we can suppose that the Roma military structures might be reoccupied on these structures. Another anomaly groups (IV and V) are appeared in the northern and western part of the investigation site. Moreover, some magnetic anomalies occasionally have a negative character (shown by thick black arrow), and it should be indicated the Roman soldier barracks. Furthermore, the anomaly in the section V has a semielliptic shape. Therefore we can define that it may derive from the archaeological structure that shows different characteristic comparing to others. The magnetic anomaly amplitudes are importantly decreased in

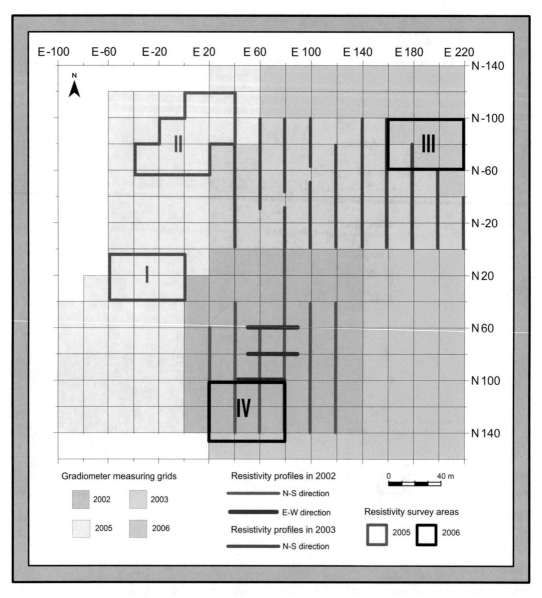

Fig. 2.6 Plan of large-scale geophysical investigations implemented between 2002 and 2006 (from Drahor et al. 2008a)

the southwestern section of the survey site, and this decreasing may result from the presence of the bedrock (limestone indicted by thick white arrow), which should be located very close to the surface.

ERT data were processed by the three-dimensional robust inversion technique to determine the subsurface archaeological structures in four different sectors investigated in 2005 and 2006. The three-dimensional resistivity inversion

results implemented from the robust approach are given as three-dimensional volumetric images together with the magnetic image in Fig. 2.8. The volumetric image of area-I was clipped from two different layers, at depths of 0.25 and 2.5 m. The archaeological features showing high resistivity were clearly determined in all depths. The volumetric image of area-II surface was also clipped from two different slices, at depths of 0.25 and 2.1 m. In this area, the archaeological

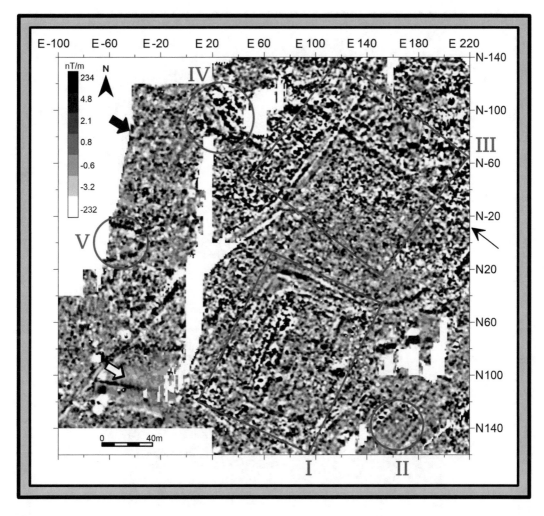

Fig. 2.7 Large-scale magnetic gradiometer image of the legion site. The five important zones were indicated by *Roman numbers* (from Drahor et al. 2008a)

features are generally observed in the northern part, and they are settled very close to the surface. The volumetric representation of area-III was completely clipped at a depth of 2.0 m. Therefore, the deeper archaeological features found in the northeastern quarter of the investigation site can also be broadly detected by this representation. The last volumetric image is produced from area-IV, and its horizontal slice was clipped from a depth of 0.75 m in order to remove the effects of ploughing in the area. Thus, the major archaeological walls and features are clearly determined in this site by ERT modelling studies. The volumetric visualization shows that structures are

found at all depths, yet are mostly situated between 1 and 2 m beneath the surface (Drahor et al. 2008a).

Large-scale integrated magnetic and ERT studies concluded that this approach gives us significant information to reconstruct the archaeological layout in the investigation site. Magnetic studies successfully identified the general outlines of the legion settlement within the investigation site. Magnetic surveys demonstrated the presence of archaeological context oriented in NE–SW and NW–SE, the buildings being arranged perpendicularly to each other. The architectural settlement plan is appeared as a regular pattern typical of a

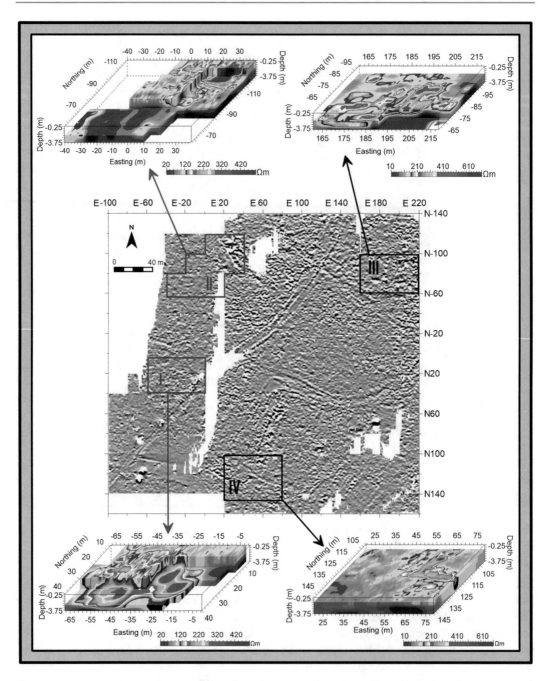

Fig. 2.8 Magnetic gradiometer image and 3D volumetric ERT representations of investigation site (from Drahor et al. 2008a)

military settlement according to magnetic surveys. ERT investigations reveal successfully in identifying the true resistivity distribution, the depths and the extensions of buried structures in the subsurface. The ERT results indicated that the archaeological features, in general, lay at shallow depths. But some of them reached down to lower strata (ca. 4 m depth). According to our 3D ERT investigations, the archaeological structures and features in the overall resistivity investigation site are of regular shape, and their orientation is nearly NE–SW and NW–SE. This result is very compatible with the magnetic outputs. Finally, we conclude that the ERT and the magnetic imaging provided matching results. This means that each method successfully determined archaeological features in the investigation site. Consequently, the integrated usage of these methods allows for a more detailed, less invasive examination of archaeological structures (Drahor et al. 2008a).

2.3.2 Amorium Archaeological Site

The Amorium archaeological site is located about 170 km southwest of Ankara. This site includes an artificial hill or höyük settled as early as from the Early Bronze Age (third millennium BC) to Phrygian periods. This site was has a major administrative and military role due to its strategic position between Syria and Constantinople in the seventh century AD. The city was the military headquarters of the Byzantine province from mid-seventh to mid-ninth century. Amorium faced repeated the Arabian attack in the Middle Age, and one of these is the well-documented siege of 838 (Lightfoot 1997). Integrated geophysical explorations in Amorium archaeological site were executed on a large area named as Middle Byzantine Military Compound. This study was aimed to detect important buried Byzantine structures. Therefore, the area divided into 10 square grids each with the dimensions of 20 by 20 m was investigated using resistivity, magnetic and self-potential methods (Fig. 2.9). The investigation site was had a flat topography. The resistivity investigation was performed using the pole–pole configuration on eight grids. The grid intervals and electrode spacing were 1×0.5 m and 1 m, respectively. The magnetic investigation was implemented using a proton magnetometer in nine grids with 1×1 m grid intervals. SP prospection was executed on the area D found between A1 and B1, where the anomalies in resistivity and magnetic prospection were generally observed in this section of the investigation site. During the SP data

collection, the gradient measurement technique was implemented by 3 m electrode spacing and 0.5×1 m grid intervals on 21 profiles directed with NW–SE using a pair of non-polarizable copper–copper sulphate electrodes (Fig. 2.9). Before the SP data acquisition, the heavy rainy weather condition affected the investigation site for 3 days. Therefore, the soil was importantly saturated for electrokinetic process owing to the capillary vertical movement of the rainwater between the soil and buried walls. Therefore, we thought that the SP survey could be useful and give some important information about the burial archaeological structures (Drahor 2004). Also we know that the SP anomaly might be obtained on burned archaeological materials and in areas that contain physical and chemical changes (Drahor et al. 1996). The investigation area was completely burned in AD 838, according to Byzantine historians. As a result of this fire, the archaeological materials in the site drastically changed compared with previous physical and chemical situations, and this event can create SP anomalies (Drahor 2004).

The magnetic, resistivity and SP data obtained from measured lines were graphed to interpret the anomalies. The geophysical anomalies obtained from theline-3 are given in Fig. 2.10. The apparent resistivity anomaly is importantly increased between 5 and 10 m of the line, while the other parts of the measured line are varied 250 and 300 Ωm, generally. This changing is showed the presence of the resistive structures in the subsurface along this line. In addition, the magnetic anomalies are similarly increased as well as resistivity anomaly in this part. This increasing might be related to the existence of magnetic materials within the archaeological structures. The amplitudes of gradient SP anomalies are changed between −10 and 25 mV/m, and they have the positive and negative anomaly forms. A large positive SP gradient anomaly reaches its maximum value at 12 m along the line. The gradient SP values have negative characteristics between 5–10 m and 13–16 m of this line. In this profile, the gradient and total SP anomalies are generally opposed to the magnetic and resistivity anomalies (Fig. 2.10). According to integrated geophysical investigations, we can conclude that the resistivity show the existence of an

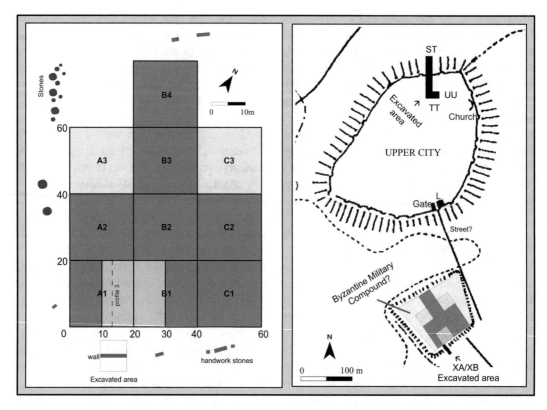

Fig. 2.9 The Amorium archaeological site map. Geophysical surveys were implemented in the Byzantine Military Compound site. Geophysical research plan is presented in *left part* of the figure. Resistivity measuring site includes A1, B1, C1, A2, B2, C2, B3 and B4 grids, while the magnetic measuring site contains A1, B1, C1, A2, B2, C2, A3, B3 and C3 grids. SP investigation was performed only on D grid (from Drahor 2004)

archaeological structure appeared in high resistive anomaly. The presence of high magnetic anomaly might also be related to burnt zones and structures that can create magnetic anomaly in high resistive zone. The total and gradient SP anomalies might be showed the physical phenomenon that could be based on the downward flow of rainwater in porous media between the buried walls and the existence of some burnt materials.

According to integrated geophysical results, the archaeologists were excavated the area D to test the geophysical outcomes. In this excavation stage, the first structural bases constructed from limestone materials were appeared very close to the surface. The following layers were built with irregular stones and bricks, and their depths are 0.5 m and 1 m, respectively. In addition, burned materials and structures are showed up at the bottom of the first structural layer. These

excavations continued, and finally the Byzantine bath with hypocaust system was found in area D (Fig. 2.11). This structure was constructed by using of limestone and Byzantine bricks. This structure consists of the polygonal structure and the circular building, which were completely filled by the stones. Furthermore, the hypocaust was found at the southern part, and it includes the thick ash layer and other burnt materials (Drahor 2004; Lightfoot and Arbel 2003).

The images of resistivity, magnetic and SP studies performed on A_1-B_1 and D areas are presented in Fig. 2.12. Also, the architectural plan of the Byzantine bath unearthed by archaeological excavation performed after the geophysical investigations. To make a successful comparison between the geophysical and excavation results, the architectural plan overlapped on the geophysical images. Therefore, the achievement of

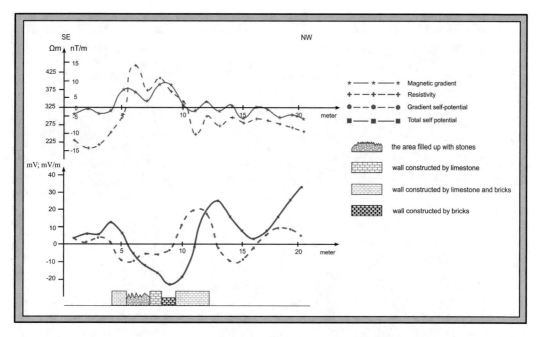

Fig. 2.10 The apparent resistivity, magnetic gradient and self-potential anomalies for line-3 of area D. The generalized section of the archaeological context revealed by excavation is given in the bottom of the sections (from Drahor 2004)

geophysical surveys was tested. The apparent resistivity image is given in Fig. 2.12a. The high resistivities (400–800 Ωm) mostly appear in the center piece of the survey site while the lower resistivities (100–300 Ωm) locate in the northeastern and southwestern parts. The very high resistivity anomalies are generally localized over the Byzantine bath building (Fig. 2.21a). The magnetic investigation was performed using gradient survey technique. The magnetic anomalies are limited between ±36 nT/m. The important positive anomalies are located in the center of area D together with the dominant negative magnetic anomalies (Fig. 2.12b). We can imply that the important resistivity and magnetic anomalies are mostly appeared with a good agreement in the middle part of the area. The gradient SP image is presented in Fig. 2.12c. The SP anomalies are ranged –40 and 20 mV/m, and important negative anomalies are found in the southern part of the excavated building while the positive anomalies are appeared in the middle part of the building.

If the archaeological excavations and the integrated results are compared together, the good harmony can be observed between them. The resistivity studies indicate the polygonal structure that was found filled with stones. Also the burnt materials and thick ash layer were found in that place showed the high magnetic values. The negative SP gradient anomalies were appeared above the tepidarium, caldarium and latrina sections of the bath, while the positive gradient anomalies were existed over and around the polygonal structures and hypocaust (Figs. 2.11 and 2.12). In addition, the horizontal projection of SP evaluation revealed that the directions and angles of the origin of each SP anomaly are a good agreement with walls and burned structures. Two different polarization angles (0° and 180°) are detected in this area. These polarizations should be indicated the presence of both burned and unburned materials. The degree of polarization angles is zero on the burned structures, while the unburned structures are shown the 180° according to excavation results. This result may be give rather interesting interpretation about the origin of SP anomalies for archaeological investigations. As a result, SP method is appearing as a successful for the

Fig. 2.11 Byzantine bath determined by integrated geophysical surveys in area D at Amorium archaeological site: (**a**) polygonal structure; (**b**) caldarium; (**c**) tepidarium (from Drahor 2004 and Lightfoot and Arbel 2003)

investigation of archaeological structures. In addition, it was encouraging to see the similarity between the excavation results and SP, magnetic and resistivity images, which emphasizes the usefulness of the integrated application of magnetic, resistivity and SP methods for the archaeological prospection.

2.3.3 Sardis Archaeological Site

Sardis was the capital city of Lydia, and it was the starting point of the western part of the royal road continuing from Sardis to Susa. Sardis is about 72 km east of İzmir, which was a well-known city of Aegean cost in antiquity. The city is located in

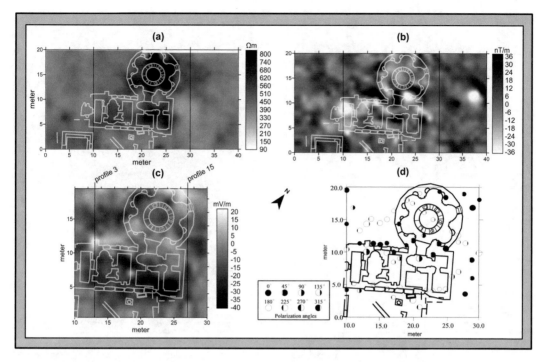

Fig. 2.12 The geophysical images in the A1-B1 and D grids and the architectural plan of the Byzantine bath unearthed by archaeological excavation. (**a**) Apparent resistivity image of A1 and B1 grids (pole-pole 1 m). (**b**) Magnetic gradient image of A1 and B1 grids. (**c**) The gradient self-potential image collected with Cu-CuSO₄ electrodes in area D. (**d**) The horizontal projection of polarization centres in area D (measured using non-polarizable electrodes). Direction: NW-SE 33°. Electrode interval: 3 m. Polarization depth: all depths (modified from Drahor 2004)

Gediz (Hermus) valley, which is one of the important grabens in western Anatolia. Also, this valley is seriously affected by earthquakes, and numerous big earthquakes have been occurred in Gediz valley since ancient times. We know that Sardis had been completely destroyed by a big earthquake in 17 AD according to historical records. The city was rather famous during the Roman period. Today, the Roman ruins such as theatre, gymnasium, stadium, etc. are noticeable on the surface. Others are covered by a very thick soil layer and there are few ruins at surface. Thus it is very hard to predict the presence of subsurface archaeological relics in this site, and this difficulty may affect the excavations. We know that the major problem can easily be solved using integrated geophysical surveys in the determination of subsurface archaeological relics. Therefore, an integrated geophysical approach was managed to define the subsurface archaeological structures and the burial plan of these structures in upper terrace of

Sardis. For this purpose magnetic, electrical resistivity tomography (ERT), VLF-R and seismic P tomography investigations were implemented in this area (Fig. 2.13). The geophysical data collected from the site were processed by various signal and image enhancement techniques, and numerous images were formed to understand the surface archaeological features in the investigation site.

The study area (about 2 ha) is very close to the stadium dating from the Roman period in Sardis. The investigation site is the terrace enclosed by the big walls in the northern, eastern and western sides. However, there are no any archaeological remains on the surface of this terrace. Therefore, an integrated geophysical survey was carried out the gridded areas (20 × 20 m) given in Fig. 2.14. Within this scope, overall area was investigated using magnetic gradiometry technique with 0.25 × 1 m grid spacing. Electrical resistivity tomography technique using pole–pole

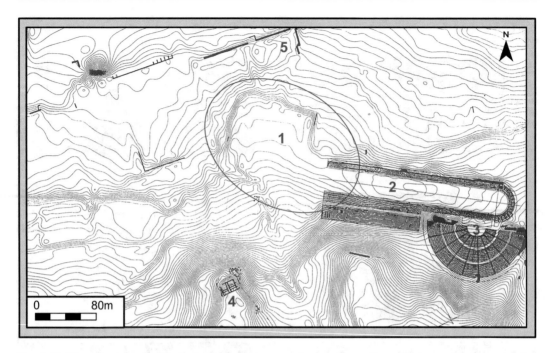

Fig. 2.13 The map shows the location of investigation area and its surroundings. The *red ellipse* demonstrates the location of geophysical investigation site. Explanation: study area (1), stadium (2), theatre (3), Byzantine fort (4), structure-A (5) (from Drahor 2006)

configuration with different dipole spacing (L = 1, 2, and 3 m) was applied in the eastern part of the site that indicates some important magnetic anomalies. The grid spacing was used as 1 × 2 m. In addition, VLF-R measurements were performed in a lesser area, which was an important anomaly group in ERT results. Measurements were conducted at 2 × 1 m grid spacing in different frequencies (19.6, and 23.4 kHz). Seismic refraction tomography investigation was also implemented along some lines running in the directions of NE–SW and SE–NW (Fig. 2.14).

The gradiometer images are represented in Fig. 2.15. The images show that magnetic contrast is very good all over the area. In these images, high magnetic anomalies are displayed in dark tone, while low magnetic anomalies are in light tone. Magnetic anomaly amplitudes in combined image are varied between ±48 nT/m (Fig. 2.15a). The principal magnetic anomalies are appeared in the northern, northeastern and southern sides of the investigation site, and these give us to be in a regular plan. In addition, the anomalies observed in the southern and southeastern parts of the area are indicated the presence of big important archaeological structures. Particularly, the anomaly encircled by a white circle is of very high magnetic value and it is wider than all other anomaly groups. Thus, we suspect the existence of an important Roman building, which might be composed of magnetic materials, such as brick, etc. In addition, some anomalies enclosed by black ellipses may be showed the streets and an antique gallery system. The gradiometry results show that there is a square settlement plan in the investigation site. Also, any interesting anomalies are not visible in middle part of this area. Therefore, we can determine that numerous archaeological features along the edges of the site in shallow depths should be existed in this area. The negative and positive anomalies showed in southeastern part of the area might be generated by the buried seats of the stadium of Sardis (Fig. 2.15b) (Drahor 2006).

In this study, resistivity data sets were inverted using 2D robust inversion technique. The robust

Fig. 2.14 The investigation plan of integrated geophysical surveys. The *grids* indicate the measured sites using by gradiometer, ERT and VLF-R. The *lines* such as RLINE1 and SLINE2 show the location of ERT and seismic refraction tomography test lines. The *solid black rectangles* demonstrate the archaeological trenches executed during the excavation season in 2002 (from Drahor 2006)

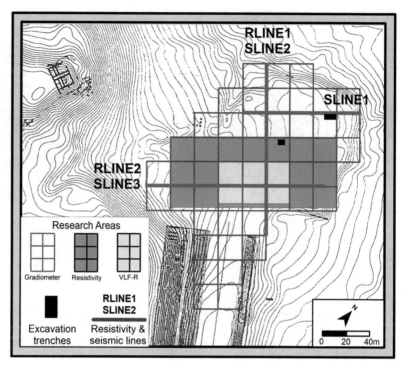

method is created to minimize the sum of the absolute value of the data misfit (Claerbout and Muir 1973). Thus, the method is not sensitive to the bad data points as much as the smoothness constrained, and therefore it produces more appropriate models for sharp structures which are similar archaeological structures. To obtain depth slices of inverted resistivity, all resistivity lines are combined to obtain the images of different depth levels. In Fig. 2.16, the ERT images four different depths (0.4, 1.3, 2.1 and 2.9 m) are presented. The first depth slice from 0.4 m shows the presence of very high resistive structure among S180–S200 and E760–E780 co-ordinates. Also, the other resistive bodies are seen in northeastern, middle and southwestern part of the investigation site. As a result, we can say that there is a good harmony between the magnetic gradient and ERT depth slice (0.4 m). In addition, this outcome reveals the existence of archaeological features found closed to the surface (Fig. 2.16a). The ERT depth slices from 1.3 and 2.1 m are displayed the presence of regular archaeological structures with high resistivity in

the south part of the investigation site. As can be seen from these slices, the buried features are commonly extended in the directions of NNE–WNW and NNW–ESE (Fig. 2.16b, c). Even though some of them extend down to 2.9 m, but the majority of archaeological features is not so much at this depth slice (Fig. 2.16d). However, the archaeological structures appeared between S180–S200 and E760–E780 co-ordinates are observed in all depth slices (Fig. 2.16a–d).

In addition, the leading aim of VLF-R investigation was to investigate the variations between the resistivities in the subsurface characteristic and to compare the results of VLF-R with other geophysical outcomes. The VLF-R investigation site was identified in the middle section of the resistivity survey site, which has high magnetic and resistivity values. The three point configuration in tie line mode was used during the VLF-R data collection. The line and measuring intervals were selected in 4 and 2 m in every 20×20 m grid, while the dipole intervals, which were perpendicular to each other, were determined in 5 m. During the investigation, two powerful radio

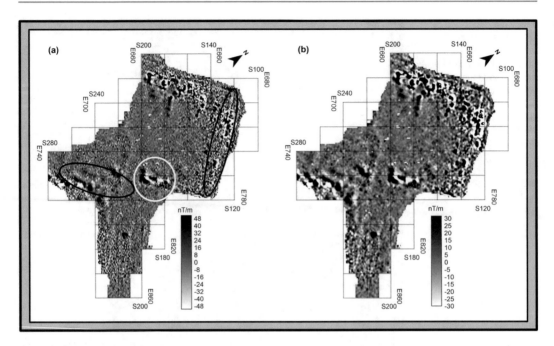

Fig. 2.15 The grey-scale magnetic images. (**a**) The combined gradiometer image after the correction process. (**b**) The low-pass filtering result (from Drahor 2006)

stations; Oxford, England (19.6 kHz) and Rhauderfehn, Germany (23.4 kHz) were used as a transmitter. The corrected data were shown in grey-scale colouring mode (Fig. 2.16e, f). The VLF-R images are showed the presence of three major anomaly groups. The best visible anomaly, which is also clearly seen in the other geophysical images, is located between S180–S200 and E760–E776 coordinates. Another group appears between S200–S220 and E740–E760, and the third one is found among S160–S180 and E740–E760 coordinates. In addition, some anomalies that have low amplitudes are visible in the images produced by two different frequencies. The primary directions of anomalies are appeared in NNE–WNW and NNW–ESE as that can be seen in other geophysical images. As a result, we can conclude that the apparent VLF-R anomalies are very similar and found in same places with other geophysical outcomes. This result shows that this technique might be effective and useful in the investigation of shallow features in archaeological prospection. Unfortunately, the technique includes some important problems, which one is the loss of transmission of VLF. This trouble

importantly prevents to constitute the large-scale images or maps obtained from VLF or VLF-R investigations of archaeological sites. Second one is the interpretation problem, which is mostly quantitative, preventing the effective interpretation (Drahor 2006). However, some studies on the inversion of VLF data showed that this problem could be solved using the inversion algorithms of VLF data (Baranwal et al. 2011; Monteiro Santos et al. 2006; Sharma and Kaikkonen 1998).

The seismic refraction tomography (SRT) is occasionally applied in archaeological prospection. The aim of this technique is to determine the subsurface archaeological features using tomographic approach to seismic P waves. In this study, the SRT data were collected along the lines of E700, E760 and S180 using a 24-channel seismic recorder. During the measuring, the interval between the geophones was selected as 1 m and the shots were implemented at next location of each geophone. A sledgehammer was also used as a seismic source. The tomographic inversion process was managed by a simulated annealing procedure including a controlled Monte-Carlo inversion. The first arrival times and survey

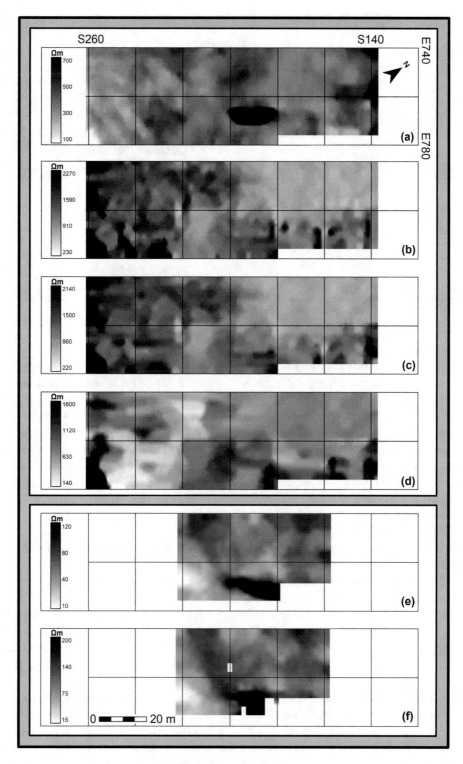

Fig. 2.16 The depth slices of ERT study: (**a**) 0.40 m, (**b**), 1.3 m, (**c**), 2.1 m and (**d**) 2.9 m. Apparent VLF-R images for two different transmitters: (**e**) 19.6 kHz transmitted from Oxford, UK and (**f**) 23.4 kHz transmitted from Rhauderfehn, Germany (modified from Drahor 2006)

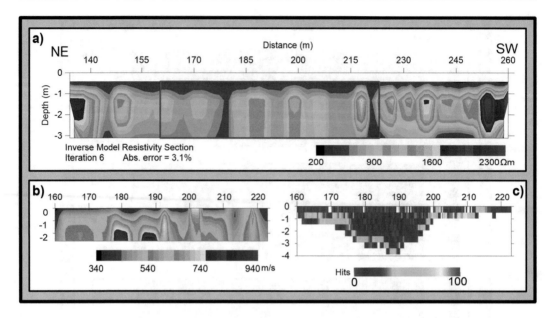

Fig. 2.17 (**a**) ERT, (**b**) SRT model sections of SLINE3, (**c**) Hits distribution of SRT model (modified from Drahor 2006)

geometry are input and no initial velocity model is needed during the modelling (Pullamma-nappallil and Louie 1994).

Figure 2.17 presents the ERT and SRT model results along the line E760. In the ERT model, we observe numerous high and medium resistive structures, which are clearly indicated the subsurface archaeological features. The structures are mostly found in very close to the surface, which are generally located from 0.6 to 2 m depths (Fig. 2.17a). SRT data were acquired on the SLINE3, and the seismic tomography model was calculated for the area indicated by the red frame. SRT model implementation was revealed the similar model output like ERT, and the vertical and horizontal variations are generally seen in the same locations with ERT model result. The archaeological structures with high P velocity are also found in shallow depths as good as resistive structures in ERT section (Fig. 2.17a, b). The hit count plot, which displays the number of seismic rays crossing each cell, is given in Fig. 2.17c. This plot indicates the distribution of seismic rays in the subsurface, and it enables to determine the reliable regions by higher hit values in every cell of seismic tomography section. As can be seen from Fig. 2.17c, show that, these values have generally medium values, except for some regions in the center of line and near the surface. This result showed that, the principle hit distribution is sufficient to define the reliable velocity regions. Consequently, the archaeological structures in shallow depths were also obtained using SRT technique. The SRT results given us impressive and similar outcomes such as ERT, magnetic and VLF-R.

2.3.4 Šapinuwa Archaeological Site

The archaeological site of Šapinuwa is located to the west of the Çekerek River on a plateau surrounded on three sides by deep valleys located in the district of Ortaköy some 53 km southeast of the city of Çorum (Fig. 2.18a). The archaeological site is enclosed with very active strike-slip fault zones such as, North Anatolian Fault, Sungurlu Fault and Kazankaya Fault zones. The North Anatolia Fault Zone is one of the very active fault zones in the earth, and this zone is caused important and very destructive earthquakes. Šapinuwa was a religious, military, and governmental centre and had close connections to Hattuşa, the capital of the Hittite

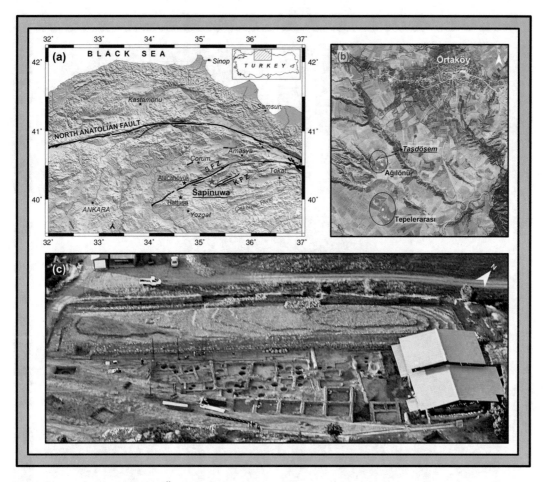

Fig. 2.18 (a) The location map of Šapinuwa (active faults are plotted according to MTA (2002) and Emre et al. (2011). *SFZ* Sungurlu Fault Zone, *KFZ* Kazankaya Fault Zone. (b) The satellite image of the site and surroundings. (c) The photo of the sacred area (Taşdöşem; the stone pavement) within the archaeological site. The holes in the photo are related to sacrificial rituals (from Drahor et al. 2015a). The base map was generated using Generic Mapping Tools (GMT) (Wessel and Smith 1995)

Empire. Archaeological finds have shown that the city functioned as an important cultic centre with a sacred area for religious rituals in the Hittite period. Today, the archaeological excavations continue in two different areas. One of them is the Tepelerarası area where excavations revealed an important official building containing more than 4000 clay tablets, dating in major parts to 1400 BC, which corresponds to the Middle Kingdom period of the Hittite Empire. The second one is the Ağılönü sacred area containing the Taşdöşem (Fig. 2.18b). The archaeological excavations in Šapinuwa have been carried out by Turkish archaeologists from Ankara University since 1990. The most important archaeological finds made at Šapinuwa so far were that of an official structure that contained an archive of Hittite cuneiform tablets and a large storage building.

The integrated geophysical investigations were applied in Tepelerarası and Ağılönü areas between 2012 and 2015 in Šapinuwa archaeological site. The main objective of this study has been to investigate and demonstrate the capability of an integrated methodological approach regarding the use of a multitude of different geophysical prospection methods applied to an archaeological

Fig. 2.19 Some photos ERT, magnetic gradiometry, seismic and GPR studies in Šapinuwa archaeological site

problem. Five different geophysical prospection techniques, namely magnetic gradiometry, electrical resistivity tomography (ERT), ground-penetrating radar (GPR), seismic refraction tomography (SRT) and multi-channel analysis of surface wave tomography (MASWT), have been comparatively tested and integrated in this study (Fig. 2.19).

2.3.4.1 The Sacred Area (Taşdöşem)

The sacred structure, known as Taşdöşem, had been found in the Hittite archaeological site of

Šapinuwa. The integrated geophysical surveys in Taşdöşem had two objectives: (1) to explore the subsurface inside this area in means of buried archaeological remains and (2) to exemplarily reveal the capabilities of the used geophysical methods. In particular, the Taşdöşem structure, with a dimension of 95 × 14 m, differs from the other excavated structures due to its specific construction character (Fig. 2.18c). The construction reaches to a depth of 1.74 m from the surface. Underneath the block of layers, another level appeared in form of a mixture of clay, agricultural

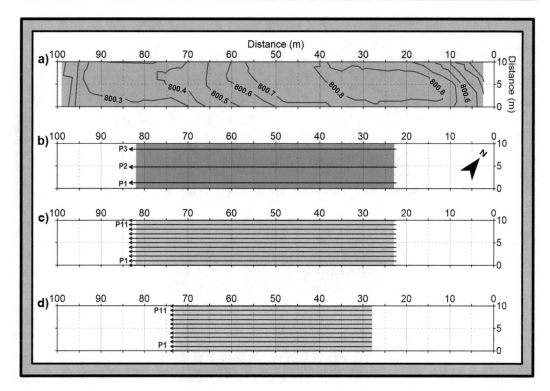

Fig. 2.20 The survey plan of the combined geophysical surveys at the Taşdöşem area. (**a**) Magnetic gradiometry with superimposed micro-topographical contours (*blue lines*), (**b**) GPR, (**c**) ERT, (**d**) SRT and MASWT survey areas (from Drahor et al. 2015a)

soil, and broken stones, down to depths of 2.3–2.4 m. The test excavation was extended to a total depth of 4.0 m, and a third level consisting of clay and hydrated lime, which was very difficult to excavate, was revealed (Drahor et al. 2015a).

In the course of two fieldwork campaigns, several ground-based geophysical prospection surveys were conducted in the Taşdöşem area, comprising an area of 95 × 10 m (Fig. 2.20). The individual surveys included magnetic gradiometry, GPR, ERT, P-wave SRT and MASWT.

Initially, geophysical survey grids and lines were established by using a highly precise GNSS system. Magnetic gradiometer measurements were carried out by parallel traverse survey of ten grids with 10 × 10 m dimensions. A 2D ERT study was performed along eleven individual profiles with 1-m crossline spacing using 60 electrodes. Due to the

dry soil condition and surface irregularities caused by small stones on the surface of the Taşdöşem, the Wenner–Schlumberger configuration was chosen for ERT measurements since it is less sensitive to noise. Furthermore, a GPR survey was conducted along 20 profiles with 0.5-m line spacing. In regard to seismic methods, 2D SRT and MASWT measurements were carried out along the ERT lines using a 24-channel seismograph and a sledgehammer source (9 kg). The used geophones of P-wave (40 Hz) and MASW (4.5 Hz) were placed 2 m apart from each other for the recording of the first arrivals and surface waves, respectively (Fig. 2.20). In the case of the P-wave seismic tomography measurements, the shots were placed near the geophones. It is known that the horizontal resolution of MASW is not as good as in the case of the seismic refraction method. The horizontal resolution is controlled by two parameters: the length of the geophone spread and the sampling interval.

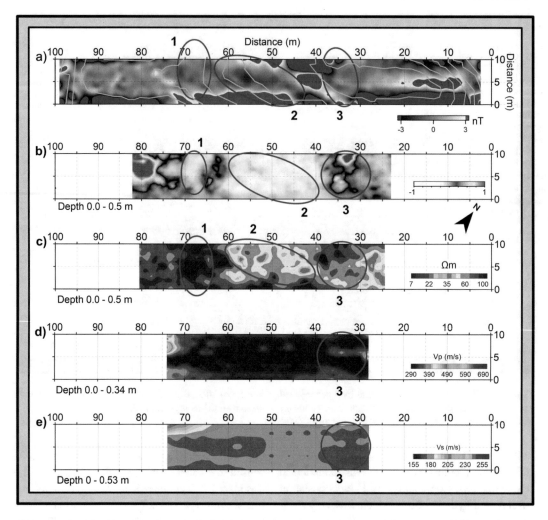

Fig. 2.21 Combined interpretation results of the shallow features. (**a**) Magnetic gradiometry, (**b**) GPR, (**c**) ERT, (**d**) SRT and (**e**) MASWT depth-slices (from Drahor et al. 2015a)

Therefore, MASWT data were collected using a fixed receiver spread configuration. In this configuration, the geophones were set up at fixed locations with 2-m geophone in-line spacing. The shots were placed between the geophones, and they were moved with an increment equal to the geophone spacing (Drahor et al. 2015a).

Figures 2.21 and 2.22 present the comparisons of the results obtained with the individual methods at shallow and intermediate depths of the Taşdöşem construction. In the magnetic data image shown in Fig. 2.21a, three important anomalies are marked with blue ellipses: (1) shows the negative anomalies in NW–SE orientation,(2) marks the distinct anomalous zone almost in E–W direction, and (3) presents the positive to negative transition zone, which is due to surface variations, as revealed by the recent agricultural activities. This variation observed in the area during the surveys is also clearly visible in the topographical contour map. In the GPR result (0–0.5 m depth slice), there is a slight correlation with the magnetic gradient image. We observe low reflection amplitudes at locations (1) and (2), whereas the high reflection amplitude appears in the zone marked with (3). The ERT slice obtained from 0 to 0.5 m depth shows an interesting result concerning the shallow part of the structure. We observe high resistivity at locations (1) and (3),

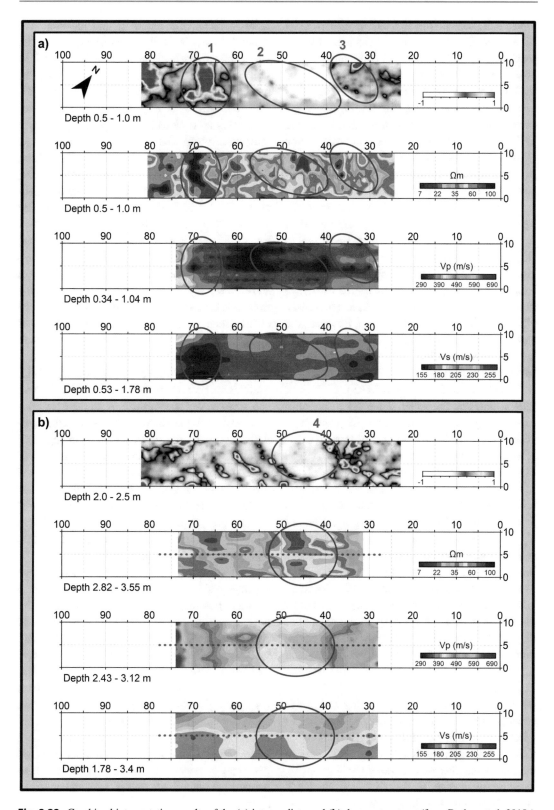

Fig. 2.22 Combined interpretation results of the (**a**) intermediate and (**b**) deeper structures (from Drahor et al. 2015a)

whereas the zone (2) displays intermediate resistivity values. The low P-wave velocities that are observed almost throughout the entire image correspond as well to negative magnetic values, indicating a correlation with the material properties in the shallow subsurface (Fig. 2.21a, d). At the same time, the low reflection amplitudes in the GPR results generally coincide with areas of low P-wave velocities (Fig. 2.21b, d). This result is one of the important conclusions of this integrated study. We imply that the attenuation caused by the shallow lossy and dispersive medium in the case of the GPR and SRT results is very important in this regard. The MASWT depth slice from 0 to 0.53 m depth shows in general low Vs velocities. However, the Vs velocities slightly decrease at the locations (1) and (3) (Fig. 2.21e). As a result, we observe a significant anomaly at the location (3) in allgeophysical data at shallow depth. However, the anomalous zones at (1) and (2) are only observed in the magnetic and ERT data, and partly in the MASWT results (Drahor et al. 2015a).

The GPR, ERT, SRT, and MASWT slices of intermediate and deeper structures are being compared in Fig. 2.22a, b. As can be seen from these slices, the anomalies in zone (1) are oriented in the NW–SE direction in all cases. The anomaly, which displayed high GPR reflection amplitudes, was found between 65 and 70 m inline coordinate, and it has a regular form in the GPR slice from 0.5 to 1 m depth. In the ERT slice of 0.5–1 m, the anomaly shows a similar character as the GPR anomaly. It has a high resistivity character, and it is located between 67 and 72 m inline coordinate. The P-wave velocity increases partly at this location, whereas Vs decreases abruptly. The character of the Vs anomaly is very similar to the anomalies displayed in the ERT and GPR slices. Thus, we think that a wall, which may be constructed of rubble or loose stones, can be found at this location. This thick wall may be situated between depth of 0.5 and 1 m, and it could be a retaining wall of this structure. The second anomalous zone (2) indicates a lossy medium in the GPR depth slice, although it shows a conductive character in the ERT slice. At the same time, this zone shows a low Vp and Vs velocity distribution in the SRT and MASWT slices. These results indicate that compacted clay material should be found in this location. The anomaly in section (3) has an arc-shaped character, and it shows clearly in the magnetic image and the GPR result at the same location. Furthermore, the changes are as well partially observed in the ERT, SRT, and MASWT slices. In the deeper part, one interesting anomalous zone (4) gives us a useful indication about the assumed burial structure within the Taşdöşem superstructure (Fig. 2.22b). This structure consists of two separate parts, displaying a geometrical character in the ERT slice (2.82–3.55 m). The structure revealed between 35 and 55 m inline coordinate consists of two different resistive bodies, which are separated from each other with a zone of intermediate resistivity. The GPR depth slice at depth of 2–2.5 m shows the lossy, absorbing character at this location. Two characteristic separated structures can be clearly seen in the MASWT slice at depth of 1.78–3.4 m. The NW part of the area shows high Vs values, whereas the southern and SE parts display low Vs velocities. This result provides us with useful information for the definition of the different characteristics of the subsurface. In the SRT slice (2.43–3.12 m), the characteristic of structure partially resembles the target structure in the ERT slice. The mentioned separation is also clearly visible in this slice. We conclude that, according to the results of this integrated geophysical study, an extensive structure has evidently been manifested within the Taşdöşem superstructure between 35 and 55 m inline coordinate of the investigation area (Drahor et al. 2015a).

This structure appears to be formed by two separate parts (red dashed line), isolated with a wall and possibly a passage. The target structure has intermediate to high resistivity values (40 and 120 Ωm) and Vs velocities (190–210 m/s), whereas it displays low-to-intermediate P-wave velocities (370–500 m/s) and a low radar reflection signature expressed by an absorbing medium (Fig. 2.22b). At the same time, a distinct positive magnetic anomaly is observed at the same location (Fig. 2.21a). Therefore, we believe that a large monumental structure, possibly covered with clayey materials, certainly is located here between depth of 2.5 and 5 m (Drahor et al. 2015a).

The integrated approach gave us a plenitude of illustrative and integrative interpretable data, compared with the simple use of single geophysical techniques. The integrated approach could simultaneously reveal relationships between the anomalies produced by conductivity, dielectricity, magnetic, and seismic (V_p and V_s) measurements. Therefore, we can possibly derive novel knowledge about subsurface features buried within an archaeological site. This study presents the usefulness of the integrated prospection approach for archaeological site investigation making use of the magnetic gradiometry, GPR, ERT, SRT and MASWT techniques. The integrated results indicate that the geophysical methods used at the archaeological site of Šapinuwa can provide archaeologists with valuable information and plans about the site and its contained structures. By making use of tomographic techniques, we can obtain various images and depth slices that will prove to be useful for the selection of targeted excavation locations, rendering the excavations more cost and time efficient for the involved archaeologists (Drahor et al. 2015a).

The integrated geophysical investigations carried out in the Taşdöşem sacred area in the archaeological site of Šapinuwa provide us with valuable information about the location of anomalies caused by presumably buried architecture. The results highlighted the reliability of the integrated prospection approach, particularly during the interpretation stage involving various physical parameters, revealing different sensitivities and resolution characteristics regarding subsurface imaging. The ERT gave us a more realistic, distinctive, and complete image than the other used techniques, in regard to quantitative and qualitative characterization of the subsurface. The GPR data did not clearly present sufficient information on the deeper archaeological structure(s), except for artificial and geological units due to the presence of surface irregularities and highly conductive material that caused a strong attenuation of the GPR signal. Magnetometry revealed information on surface irregularities and the covering soil regarding their magnetic properties, but it failed to provide answers concerning deeper structures. MASWT surprisingly generated more information on the shallow and deeper archaeological structures in the Taşdöşem construction. This is an important conclusion concerning the detection ability of this technique regarding archaeological structures. According to the MASWT results, a wall-type structure located in the shallow part of the area has evidently been determined, whereas the fulfilling results have been concerned with the subsurface characteristics in deeper part of the Taşdöşem in MASWT slicing representations. Seismic P-wave tomography presented a reasonable result concerning the archaeological remains in the subsurface. The observed intermediate V_p velocities could correspond to an embedded structure, such as a burial chamber filled with air. This assumption has also been strengthened by the measurement of high induced resistivity values (Drahor et al. 2015a).

This study revealed that the integrated approach of a combination of geophysical prospection methods enabled a better definition of the position, localisation, depth, thickness, extension, and physical characteristics of buried archaeological structures as well as the geological substratum at archaeological sites comparable to that of Šapinuwa.

2.3.4.2   City Area (Tepeleraras*ı*)

Tepeleraras*ı* is a highly problematic area due to three important causes for geophysical surveys. These are; the influences of major earthquakes, which were constituted the complex soil distribution, had been affected the city during Hittite times. This complication may directly influence the results of geophysical investigations. The archaeological context in the subsurface of the city was extremely affected by the intense and extensive fire possibly occurred from the big Earthquakes. This effect may have caused confusing magnetic anomalies in the area. In addition, the archaeological structures are extremely close to the surface. Therefore, the archaeological context has importantly been affected from the intensive modern agricultural activity, which may cause spurious undesirable anomalies in geophysical surveys. Thus the integrative geophysical approach was here preferred to explore the buried archaeological context and to reveal the city plan. To this purpose, magnetic gradiometry, Ground

Penetrating Radar (GPR), Electrical Resistivity Tomography (ERT), Seismic Refraction Tomography (SRT) and Multichannel Analysis of Surface Wave Tomography (MASWT) techniques were used in the Tepelerarası area of Šapinuwa archaeological site (Fig. 2.23).

Gradiometric magnetic data were collected in all area showed by grey shading in Fig. 2.23. This data were processed using standard correction and evaluation techniques that apply in magnetic gradiometric surveys in archaeological investigations. A great number of traces of archaeological structures found much closed to the surface are clearly observed in this image (Fig. 2.24). In addition, the effects of agricultural cultivations are also seen in this image, mostly in the northern and northeastern part of the investigation area. These effects are dominantly masked the observable magnetic anomalies. In any case, the important anomaly groups that show the big structures in the subsurface are also clearly appeared in the image. The significant anomalies are enclosed by red circles and ellipses in image presented in Fig. 2.24. The burning areas are evidently observed in the image, and they are enclosed by yellow circles and ellipses. In addition, the field drainages and the borders are clearly displayed in magnetic image as positive and negative anomalies according to their geometrical properties. The area encircled by orange ellipse is very complex and there is not any interpretable anomaly in this area. The burned archaeological structures and other materials are observed as positive magnetic anomalies, while the non-magnetic materials such as the walls produced by limestone and similar materials are showed as negative magnetic anomalies in this image. According to regular positive and negative consecutive magnetic anomalies, the area might be has a systematically city plan. This conclusion is highly important result to understand the architectural plan of the Hittite cities (Drahor et al. 2015b).

GPR were obtained by blue, yellow and orange colour frames that show the investigation sites explored in 2013, 2014 and 2015, respectively (Fig. 2.23). Overall GPR data were combined and then processed by standard GPR processing stages. After that the GPR interpretation of whole area was carried out using different depth slices. There important depth slices are given in Fig. 2.25. First depth slice of GPR is shown in Fig. 2.25a, and it displays the region showed between 22 and 33 cm under the surface. In this depth slice, the archaeological structures are almost appeared in encircled area with blue circle. In this area, the radar reflection traces from the subsurface structures demonstrated by dark tones are clearly observed in this area. The high reflective anomalies reveal that the archaeological structures found in very shallow depths are oriented in NW-SE and NE-SW direction. This orientation is suited the typical Hittite city plan in Šapinuwa archaeological site. The GPR anomalies show that the important archaeological structures should be found in near surface depths in this area. In this depth slice, the yellow circle and ellipse is showed the results of GPR study performed in 2014. The GPR anomalies are observable in this slice with similar orientations, but they have lower amplitudes. Also, the anomaly complexity is more visible in this image. The red ellipses shows the important anomalies obtained from 2015 GPR surveys. In the southern part of the investigated area, the GPR reflections have lower amplitudes than the others. However, some thin traces are also observed with similar orientations in this image. Moreover, the GPR anomalies in the eastern part are more observable than the others, and they have also high amplitudes. Second depth slice indicates the region of 54–65 cm depths. In this depth slice, the GPR signals attenuated in the northern part are enclosed by blue circle. Also similar attenuation is observed in other parts of the GPR image, except for the southern part of the investigation area. The visibility is increased in this area, and some regular traces with high amplitudes are appeared (Fig. 2.25b). Therefore, we can determine that some regular structures are observed in this part, and they should show the existence of some important archaeological relics. The last slice is related with the region found between 97 and 108 cm depths. In this depth slice, the amplitudes of the anomalies observed in northern part of the area are importantly decreased.

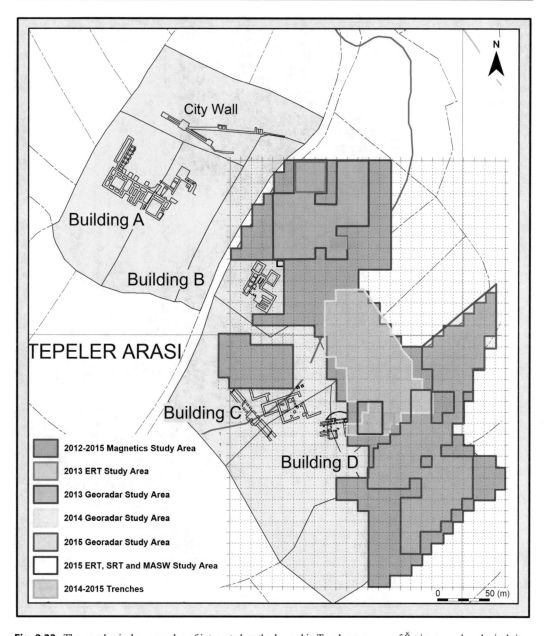

Fig. 2.23 The geophysical survey plan of integrated methods used in Tepelerarası area of Šapinuwa archaeological site

However, some regular reflection traces are also shown. These traces should be related with the bottom of this archaeological layer. However, some GPR traces with the similar orientation and the regular form are also observed in the area enclosed with yellow circle. This result shows the presence of archaeological structures in this depth. Also the similar situation is clearly seen in the southern part of the investigation site (Fig. 2.25c).

The ERT, SRT and MASWT studies were carried out in some localities in the investigation site of 2014. These investigation localities are presented in Fig. 2.23. The results of these studies will be given in following section together with the archaeological interpretation obtained from the integrated results.

Fig. 2.24 The large-scale magnetic image of Tepelerarası area in Šapinuwa archaeological site. The data were corrected and processed standard signal and image processing techniques

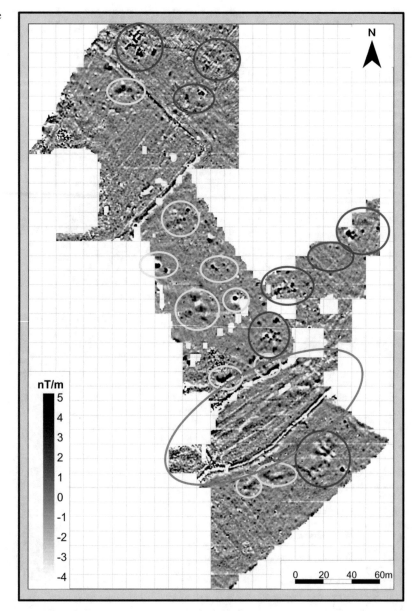

2.4 Archaeological Interpretations and Excavations According to Results of Integrated Investigations

The archaeological interpretation is thoroughly described the subsurface archaeological context according to geophysical results before the excavations. It's succeeding depends on the selection and number of the methods as well as the measuring procedures and quality, integrated GPS usage, soil conditions and the unwanted noises created by the presence of buried actual metal materials, ploughing, forbidden excavation, earthquake damage and etc. The integrated usage of geophysical prospection methods is importantly increased the success of true results in archaeological prospection. As we know, the

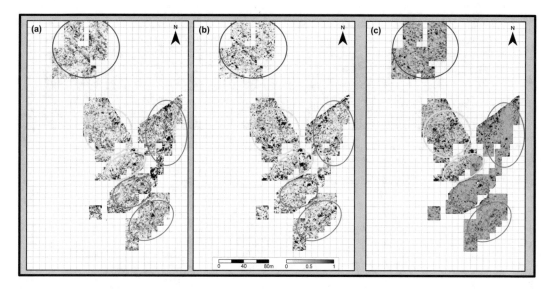

Fig. 2.25 The corrected and processed large-scale GPR depth slices of Tepelerarası area in Šapinuwa archaeological site. (**a**) 22–33 cm, (**b**) 54–65 cm, (**c**) 97–108 cm

used geophysical method in integrated approach gives us different information about the subsurface characterization of investigated archaeological sites. Therefore, we determine the characterization of burial archaeological context according to magnetic properties, electrical potential, electromagnetic wave propagation and seismic velocity changes. Thus, the materials that contain magnetic and/or non-magnetic properties, the structures with high, medium and low resistivity, the reflection characteristics of reflected layer, the seismic velocity features of buried archaeological relics and their surrounds might be obtained in detail. This elaborative approach could be enabled to constitute of conceptual model and detailed plan of investigated archaeological site. Therefore, the archaeologists could be constituted more detailed and high speed excavation plans. In addition, the direct determination of optimum excavation locations using integrated geophysical surveys is decreased the extensive excavation procedure in archaeology. Thus the damage caused by speed excavation process might be importantly decreased in determination of archaeological context. Ultimately the archaeologists will be studied in limited excavation areas determined by integrated solutions of geophysical methods, and so that they could also

be obtained the maximum information about the subsurface characterization of investigation sites. To better understand the usefulness of this approach, the results of integrated geophysical surveys and the related excavation outcomes obtained three archaeological sites in Turkey will be presented in following section.

2.4.1   Legion IV Scythica in Zeugma

In this study an archaeological interpretation was performed based on magnetic gradiometry and ERT surveys. Thus, the settlement plan was implemented according to detailed archaeological interpretation obtained by geophysical results (Fig. 2.26). The archaeological interpretation gives us very regular and systematic settlement plan about the Military installation site and its surrounds. In the settlement plan, some insulae with NE–SW and NW–SE orientations are shown in the northeastern part of the investigation site, and they are separated from each other by side and main streets perpendicular to each other. In the southeastern part of the area, the archaeological structures have two different orientations (WNW–ESE and WSW–ENE). The traces of archaeological features constituted from integrated geophysical results

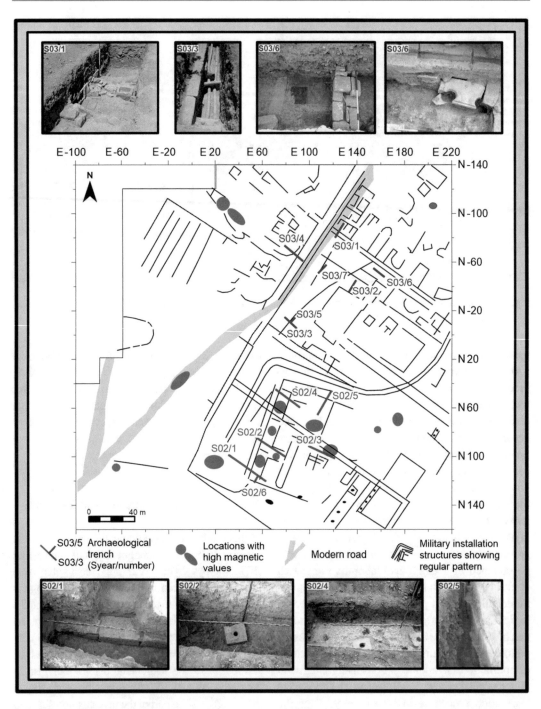

Fig. 2.26 Archaeological interpretation of large-scale geophysical data. Diagram shows the drawing of buried military structures and the locations of high magnetic anomalies in the investigation area. The photos taken from excavation trenches are displayed in upper and bottom section (modified from Drahor et al. 2008a)

have a very regular plan in this section, and it should be showed the plan of a typical administrative building of the Roman legionary camp area.

But some differences originated from the directions of the buried structures can also see in the plan. These differences reveal that the presence of two different legion habitations, which may remain from different periods, should exist in the southern section of the survey site. Trial archaeological trenching studies verified and clarified the geophysical results in the area. The stratigraphy obtained from different trenches shows the existence of different periods inside the construction. In particular, the significant amount of Roman military equipment (arrow- and spearheads, armour scales, mail shirts, etc.) as well as five coins, point towards a date between the end of the first and the middle of the third century AD. The military equipment from the trenches as well as the numerous stamped tiles from the surface of the area investigated proves the military character of these archaeological remains (Drahor et al. 2008a).

2.4.2 Sardis Archaeological Site

The geophysical results of integrated explorations in Sardis archaeological site are well correlated among the used techniques. This study proved the importance and role of integrated geophysics in rather complex archaeological sites such as Sardis. Overall integrated results are compared in Fig. 2.26. The gradiometer study shows that this area could be an agora of Sardis due anomaly characteristics of gradiometer results and its dimensions (app. 100 × 100 m) (Fig. 2.27a). ERT results performed in the middle part of the investigation area revealed the magnetic gradiometer results (Fig. 2.27b, c). As can be seen from the figure, a good harmony between the ERT and the magnetic images is seen obviously. ERT model and depth slice enabled the definition of burial depths and true resistivities of the structures. In ERT depth slices, the lateral extensions of the buried structures are oriented in N–S and E–W directions (Fig. 2.27c). The VLF-R investigation provided unexpectedly informative and supportive results, which were clearly supported by the results of other geophysical investigations (Fig. 2.27d, e). The results of VLF-R studies pointed out that this technique might be useful to detect the subsurface features in archaeological sites. Moreover, seismic refraction tomography investigation implemented on the

test lines revealed that technique could give useful outcomes in archaeological site investigations. Particularly, the SRT model section of SLINE3 was given very compatible results with other used geophysical techniques (Fig. 2.27f). This result shows the availability of SRT application in archaeology. The results of Sardis investigations presented that the major archaeological structures might be mostly found in shallow depths. In addition, a good comparison is observed for overall applied geophysical techniques, and the results of them clearly support each other. Therefore, we recommend that the integrated application of geophysics is highly simplified the archaeological interpretation in means of excavation strategy (Drahor 2006).

After the integrated geophysical survey, an archaeological test excavation was conducted on two trenches in 2002. The first trench was located at the edge of terrace, on the north side and towards its west, Trench F 55 02.1; the other one is in the central part of the terrace, Trench F 55 02.2 (Fig. 2.28). The dimensions of first trench were 4 × 10 m (at E701–705/S120–130 on the grid). The trench contained a variety of occupation surfaces, walls, terracotta pipe conduits, fills and dumps. Due to limited exposure, their functional significance is unclear. The principal architectural feature of the trench is a fieldstone and mortar wall (Fig. 2.28a). Oriented east–west, the wall is 1.80 m thick and stands exposed to a height of 4 m on its north side. The bottom of the wall was not located. Trench F 55 02.2 was 4 × 4 m (at E722–726/S167–171 on the grid). Eleven stratified fills were recognized in excavation. They contained varying quantities of ceramic items (pottery, lamps, terracottas), bones and coins (35 total). The upper five fills, totalling 1.36–1.80 m thick, were deposited in the seventh century AD. Upper fills probably were deposited by erosion. Three walls of modest size and simple construction (fieldstone, tile, mud mortar) appeared in the upper part of the trench, and are probably from the fifth or sixth century AD. The two fissures, which extend through the trench in a roughly east–west orientation, have a maximum width of 10 cm and were exposed to a depth of 2.5 m (including a depth of at least 1 m

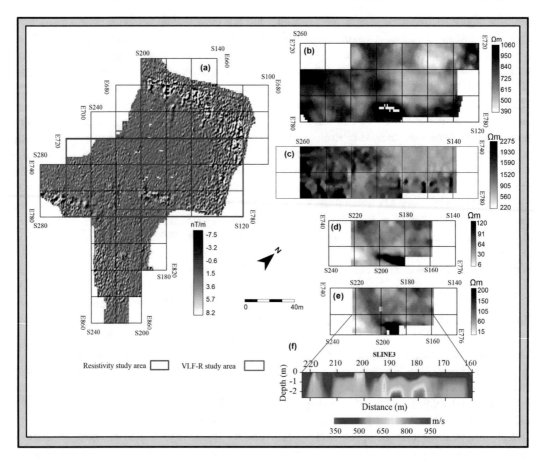

Fig. 2.27 The comparative figure indicates the overall geophysical results. (**a**) Gradiometer relief image. (**b**) Apparent resistivity image of L = 2 m. (**c**) The depth slice at 1.3 m depth of ERT study. The images of VLF-R data (**d**) for 19.6 kHz, (**e**) for 23.4 kHz. (**f**) The SRT model section of SLINE3. The investigation areas of ERT and VLF-R are indicated the *blue* and *red rectangles* on the gradiometer image, respectively (modified from Drahor 2006)

below the lowest excavated level). They evidently came into existence in the seventh century AD, and may have been the result of earthquake (Fig. 2.28b) (Greenewalt 2003). The excavation results showed that the integrated geophysical approaches would be very useful in obtaining satisfactory and more informative results from archaeological sites (Drahor 2006).

2.4.3   Šapinuwa Archaeological Site

The best comparison of the results of integrated geophysical explorations in Šapinuwa archaeological site is presented in Fig. 2.29. The well correlated

results among the used geophysical techniques are obtained in this study. The area was selected to test the results of integrated geophysical techniques (magnetic scanning, GPR, ERT, SRT and MASWT). Therefore, this study proved the importance and role of integrated geophysics in very complex archaeological areas such as Šapinuwa. The overall integrated results of geophysical investigations are given in Fig. 2.29 together with excavation result. The investigation plan of the test site is presented in Fig. 2.23. The magnetic gradiometry result of this site and its environs are clearly observed in Fig. 2.29a. There are two important anomalous zones in this area. One of them is enclosed with red circle, and the anomalies have

Fig. 2.28 Archaeological trenches conducted by excavations in 2002. The archaeological structures found in very shallow depths. (**a**) Trench F55 02.1, (**b**) Trench F55 02.2. The trenches contained a variety of occupation surfaces, walls, terracotta pipe conduits, fills and dumps. The fissures, which extend through the trench in a roughly east–west orientation, have a maximum width of 10 cm and they may have been the result of an important earthquake according to archaeologist (from Drahor 2006)

generally complicated image that shows positive and negative magnetic complexity. However, the direction of these anomalies is oriented in NE-SW and NW-SE. Another anomaly has a positive character with NE-SW direction, and it is enclosed with yellow ellipse (Fig. 2.29a). According to magnetic results, this structure should be consisted of burnt materials such as mud brick or similar any material. In addition, the first anomaly group could be showed some shallow structures that have extremely complex character including burnt materials. The image of GPR study performed in this site is given in Fig. 2.29b. The first structural group enclosed with red circle is slightly visible in GPR depth slice taken from 22 to 33 cm depths. The extensions of radar anomalies with high reflection are generally similar together with magnetic anomalies. Second anomaly enclosed with yellow ellipse has a high reflective character, but its form is different from the magnetic anomaly. Therefore we can think that the structure might be showed a fragmental character. The area was investigated with 2D ERT technique, and the collected data were processed by 3D robust inversion technique. Therefore the images from different depth slices were constituted to investigate the subsurface

archaeological structures (Fig. 2.29c). The first slice is displayed the section between 0 and 25 cm depths. The resistivity is changed among 13 and 70 Ωm. High resistivities are found in certain locations, and the anomalies are generally oriented NW-SE and NE-SW directions as well as magnetic and GPR images. The high resistivities encircled by red circle have very good harmony with magnetic image. Therefore, we can think that the archaeological structures are mostly buried in very near surface. The following slice from 25 to 50 cm is also similar with the first slice. The third slice from 50 to 110 cm is generally similar with previous slice, but the resistivities are slightly decreased in some locations. In the last slice, the resistivities are importantly decreased, and the anomaly forms are considerably changed. This result shows that the archaeological structures in the investigation site are mostly located between 0 and 1 m depths (Fig. 2.29c). In 2015 investigations, the SRT and MASWT studies were carried out in same area. The tomographic results of line 20 measured during these surveys are presented in Fig. 2.29e, f. To compare these results with ERT, the 2D ERT model section is also given in Fig. 2.29d. In the ERT model, the resistivities are generally changed

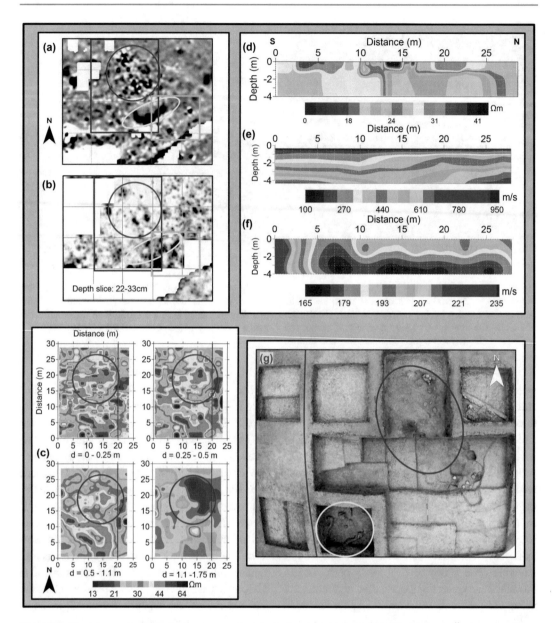

Fig. 2.29 The integrated geophysical and excavation results in a test site in Tepelerarası region in Šapinuwa archaeological site. The *red line* shows the Line 20 in the investigation site

between 13 and 40 Ωm, and they have medium resistivity value for Šapinuwa archaeological site. The medium resistivities are almost located in 0–1 m depths except for the results obtained from 0 to 13 m of the line, which are continued to deeper parts of the investigation line. The low resistivities are changed among 13–21 Ωm, and they should display the unburned mud-brick depositions with high water content. The SRT section is given in

Fig. 2.29e. The seismic P waves are varied between 100 and 700 m/s, and low velocities (100–300 m/s) are generally situated from the surface to 1 m depth. The medium velocities (300–500 m/s) are shown similar characteristic with low velocities. The low and medium velocities go down deep in southern part of the line. The high P wave velocities (500–800 m/s) are closer to the surface in the northern part, and the variation among low and medium velocities of P wave are moderately in accordance with the resistivity changes. The P wave results indicate that the archaeological structures are mostly extended to deeper part of the area, and their seismic P velocities have low and medium values due to the complex and porous character of the burial materials. The MASWT section is presented in Fig. 2.29f. The S values calculated from MASW data are changed between 156 and 240 m/s. The model has generally low S velocity character except for southern part of the line. This increasing should be related to an archaeological structure appeared in ERT, GPR and magnetic gradiometry results. In MASWT section, the medium S velocities are extended 2 m depths, and then the velocities are decreased in the deeper part of the line. In addition, the low velocities appeared between 4 and 10 m of the line are continued to end of the section. After the integrative geophysical survey, the archaeologists performed an archaeological excavation in area with 20 × 15 m dimensions in the test site. The latest photo of the excavation site is given in Fig. 2.29g. During the excavation, a very solid mud brick base structure was explored from 20 cm to 1 or 1.5 m depths in different regions. Some small wall fragments and stones are appeared in between 1 and 2 m depths of the excavated area encircled by yellow circle. This result is rather compatible with the geophysical conclusions. In addition, a burned zone enclosed with blue ellipse is revealed in 1 m depth approximately. The effect of this zone is shown a small positive anomaly in magnetic image, while it is appeared as negative amplitude in GPR result (Fig. 2.29a, b, g).

This study indicated that the integrated approach of different geophysical techniques in archaeological sites at Šapinuwa enabled better definition of position, localisation, depth, thickness, extension and the physical characteristics of buried archaeological structures and their geological context. Therefore, we can conclude that, magnetic gradiometry provides valuable knowledge about the shallow archaeological remains and the covering soil by means of their magnetic properties; ERT enables determination of the location, extension, depth, thickness and electrical characteristics of subsurface features including geological units; GPR revealed the presence of archaeological remains; seismic P tomography supplies additional information about the buried archaeological structures; MASWT detects the lateral and vertical Vs velocity changes of archaeological context. As a consequence, the integrated approach of different geophysical techniques provides an important tool for the archaeologists to inform their detailed and extensive excavation.

2.5 Conclusions

This chapter addresses the capability of integrated applications of geophysical methods and their usefulness in non-destructive site investigation of archaeological sites. In archaeological sites, soil has extremely importance, and it shows majorly complex character with a very large variability. Its characterisation is required the integrated approach of geophysical investigation for identification of archaeological sites. The integrated geophysical investigations carried out in Roman Legionary Camp sites; Satala and Zeugma, Amorium Byzantine Military Compound, Sardis and Šapinuwa archaeological sites provide us with valuable information about the location of anomalies caused by presumably buried architecture of archaeological sites. The results highlighted the reliability of the integrated prospection approach, particularly during the interpretation stage involving various physical parameters, revealing different sensitivities and resolution characteristics regarding subsurface imaging. These studies showed that magnetometry studies revealed information on surface irregularities and the covering soil regarding their magnetic properties, but it failed to provide answers concerning deeper structures. Particularly this method can give us valuable results to determine the burned zones and structures. The GPR generally gave sufficient information from the surface to the deeper parts, except for artificial and geological units due to the presence of surface irregularities and highly conductive material that caused a strong attenuation of the GPR signal. The ERT presented us a more realistic, distinctive, and complete images to define the quantitative and qualitative characterization of the subsurface together with other used techniques. The Seismic P-wave tomography presented a reasonable result concerning the archaeological remains in the subsurface. The SP application introduced a successful result in investigation of archaeological structures. Particularly, the burnt structures and some wall artefacts could be determined after SP studies, and the results supported to other geophysical applications. VLF-R given us productive results located in near surface structures in archaeological sites. MASWT surprisingly generated more information on the shallow and deeper archaeological structures in the investigation sites. This is an important conclusion concerning the detection ability of this technique regarding archaeological structures. According to the MASWT results, a wall-type structure located in the shallow part in Taşdöşem, Šapinuwa has evidently been determined.

As a final conclusion, it should determine that, integrative research applications of geophysics in archaeological sites should increase using multimethodical and more systematically approaches by many researchers.

Acknowledgements The author acknowledges the GEOIM LTD Company for supporting of some data and results. Also I am grateful to Atilla Ongar, who is a research assistant in Dokuz Eylül University in Department of Geophysics, assisted during preparing of overall figures.

References

Arato A, Piro S, Sambuelli L (2015) 3D inversion of ERT data on an archaeological site using GPR reflection and 3D inverted magnetic data as a priori information. Near Surf Geophys 13:545–556

Baranwal VC, Franke A, Börner RU, Spitzer K (2011) Unstructured grid based 2-D inversion of VLF data for models including topography. J Appl Geophys 75:363–372

Cardarelli E, Di Filippo G (2004) Integrated geophysical surveys on waste dumps: evaluation of physical parameters to characterize an urban waste dump (four case studies in Italy). Waste Manag Res 22:390–402

Cardarelli E, Di Filippo G (2009) Integrated geophysical methods for the characterisation of an archaeological site (Massenzio Basilica–Roman forum, Rome, Italy). J Appl Geophys 68:508–521

Claerbout JF, Muir F (1973) Robust modeling with erratic data. Geophysics 38:826–844

Dabas M, Hesse A, Tabbagh J (2000) Experimental resistivity survey at Wroxeter archaeological site with a fast and light recording device. Archaeol Prospect 7:107–118

De Domenico D, Giannino F, Leucci G, Bottari C (2006) Integrated geophysical surveys at the archaeological site of Tindari (Sicily, Italy). J Archaeol Sci 33:961–970

Diamanti N, Tsokas G, Tsourlos P, Vafidis A (2005) Integrated interpretation of geophysical data in the archaeological site of Europos (northern Greece). Archaeol Prospect 12:79–91

Drahor MG (2004) Application of the self-potential method to archaeological prospection: some case studies. Archaeol Prospect 11:77–105

Drahor MG (2006) Integrated geophysical studies in the upper part of Sardis archaeological site, Turkey. J Appl Geophys 59:205–223

Drahor MG (2011) A review of integrated geophysical investigations from archaeological and cultural sites under encroaching urbanisation in Izmir, Turkey. Phys Chem Earth 36:1294–1309

Drahor MG, Akyol AL, Dilaver N (1996) An application of the self-potential (SP) method in archaeogeophysical prospection. Archaeol Prospect 3:141–158

Drahor MG, Göktürkler G, Berge MA, Kurtulmuş TÖ, Tuna N (2007) 3D resistivity imaging from an archaeological site in south-western Anatolia, Turkey: a case study. Near Surf Geophys 5:195–201

Drahor MG, Berge MA, Kurtulmuş TÖ, Hartmann M, Speidel MA (2008a) Magnetic and electrical resistivity tomography investigations in a Roman Legionary camp site (Legio IV Scythica) in Zeugma, southeastern Anatolia, Turkey. Archaeol Prospect 15:159–186

Drahor MG, Kurtulmuş TO, Berge MA, Hartmann M, Speidel MA (2008b) Magnetic imaging and electrical resistivity tomography studies in a Roman Military installation found in Satala archaeological site from northeastern of Anatolia, Turkey. J Archaeol Sci 35:259–271

Drahor MG, Berge MA, Öztürk C, Alpaslan N, Ergene G (2009) Integrated usage of geophysical prospection techniques in Höyük (tepe, tell)-type archaeological settlements. ArchaeoSciences, revue d'archéométrie 33 (suppl):291–293

Drahor MG, Berge MA, Öztürk C (2011) Integrated geophysical surveys in the Agios Voukolos church, Izmir, Turkey. J Archaeol Sci 38:2231–2242

Drahor MG, Berge MA, Öztürk C, Ortan B (2015a) Integrated geophysical investigations at a sacred Hittite Area in Central Anatolia, Turkey. Near Surf Geophys 13:523–543

Drahor MG, Öztürk C, Ortan B, Berge MA, Ongar A (2015b) A report on integrated geophysical investigation in the Šapinuva archaeological site in Central Anatolia of Turkey. Geoim LTD. 2015ARKEO1-05, 90 p (in Turkish)

Emre Ö, Duman TY, Özalp S, Elmacı H, Olgun Ş (2011) 1/250,000 Scale active fault map series of Turkey, Çorum (NK 36-16) quadrangle. General Directorate of Mineral Research and Exploration, Ankara, Turkey

Gaffney CF, Gater JA, Linford P, Gaffney VL, White R (2000) Large-scale systematic fluxgate gradiometry at the Roman city of Wroxeter. Archaeol Prospect 7:81–99

Gaffney V, Patterson H, Piro S, Goodman D, Nishimura Y (2004) Multimethodological approach to study and characterize Forum Novum (Vescovio, Central Italy). Archaeol Prospect 11:201–212

Gaffney C, Harris C, Pope-Carter F, Bonsall J, Fry R, Parkyn A (2015) Still searching for graves: an analytical strategy for interpreting geophysical data used in the search for "unmarked" graves. Near Surf Geophys 13:557–569

Greenewalt CH (2003) Archaeological exploration of Sardis, 2002 field season. Report to the General Directorate of Monuments and Museums, Ministry of Culture, Republic of Turkey

Hartmann M, Speidel MA (2003) The Roman army at Zeugma: recent research results. In: R. Early et al., Zeugma: Interim Reports. J Roman Archaeol 51 (suppl):101–126

Kvamme KL (2006) Integrating multidimensional archaeological data. Archaeol Prospect 13:57–72

Leucci G, Negri S (2006) Use of ground penetrating radar to map subsurface archaeological features in an urban area. J Archaeol Sci 33:502–512

Leucci G, Greco F, De Giorgi L, Mauceri R (2007) Three-dimensional image of seismic refraction tomography and electrical resistivity tomography survey in the castle of Occhiolà (Sicily, Italy). J Archaeol Sci 34:233–242

Lightfoot CS (1991) Archaeological surveys at Satala in 1989. In: The proceeding of VIIIth meeting of research results of symposium of excavation, survey and archaeometry, Ankara, Turkey, May 28 and June 1, 1990, pp 299–309 (In Turkish)

Lightfoot CS (1997) Excavations at Amorium in 1996. Bull Br Byzantine Stud 23:39–49

Lightfoot CS, Arbel Y (2003) Amorium excavation 2001. In: XXIVth symposium on excavation results proceedings, pp 521–532 (In Turkish)

Linford N, Linford P, Payne A (2015) Chasing aeroplanes: developing a vehicle-towed caesium magnetometer array to complement aerial photography over three recently surveyed sites in the UK. Near Surf Geophys 13:623–631

Matera L, Noviello M, Ciminale M, Persico R (2015) Integration of multisensor data: an experiment in the archaeological park of Egnazia (Apulia, Southern Italy). Near Surf Geophys 13:613–621

Monteiro Santos FA, Mateus A, Figueiras J, Gonçalves MA (2006) Mapping groundwater contamination around a landfill facility using the VLF-EM method—a case study. J Appl Geophys 60:115–125

MTA (2002) 1/500.000 Scale Geological Maps of Turkey, 7, General Directorate of Mineral Research and Exploration (MTA), Ankara, Turkey

Neubauer W, Eder-Hinterleitner A (1997) Resistivity and magnetic of the Roman town Carnuntum, Austria: an example of combined interpretation of prospection data. Archaeol Prospect 4:179–189

Papadopoulos N, Sarris A, Yi MJ, Kim JH (2009) Urban archaeological investigations using surface 3D ground penetrating radar and electrical resistivity tomography methods. Explor Geophys 40:56–68

Papadopoulos NG, Sarris A, Salvi MC, Dederix S, Soupios P, Dikmen U (2012) Rediscovering the small theatre and amphitheatre of ancient Ierapytna (SE Crete) by integrated geophysical methods. J Archaeol Sci 39:1960–1973

Piro S, Mauriello P, Cammarano F (2000) Quantitative integration of geophysical methods for archaeological prospection. Archaeol Prospect 7:203–213

Pullammanappallil SK, Louie JN (1994) A generalized simulated annealing optimization for inversion of first arrival times. Bull Seismol Soc Am 84(5):1397–1409

Sarris A, Papadopoulos NG, Agapiou A, Salvi MC, Hadjimitsis DG, Parkinson WA et al (2013) Integration of geophysical surveys, ground hyperspectral measurements, aerial and satellite imagery for archaeological prospection of prehistoric sites: the case study of Vésztő-Mágor Tell, Hungary. J Archaeol Sci 40:1454–1470

Seren S, Eder-Hinterleitner A, Neubauer W, Groh S (2004) Combined high resolution magnetics and GPR surveys of the roman town of Flavia Solva. Near Surf Geophys 2:63–68

Sharma SP, Kaikkonen P (1998) Two-dimensional non-linear inversion of VLF-R data using simulated annealing. Geophys J Int 133:649–668

Simon F-X, Kalayci T, Donati JC, Cuenca Garcia C, Manataki M, Sarris A (2015) How efficient is an integrative approach in archaeological geophysics? Comparative case studies from Neolithic settlements in Thessaly (Central Greece). Near Surf Geophys 13:601–611

Simyrdanis K, Papadopoulos N, Kim JH, Tsourlos PI, Moffat I (2015) Archaeological investigations in the shallow seawater environment with electrical resistivity tomography. Near Surf Geophys 13:601–611

Speidel MA (1998) Legio IV Scythica: its movements and men. In The Twin Towns of Zeugma on the Euphrates. Rescue work and historical studies, Kennedy D (ed.). J Roman Archaeol 27(suppl):163–204

Tsokas GN, Giannopoulos A, Tsourlos P, Vargemezis G, Tealby JM, Sarris A, Papazachos CB, Savopoulou T (1994) A large scale geophysical survey in the archaeological site of Europos (N. Greece). J Appl Geophys 32:85–98

Tsokas GN, Tsourlos PI, Vargemezis G, Novack M (2008) Non-destructive electrical resistivity tomography for indoor investigation: the case of Kapnikarea church in Athens. Archaeol Prospect 15:47–61

Tsokas GN, Kim JH, Tsourlos PI, Angistalis G, Vargemezis G, Stampolidis A, Diamanti N (2015) Investigating behind the lining of the Tunnel of Eupalinus in Samos (Greece) using ERT and GPR. Near Surf Geophys 13:571–583

Vafidis A, Economou N, Ganiatsos Y, Manakou M, Poulioudis G, Sourlas G et al (2005) Integrated geophysical studies at ancient Itanos (Greece). J Archaeol Sci 32:1023–1036

Wagner J (1976) Seleukeia am Euphrat/Zeugma. Karte II

Wessel P, Smith WHF (1995) New version of the generic mapping tools released. EOS Trans Am Geophys Union 76:329

Wilken D, Wunderlich T, Stümpel H, Rabbel W, Pašteka R, Erkul E, Papčo J, Putiška R, Krajňák M, Kušnirák D (2015) Case history: integrated geophysical survey at Katarínka Monastery (Slovakia). Near Surf Geophys 13:585–599

Application of Tensorial Electrical Resistivity Mapping to Archaeological Prospection

3

Mihály Varga, Attila Novák, Sándor Szalai, and László Szarka

Abstract

In an archaeological site (Pilisszentkereszt Cistercian Monastery, Hungary) we carried out 3D tensorial geoelectric mapping measurements. We applied the well known tensorial form of Ohm's differential law, where a 2×2 resistivity tensor relates the horizontal current density vector and the corresponding electric field vector. In the DC apparent resistivity tensor there are three independent rotational invariants, and we defined two alternative sets. In the field two perpendicular AB directions were used, and $16 \cdot 15 = 240$ potential electrodes (with an equidistant space of $\Delta_x = \Delta_y = 50$ cm) were put in the central (nearly squared, 7.5 m $\cdot$ 7 m) area between the current electrodes. Due to a four-channel measuring system, it was possible to determine both components of a horizontal electric vector at the same time. The time needed to measure all potential differences between the neighbouring potential electrodes (thus to obtain $15 \cdot 14 = 210$ resistivity tensors), was about 40 min. The tensorial results are shown together with the results of traditional measurements. Man-made origin anomalies as a subsurface channel, building remnants, a furnace and an ancient road have been discovered and described. In field conditions, any resistivity estimation provides reliable information about the subsurface (both the tensor invariants and the traditional mean values). At the same time, the multidimensional (2D and 3D) indicators proved to be informative only in case of significant subsurface inhomogeneities.

Keywords

Tensor $\cdot$ Resistivity $\cdot$ Mapping $\cdot$ Archaeology

3.1 Introduction

The Cistercian Monastery in Pilisszentkereszt (30 km N of Budapest, Hungary) was founded by Béla III of Hungary in 1184. The buildings were destroyed during the Ottoman period, only some wall remnants and ruins of the cathedral pillars still can be seen on the surface. (For a more detailed historical description see Gerevich 1977). In co-operation with the Archaeological Institute of Hungarian Academy of Sciences, we

M. Varga
KBFI-Triász Kft., Budapest, Hungary

University of West-Hungary, Sopron, Hungary
e-mail: varga1m@digikabel.hu

A. Novák · S. Szalai (✉) · L. Szarka
Geodetic and Geophysical Research Institute of the Hungarian Academy of Sciences, Sopron, Hungary

University of West-Hungary, Sopron, Hungary
e-mail: novak@ggki.hu; szalai@ggki.hu; szarka@ggki.hu

© Springer International Publishing AG, part of Springer Nature 2019
G. El-Qady, M. Metwaly (eds.), *Archaeogeophysics*, Natural Science in Archaeology,
https://doi.org/10.1007/978-3-319-78861-6_3

carried out a geophysical exploration around the monastery, in order to be able to answer if there had been any other establishments (royal- and church buildings, built even before the Cistercian monastery) there.

Geoelectric methods have been widely applied in archaeology (Diamanti et al. 2005). Instead of the routinely applied pole-pole methods, recently proposed in the tomographic context by Papadopoulos et al. (2006), we developed a mapping technique, based on an old, the so-called "potential gradient mapping" method (PM, where potential gradients are measured in a central area, between two distant current electrodes), widely applied a few decades ago in Hungarian bauxite exploration (Majkuth et al. 1973; Simon 1974; Kakas 1981, based on the theoretical description by Kunetz 1966). The scheme of electrodes, shown in Fig. 3.1, is the same for both the traditional and the new, tensorial measurements.

Due to an available four-channel instrument (MES-04, produced by KBFI-Triász Kft., Hungary) in 2003, it was possible to measure both horizontal components of the electric field simultaneously. This setup is similar to those by Cammarano et al. (2000), Di Fiore et al. (2002), and Futterer (2000).

In this paper, on basis of theoretical-, numerical modelling- and field results, we summarize the main experiences about the tensorial resistivity mapping in near-surface (archaeological) prospection. We also provide a comparison between the tensorial and the traditional results.

In Sect. 3.2, we derive the tensor elements and define two alternative sets of independent invariants of the apparent resistivity tensor. In Sect. 3.3, we describe the technical details of the mapping method. In Sect. 3.4, the field results are discussed.

3.2   Theory

On the horizontal surface the apparent resistivity tensor $\underline{\underline{\rho}}$ relates the horizontal electric field $\mathbf{E}$ and the current density vector $\mathbf{j}$ by Ohm's differential law (Bibby 1977):

$$\mathbf{E} = \underline{\underline{\rho}}\,\mathbf{j},\ \text{that is}\ \begin{bmatrix} E_x \\ E_y \end{bmatrix} = \begin{bmatrix} \rho_{xx} & \rho_{xy} \\ \rho_{yx} & \rho_{yy} \end{bmatrix} \begin{bmatrix} j_x \\ j_y \end{bmatrix} \quad (3.1)$$

3.2.1   Determination of the Tensor Elements

In order to compute the elements of tensor $\underline{\underline{\rho}}$, at first $\mathbf{j}$ and $\mathbf{E}$ are to be determined. Having current electrodes A and B and a current intensity I (+I: A, –I: B), the horizontal current density vector at a measuring site (where its coordinates are determined by $\mathbf{r}_A$ and $\mathbf{r}_B$, two vectors pointing from the corresponding current electrode to the measuring point), is given as:

$$\mathbf{j}_{AB} = \frac{I}{2\pi}\left(\frac{\mathbf{r}_A}{r_A^3} - \frac{\mathbf{r}_B}{r_B^3}\right) \quad (3.2)$$

The components of the electric field vector at the surface are determined from the voltages ΔU_x and ΔU_y, measured between closely spaced potential electrodes as

$$E_x \sim -\Delta U_x/\Delta x \ \text{and}\ E_y \sim -\Delta U_y/\Delta y \quad (3.3)$$

where the negative sign refers to the negative gradient in the definition $\mathbf{E} = -\,grad\,\Delta U$, and Δx, Δy are the distances between the neighbouring electrodes in x and y directions.

Applying one single pair of current electrodes, one current density vector and one electric field vector are obtained. In order to be able to determine all elements of the resistivity tensor, at least two current directions are needed. In that case just as many (four) equations are obtained, as it is the number (four) of the elements in the resistivity tensor. If indices 1 and 2 refer to the applied AB-s, then the four equations are

$$E_{x1} = \rho_{xx}j_{x1} + \rho_{xy}j_{y1} \quad (3.4a)$$

$$E_{x2} = \rho_{xx}j_{x2} + \rho_{xy}j_{y2} \quad (3.4b)$$

Fig. 3.1 The layout of the tensorial resistivity mapping is the same as that of the traditional potential gradient mapping (The actual parameters for the archaeological application is summarised in Sect. 3.3.)

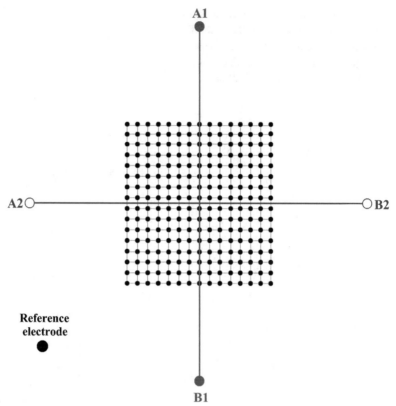

$$E_{y1} = \rho_{yx} j_{x1} + \rho_{yy} j_{y1} \qquad (3.4c)$$

$$E_{y2} = \rho_{yx} j_{x2} + \rho_{yy} j_{y2} \qquad (3.4d)$$

Then the apparent resistivity tensor (Bibby 1986, where some elements were misprinted) is given as

$$\underline{\rho} = \frac{1}{(j_{x1} j_{y2} - j_{y1} j_{x2})} \begin{bmatrix} E_{x1} j_{y2} - E_{x2} j_{y1} & E_{x2} j_{x1} - E_{x1} j_{x2} \\ E_{y1} j_{y2} - E_{y2} j_{y1} & E_{y2} j_{x1} - E_{y1} j_{x2} \end{bmatrix}$$

$$(3.5)$$

Once the resistivity tensor is obtained, we recommend, especially in archaeological studies, applying its rotational invariants. As known, rotational invariants do not depend on the direction of the applied current, thus rotational-invariant based apparent resistivity maps are expected to provide anomalies, reflecting the true geometry of the subsurface inhomogeneities.

Szarka and Menvielle (1997) showed that in the magnetotelluric impedance tensor, having

2 × 2 complex elements, the number of independent invariants is seven, and the eighth parameter is the measuring direction itself. Consequently, the DC resistivity tensor with 2 × 2 real elements can be transformed into three independent rotational invariants and into the measuring direction.

In the following sub-sections we define two alternative sets of independent invariants: (1) three independent resistivity estimations (determinant, sum of squares of elements and trace); (2) a further resistivity estimation, a 2D- and a 3D indicator.

3.2.2   A Set of Independent Tensor Invariants

It is possible to define an infinite number of alternative sets of three independent rotational invariants. In the simplest, mathematical approach of the DC resistivity tensor, the determinant ("det"), the sum of the square of the tensor element ("ssq"), and the trace (the sum of the main elements, denoted as "trace") are the three independent invariants. The corresponding resistivity definitions are as follows:

$$\rho_{\text{det}} = \left(\det\underline{\underline{\rho}} \right)^{1/2}$$

$$= \left(\rho_{xx}\rho_{yy} - \rho_{xy}\rho_{yx} \right)^{1/2} \tag{3.6a}$$

$$\rho_{\text{ssq}} = \left(1/2 \, \text{ssq}\underline{\underline{\rho}} \right)^{1/2}$$
$$= \left[1/2 \left(\rho_{xx}^2 + \rho_{xy}^2 + \rho_{yx}^2 + \rho_{yy}^2 \right) \right]^{1/2} \tag{3.6b}$$

$$\rho_{\text{trace}} = 1/2 \left(\text{trace}\underline{\underline{\rho}} \right)$$
$$= 1/2 \left(\rho_{xx} + \rho_{yy} \right) \tag{3.6c}$$

Any further invariant, e.g. the well-known dif (half difference of the xy and yx elements)

$$\text{dif}\,\underline{\underline{\rho}} = \frac{1}{2} \left(\rho_{xy} - \rho_{yx} \right) \tag{3.7}$$

is no more independent, since it can be easily demonstrated that

$$4\,\text{trace}^2\underline{\underline{\rho}} + 4\,\text{dif}^2\underline{\underline{\rho}} = \text{ssq}\underline{\underline{\rho}} + 2\,\det\underline{\underline{\rho}} \tag{3.8}$$

A wide variety of possible tensorial resistivity definitions is provided by Szarka et al. (2005).

3.2.3 The WAL Invariants

Of course, any other independent invariants could be alternatively used. The set of the so-called WAL invariants in magnetotellurics (after the initials of Weaver, Agarwal and Lilley, the co-authors of the paper by Weaver et al. 2000), provides not only a 1D resistivity estimation, but also a 2D- and a 3D indicator about the subsurface.

We decompose the DC apparent resistivity tensor into a symmetric tensor and an asymmetric one:

$$\underline{\underline{\rho}} = \begin{bmatrix} \rho_{xx} & \rho_{xy} \\ \rho_{yx} & \rho_{yy} \end{bmatrix}$$
$$= \frac{1}{2} \begin{bmatrix} \rho_{xx} - \rho_{yy} & \rho_{xy} + \rho_{yx} \\ \rho_{xy} + \rho_{yx} & \rho_{yy} - \rho_{xx} \end{bmatrix} + \frac{1}{2} \begin{bmatrix} \rho_{xx} + \rho_{yy} & \rho_{xy} - \rho_{yx} \\ \rho_{yx} - \rho_{xy} & \rho_{xx} + \rho_{yy} \end{bmatrix} \tag{3.9}$$

On basis of the papers by Bibby (1977, 1986) we apply

$$\sin 2\alpha = \frac{\rho_{xx} + \rho_{yx}}{\sqrt{\left(\rho_{xy} + \rho_{yx} \right)^2 + \left(\rho_{yy} - \rho_{xx} \right)^2}} \quad \text{and}$$

$$\sin 2\beta = \frac{\rho_{xy} - \rho_{yx}}{\sqrt{\left(\rho_{yx} - \rho_{xy} \right)^2 + \left(\rho_{xx} + \rho_{yy} \right)^2}} \tag{3.10}$$

In such a way we write:

$$\underline{\underline{\rho}} = \frac{1}{2}\sqrt{\left(\rho_{xy} + \rho_{yx} \right)^2 + \left(\rho_{yy} - \rho_{xx} \right)^2}$$
$$\begin{bmatrix} \cos 2\alpha & \sin 2\alpha \\ \sin 2\alpha & -\cos 2\alpha \end{bmatrix}$$
$$+ \frac{1}{2}\sqrt{\left(\rho_{yx} - \rho_{xy} \right)^2 + \left(\rho_{xx} + \rho_{yy} \right)^2}$$
$$\begin{bmatrix} \cos 2\beta & \sin 2\beta \\ -\sin 2\beta & \cos 2\beta \end{bmatrix} \tag{3.11}$$

Since

$$\frac{1}{2}\sqrt{\left(\rho_{xy} + \rho_{yx} \right)^2 + \left(\rho_{yy} - \rho_{xx} \right)^2}$$
$$= \frac{1}{2}\sqrt{\text{ssq}\underline{\underline{\rho}} - 2\det\underline{\underline{\rho}}} = \Pi_1 \tag{3.12a}$$

$$\frac{1}{2}\sqrt{\left(\rho_{yx} - \rho_{xy} \right)^2 + \left(\rho_{xx} + \rho_{yy} \right)^2}$$
$$= \frac{1}{2}\sqrt{\text{ssq}\underline{\underline{\rho}} + 2\det\underline{\underline{\rho}}} = \Pi_2 \tag{3.12b}$$

therefore

$$\underline{\underline{\rho}} = \Pi_1 \begin{bmatrix} \cos 2\alpha & \sin 2\alpha \\ \sin 2\alpha & -\cos 2\alpha \end{bmatrix}$$
$$+ \Pi_2 \begin{bmatrix} \cos 2\beta & \sin 2\beta \\ -\sin 2\beta & \cos 2\beta \end{bmatrix}, \tag{3.13}$$

where $\quad \sin 2\alpha = \dfrac{\rho_{xy}+\rho_{yx}}{\sqrt{\mathrm{ssq}\underline{\underline{\rho}}-2\det\underline{\underline{\rho}}}}\quad$ and

$$\sin 2\beta = \dfrac{\rho_{xy}-\rho_{yx}}{\sqrt{\mathrm{ssq}\underline{\underline{\rho}}+2\det\underline{\underline{\rho}}}}.$$

In Eq. (3.13) Π_1, Π_2, and β are invariants, α is not an invariant, since $\rho_{xy}+\rho_{yx}$ is not invariant. In such a way, instead of the seven WAL invariants in magnetotellurics (I_1–I_7), in case of the DC resistivity tensor only I_1, I_3 and I_5 exist. They are as follows.

1. An estimation of the invariant resistivity I_1, which we prefer to denote as I_{1D}

$$I_{1D} = I_1 = \Pi_2 = \frac{1}{2}\sqrt{\mathrm{ssq}\underline{\underline{\rho}} + 2\det\underline{\underline{\rho}}} \quad (3.14)$$

The corresponding apparent resistivity is the root mean square of ρ_{ssq} and ρ_{det}, since

$$\rho_{I_1} = I_1 = \sqrt{\frac{\rho_{\mathrm{ssq}}^2 + \rho_{\mathrm{det}}^2}{2}} \quad (3.15)$$

2. The two-dimensional anisotropy, I_{2D}, obtained from the magnetotelluric parameter I_3 as

$$I_{2D} = I_3 = \frac{\Pi_1}{\Pi_2} = \sqrt{\frac{\mathrm{ssq}\underline{\underline{\rho}} - 2\det\underline{\underline{\rho}}}{\mathrm{ssq}\underline{\underline{\rho}} + 2\det\underline{\underline{\rho}}}}. \quad (3.16)$$

3. The measure of the three-dimensionality, I_{3D}, corresponding to the original I_5 by Weaver et al. (2000):

$$I_{3D} = I_5 = \sin 2\beta = \frac{\rho_{xy} - \rho_{yx}}{2\Pi_2} = \frac{\mathrm{dif}\underline{\underline{\rho}}}{\Pi_2} \quad (3.17)$$

I_{2D} and I_{3D} are dimensionless numbers, with absolute values between 0 and 1. In one-dimensional case both of them are zero. Along two-dimensional bodies I_{3D} remains zero,

but I_{2D} will differ from zero. Over three-dimensional inhomogeneities both of them are non-zero.

3.2.4 Relation Between the Traditional- and the Tensorial Versions

In the so-called "potential gradient mapping" (PM), extensively applied in Hungarian bauxite exploration a few decades ago (Majkuth et al. 1973; Simon 1974; Kakas 1981), the potential differences were measured over a central area, between two distant current electrodes. Two current directions were applied, and the corresponding equations are as follows.

$$E_{x1} = \rho_{xx}^{PM} j_{x1} \quad (3.18a)$$

$$E_{y2} = \rho_{yy}^{PM} j_{y2} \quad (3.18b)$$

As it was found by analogue modelling experiments (Szarka 1984, 1987), various mean values, obtained from the two measurements provided satisfactory image about sinks and horsts in a high-resistivity bottom. The geometric mean, the root mean square and the arithmetic mean are in close relationship with the determinant, sum of square and trace (Eqs. 3.6a–c), as follows:

$$\rho_{\mathrm{det}}^{PM} = \left(\rho_{xx}^{PM}\rho_{yy}^{PM}\right)^{1/2} \sim \left(\det\underline{\underline{\rho}}\right)^{1/2} \quad (3.19a)$$

$$\rho_{\mathrm{ssq}}^{PM} = \left[1/2\left(\rho_{xx}^{PM}\right)^2 + 1/2\left(\rho_{yy}^{PM}\right)^2\right]^{1/2}$$
$$\sim \left(1/2\,\mathrm{ssq}\underline{\underline{\rho}}\right)^{1/2} \quad (3.19b)$$

$$\rho_{\mathrm{trace}}^{PM} = 1/2\left(\rho_{xx}^{PM} + \rho_{yy}^{PM}\right)$$
$$\sim 1/2\left(\mathrm{trace}\underline{\underline{\rho}}\right) \quad (3.19c)$$

In this way, Eqs. (3.4a–d) represent a full tensorial extension of equations (3.19a–c).

3.2.5   Synthetic Example

The main characteristics of all possible parameters (traditional and full tensor-derived ones) are shown in a numerical example. The high-resistivity frame, shown in Fig. 3.2, is a model of the foundations of a building. The results, summarized in Fig. 3.2, are as follows.

- Rows 1 and 2: The traditional ρ_{xx}^{PM} and ρ_{yy}^{PM} maps, as well as the tensorial ρ_{xx} and ρ_{yy} elements of the tensor indicate that the method is sensitive to inhomogeneities, perpendicular to the current flow.
- Rows 3 and 4: There is no significant difference among the six resistivity maps (the traditional mean values $\rho_{\mathrm{det}}^{PM}, \rho_{\mathrm{ssq}}^{PM}$ and $\rho_{\mathrm{trace}}^{PM}$ and the rotational invariants $\rho_{\mathrm{det}}, \rho_{\mathrm{ssq}}$ and ρ_{trace}).
- Rows 5: Among the WAL invariants ρ_{I_1} is practically the same as $\rho_{\mathrm{det}}, \rho_{\mathrm{ssq}}$ and ρ_{trace}. The multi-dimensional indicators I_{2D} and I_{3D} are relative numbers, shown in percents. They differ from zero in case of two or three-dimensional inhomogeneities.

If the square frame model is at depth, and/or the resistivity contrast is less, all anomalies are of course less significant.

In Fig. 3.3 the WAL invariants are shown in presence of $\pm 3\%$ random noise, added to the measured electric field. The multidimensional indicators are more sensitive to noise than ρ_{I_1}.

3.3   Field Realization

We developed the near-surface tensorial version of the PM resistivity mapping technique. In Fig. 3.1 the field layout is shown. In this section the technical parameters are specified.

3.3.1   MN Distance, Mapping Area, AB Distance

It is known from various modelling studies that the lateral resolution depends on the distance between the potential electrodes M and N. We selected MN = 50 cm, a usual value in archaeological studies. The instruments allowed an easy handling of about 240 electrode positions, so it was an evident choice to define a nearly squared measuring area with $16 \cdot 15 = 240$ electrodes, covering an area of 7.5 m · 7 m.

In field conditions very precise electrode positions can not be guaranteed. The surface inhomogeneities may cause sometimes up to 5 cm horizontal displacements, and the soil hardness led to about 10–15 cm depth differences among the electrodes. Sporadic errors could be easily eliminated, and a special effort was made to avoid parallel shifts of neighbouring measuring lines.

Over a given mapping area and fixed AB electrodes, the depth of investigation varies as a function of the AB separation, and it also may vary from site to site. Near-surface investigations require a short AB distance, while the homogeneity of the depth of investigation within the measuring area would require an as large AB distance as possible. One should find a compromise between the two confronting requirements.

On basis of theoretical depth of investigation by Roy and Apparao (1971), Bhattacharya and Dutta (1982) and Szalai (2000), for a measuring area of 7.5 m · 7 m we selected AB = 15 m.

3.3.2   Measuring Process

After placing the fixed electrodes 15 m away one from the other, and connecting the potential electrodes by using multicore cables, the whole measuring process is computer-controlled. All ΔU_x and ΔU_x potential differences are determined: at first due to current electrodes A_1B_1 then due to A_2B_2, as follows. By using the four-channel instrument, at the same time four potential values at electrode positions P_i, P_{i+1}, P_{i+15}, and at the reference electrode are measured. ("i" had altogether 210 values of the total 240 ones; electrodes "i" and "i + 1" are in the same row; electrodes "i" and "i + 15" are in the same column).

Thus both components of horizontal the electric field vector (shown in Eq. 3.3) were

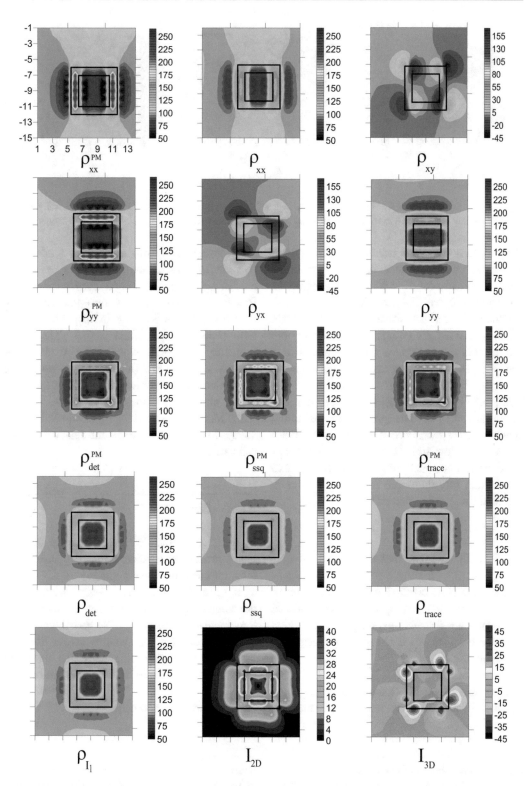

Fig. 3.2 3D numerical results over a rectangular high-resistivity square frame model. Host resistivity: 100 ohmm; parameters of the frame: resistivity: 1000 ohmm, burial depth: 0 m, wall thickness: 0.5 m, vertical thickness: 1 m. *Rows 1–2*: traditional potential gradient (PM) maps $\left(\rho_{xx}^{PM}, \rho_{yy}^{PM}\right)$, and maps of the *xx, xy,*

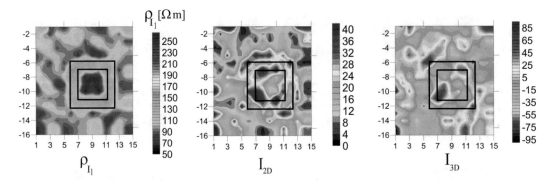

Fig. 3.3 The DC versions of the WAL invariants (ρ_{I_1}, I_{2D} and I_{3D}) over the same model as shown in Fig. 3.1, in presence of $\pm 3\%$ random noise, added to the electric field. ρ_{I_1} is given in ohm-meters; I_{2D} and I_{3D} are given in percents

determined at the same instant. The current density vector values are calculated simply from the current intensity (of 50–500 mA, depending on soil condition), by using Eq. (3.2).

Routinely, the net measuring time to obtain $15 \cdot 14 = 210$ resistivity tensors is about 40 min. During this time, the next 240 electrodes over the adjacent measuring area can be easily arranged. Having at least three people in the field, usually 8–12 adjacent maps per day can be measured. The data processing (assuming that everything was correctly done in the field) takes much less time. So far 239 such maps have been completed, thus we have $239 \cdot 210$ (more than fifty thousand) resistivity tensors in the field around the monastery.

3.3.3 Data Interpretation

The large area is covered with a succession of adjacent mapping areas. By computing rotational invariant maps for each mapping area, and tying them together, we expect to see the true geometry of buried targets.

In case of one single AB length, of course it is not possible to get information about the depth variation of the resistivity, and it would take too

much time to carry out a sounding at each site. Nevertheless, there are some depth estimation methods. One of them is to carry out a spatial frequency analysis of the observed resistivity anomalies, as it is done e.g. in magnetic exploration (Spector and Grant 1970). The high-frequency anomalies would refer to shallower effects, while the low-frequency ones can origin from relatively deeper regions. Another approach is the probability tomography method (Patella 1997; Mauriello et al. 1998; Mauriello and Patella 1999). In this paper we do not deal with depth estimation.

3.4 Results

In Fig. 3.4 all the 239 maps (namely the ρ_{det} maps) are shown together with the plan-view of the monastery and the topography map.

In spite of the fact that the resistivity maps are not completely free of measuring errors and positioning inaccuracies (e.g., the resistivity values are not perfectly continuous across the the boundaries of neighbouring maps), the integrated resistivity map is really informative. Among the variations of natural origin, there are a lot of evidently artificial anomalies.

Fig. 3.2 (Continued) yx and yy elements of the resistivity tensor (ρ_{xx}, ρ_{xy}, ρ_{yx} and ρ_{yy}). *Row 3*: PM mean values $\left(\rho_{det}^{PM}, \rho_{ssq}^{PM}, \text{and } \rho_{trace}^{PM} \right)$. *Row 4*: the first set of independent invariants (ρ_{det}, ρ_{ssq}, ρ_{trace}). *Row 5*: the DC versions of the WAL invariants (ρ_{I_1}, I_{2D} and I_{3D}). The x direction is horizontal; the resistivities are given in ohm-meters

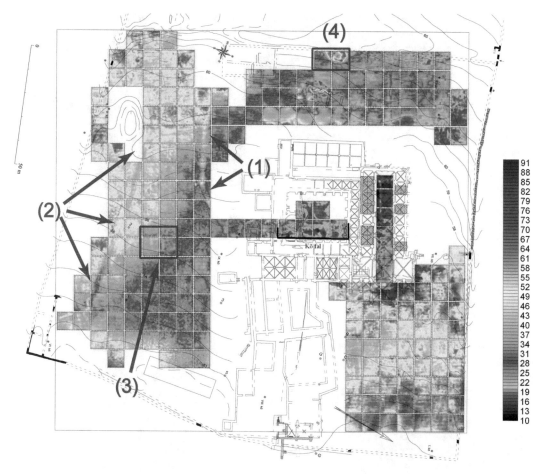

Fig. 3.4 239 apparent resistivity maps (given in ohm-meters), together with the map of the monastery complex

The following features—among others—are obviously of man-made origin:

1. The anomaly, bending towards the ambulatory must be a subsurface channel, carrying water from an artificial lake to the monastery buildings (fountain, kitchen, water closet, etc.)
2. The diagonal lineament, bending in the upper part toward right, coincides with the present lane. It is of low resistivity. The very high resistivities to the left and above the diagonal lane are most probably due to buried building remnants. The diagonal anomaly might have been an ancient road.
3. A relatively high-resistivity square-frame like anomaly with a dimension of 6 m·6 m is probably a remnant of some building foundation.
4. A horse-shoe like high resistivity anomaly might have been a furnace.

We discuss in details the resistivity mapping results over anomalies (3) and (4).

3.4.1 Square Frame

Field results over the square frame are shown in Fig. 3.5, in the same order as the modelling results are shown in Fig. 3.3. The imaging properties of the independent resistivity invariants (det, ssq, trace, the I_{1D}-based resistivity) are very similar to each other. The PM-based resistivities are also very similar. All of them have the same features as the numerical modelling results.

At the same time, the multidimensional indicators I_{2D} and I_{3D} obtained from the field measurements are completely useless. The corresponding signals are too small. Consequently the measuring errors and various noises

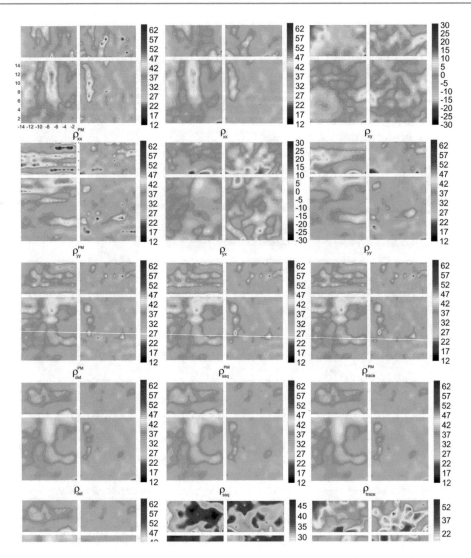

Fig. 3.5 Field results over anomaly (3). *Rows 1–2*: traditional potential gradient (PM) maps $\left(\rho_{xx}^{PM}, \rho_{yy}^{PM}\right)$, and maps of the *xx*, *xy*, *yx* and *yy* elements of the resistivity tensor (ρ_{xx}, ρ_{xy}, ρ_{yx} and ρ_{yy}). *Row 3*: PM mean values $\left(\rho_{det}^{PM}, \rho_{ssq}^{PM}, \text{and } \rho_{trace}^{PM}\right)$. *Row 4*: the first set of independent invariants (ρ_{det}, ρ_{ssq}, ρ_{trace}). *Row 5*: the DC versions of the WAL invariants (ρ_{I_1}, I_{2D} and I_{3D}). The *x* direction is horizontal. The resistivities are given in ohm-meters; I_{2D} and I_{3D} are given in percents

suppress the physical effect, shown in the numerical study.

3.4.2 Furnace

The furnace produces a much stronger anomaly. Again, all resistivity anomalies proved to be very similar to each other. In Fig. 3.6 the measured elements of the resistivity tensor and the three WAL invariants are shown. Figure 3.7 shows the same for the corresponding modelled quantities. Over the furnace even the measured I_{2D} and I_{3D} are meaningful: they are not the same as in the numerical modelling study, but somewhat remind to them. These multidimensional maps seem to indicate the deviation of the subsurface structure from the assumed model.

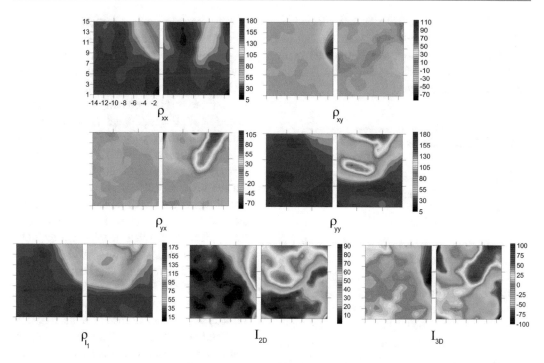

Fig. 3.6 Field results over the assumed furnace. *Rows 1–2*: Maps of ρ_{xx}, ρ_{xy}, ρ_{yx} and ρ_{yy} elements of the resistivity tensor. *Row 3*: the DC versions of the WAL invariants (ρ_{I_1}, I_{2D} and I_{3D}). The x direction is horizontal. The resistivities are given in ohm-meters; I_{2D} and I_{3D} are given in percents

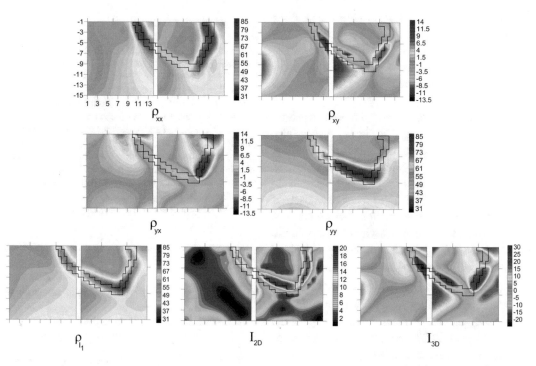

Fig. 3.7 Numerical results over the furnace model. Host resistivity: 30 ohmm; parameters of the furnace model: resistivity: 3000 ohmm, burial depth: 0.5 m, wall thickness: 0.5 m, vertical thickness: 1 m. *Rows 1–2*: Maps of ρ_{xx}, ρ_{xy}, ρ_{yx} and ρ_{yy} elements of the resistivity tensor. *Row 3*: the DC versions of the WAL invariants (ρ_{I_1}, I_{2D} and I_{3D}). The x direction is horizontal. The resistivities are given in ohm-meters; I_{2D} and I_{3D} are given in percents

Even in this case, in order to detect the plan-view geometry of the buried furnace remnants, it would have been enough to measure the main components, as it is done with the traditional potential gradient mapping.

3.5 Conclusions

The theoretical, practical (field), methodological and archaeological conclusions of this work are as follows:

1. **Theoretical**. In order to obtain all possible DC resistivity information about the true plan-view geometry of subsurface targets, we propose the use of tensor invariants. Two alternative sets of three independent tensor-invariants are derived. The first set consists of the determinant, the sum of square of elements and the trace; the second one consists of a further resistivity estimator, a 2D- and a 3D indicator. As shown by numerical models, any of the ρ_{det}, ρ_{ssq}, ρ_{trace} and ρ_{I_1} maps (and also the simple ρ_{det}^{PM}, ρ_{ssq}^{PM} and ρ_{trace}^{PM} maps) give information about the true plan-view geometry. The multidimensional indicators provide additional information: I_{2D} about the sides (i.e. the "strike directions"), I_{3D} about the corners.
2. **Field experiences**. The proposed tensorial resistivity mapping has been successfully realized: more than fifty thousand rotational invariant-based apparent resistivity values have been obtained in the field. Man-made features are really easy to recognise. Each of the four resistivity estimations (all members of the first set and the resistivity estimator in the second set) provide practically identical results. It is remarkable that the simple ρ_{det}^{PM}, ρ_{ssq}^{PM} and ρ_{trace}^{PM} maps are equally good. Moreover, the multidimensional invariant indicators (based on the so-called null-components, Szalai et al. 2002) are useful only in case of very significant anomalies.
3. **Methodological**. As we see, the measurements could have done twice quicker, if only the components parallel to the main current flow are measured. According to the modelling results, the multidimensional invariants have clear physical meaning, but their benefits can be exploited only in case of much more precise geometry of the electrodes in the field. The traditional version is a good solution for an everyday use. If a more precise geometry (e.g., by a fixed frame) can be fixed, it would be worth measuring the full tensor.

As the resistivity tensor shows deficiency in investigating the depth of the structures, it is advised to accomplish tensorial surveys with the application of other geophysical methods. A few soundings or ERT profiles at each anomalous feature might also be useful.

4. **Archaeological**. Man-made origin anomalies as a subsurface channel, building remnants, a furnace and an ancient road have been discovered and described. The excavations have not started yet, but the geophysical anomalies have already inspired a lot of new ideas for the archaeologists.

Acknowledgements Theoretical part of the work was started out already in frames of projects TS408048 (2002–2004) and T37694 (2002–2005) of the Hungarian Research Fund; various aspects were elaborated in frames of projects T049604 (2005) and NI 61013 (started in 2006). The field measurements were sponsored by the Archaeological Institute of the Hungarian Academy of Sciences (where the leader of project Medium Regni was Dr. Elek Benkő). J. Túri, A. Kovács and several students (first of all Zs. Pap and A. Károlyi) took part in the field measurement or/and the data processing. M. Varga and A. Novák are PhD students at the University of West-Hungary, Sopron. Comments by A. Ádám (GGRI HAS), the referees (G. Tsokas, P. Mauriello) and the Associate Editor (P. Tsourlos) are also acknowledged.

References

Bhattacharya BB, Dutta I (1982) Depth of investigation studies for gradient arrays over homogeneous isotropic half-space. Geophysics 47:1198–2003

Bibby HM (1977) The apparent resistivity tensor. Geophysics 42:1258–1261

Bibby HM (1986) Analysis of multiple-source bipole-quadripole resistivity surveys using the apparent resistivity tensor. Geophysics 51:972–983

Cammarano F, Di Fiore B, Mauriello P, Patella D (2000) Examples of application of electrical tomographies and radar profiling to cultural heritage. Annali di Geophysica 43:309–324

Diamanti N, Tsokas G, Tsourlos P, Vafidis A (2005) Integrated interpretation of geophysical data in the archaeological site of Europos (northern Greece). Archaeol Prospect 12:79–91

Di Fiore B, Mauriello P, Monna D, Patella D (2002) Examples of application of tensorial resistivity probality tomography to architectonic and archaeological targets. Ann Geophys 45:417–429

Futterer B (2000) Tensorgeoelektrik in der Anwendung. Insttut für Geophysik, TU Bergakademie Freiberg (Supervisor: R.U. Börner), Student Research Project. http://www.geophysik.tu-freiberg.de/spitzer/Download_Store/Publications/Students-Research-Project/Futterer-Studienarbeit.pdf

Gerevich L (1977) Pilis Abbey a cultural center. Acta Archaeologica Academiae Scientiarum Hungaricae 29:155–198

Kakas K (1981) DC potential mapping (PM). In: Annual report of the Eötvös Loránd Geophysical Institute of Hungary for 1980, pp 163–165

Kunetz G (1966) Principles of direct current resistivity prospecting. Borntraeger, Berlin-Nikolassee, pp 31–33

Majkuth T, Ráner G, Szabadváry L, Tóth CS (1973) Results of methodological developments in the Transdanubian Central Range: direct exploration of bauxite bearing structures near Bakonyoszlop (in Hungarian). Report of the Eötvös Loránd Geophysical Institute, Budapest

Mauriello P, Patella D (1999) Resistivity imaging by probability tomography. Geophys Prospect 47:411–429

Mauriello P, Monna D, Patella D (1998) 3D geoelectric tomography and archaeological applications. Geophys Prospect 46:543–570

Papadopoulos NG, Tsourlos P, Tsokas GN, Sarris A (2006) Two-dimensional and three-dimensional resistivity imaging in archaeological site investigation. Archaeol Prospect 13:163–181

Patella D (1997) Introduction to ground surface self-potential tomography. Geophys Prospect 45:653–681

Roy A, Apparao A (1971) Depth of investigation in direct current methods. Geophysics 36:943–959

Simon A (1974) Theory of potential mapping and its processing methods (in Hungarian). Report of the Eötvös Loránd Geophysical Institute, Budapest

Spector A, Grant FS (1970) Statistical models for interpreting aeromagnetic data. Geophysics 35:293–302

Szalai S (2000) About the depth of investigation of different D.C. dipole-dipole arrays. Acta Geodaetica et Geophysica Hungarica 35:63–73

Szalai S, Szarka L, Prácser E, Bosch F, Müller I, Turberg P (2002) Geoelectric mapping of near-surface karstic fractures by using null-arrays. Geophysics 67:1769–1778

Szarka L (1984) Analogue modelling of DC mapping methods. Acta Geod Geophys Mont Hung 19:451–465

Szarka L (1987) Geophysical mapping by stationary electric and magnetic field components: a combination of potential gradient mapping and magnetometric resistivity (MMR) methods. Geophys Prospect 35:424–444

Szarka L, Menvielle M (1997) Analysis of rotational invariants of magnetotelluric impedance tensor. Geophys J Int 129:133–142

Szarka L, Ádám A, Menvielle M (2005) A field test of imaging properties of rotational invariants of the magnetotelluric impedance tensor. Geophys Prospect 53:325–334

Weaver JT, Agarwal AK, Lilley FEM (2000) Characterization of the magnetotelluric tensor in terms of its invariants. Geophys J Int 141:321–336

Combined Seismic Tomographic and Ultra-Shallow Seismic Reflection Study of an Early Dynastic Mastaba, Saqqara, Egypt

4

Mohamed Metwaly, Alan G. Green, Heinrich Horstmeyer, Hansruedi Maurer, and Abbas M. Abbas

Abstract

Mastabas were large rectangular structures built for the funerals and burials of the earliest Pharaohs. One such mastaba was the basic building block that led to the first known stone pyramid, the >4600-year old Step Pyramid within the Saqqara necropolis of Egypt. We have tested a number of shallow geophysical techniques for investigating in a non-invasive manner the subsurface beneath a large Early Dynastic mastaba located close to the Step Pyramid. After discovering that near-surface sedimentary rocks with unusually high electrical conductivities precluded the use of the ground-penetrating radar method, a very high-resolution seismic data set was collected along a profile that extended the 42.5 m length of the mastaba. A sledgehammer source was used every 0.2 m and the data were recorded using a 48-channel array of single geophones spaced at 0.2 m intervals. Inversions of the direct- and refracted-wave traveltimes provided P-wave velocity tomograms of the shallow subsurface, whereas relatively standard processing techniques yielded a high-fold (50–80) ultra-shallow seismic reflection section. The tomographic and reflection images were jointly interpreted in terms of loose sand and friable limestone layers with low P-wave velocities of 150–650 m/s overlying consolidated limestone and shale with velocities >1500 m/s. The sharp contact between the low and high velocity regimes was approximately horizontal at a depth of ~2 m. This contact was the source of a strong seismic reflection. Above this contact, the velocity tomogram revealed moderately high velocities at the surface location of a friable limestone outcrop and two low velocity blocks that probably outlined sand-filled shafts. Below the contact, three regularly spaced low velocity blocks likely represented tunnels and/or subsurface chambers.

M. Metwaly (✉)
National Research Institute of Astronomy and Geophysics, Helen, Cairo, Egypt

Department of Archaeology, College of Tourism and Archaeology, King Saud University, Riyadh, Saudi Arabia

A. G. Green · H. Horstmeyer · H. Maurer
Institute of Geophysics, ETH-Hoenggerberg, Zurich, Switzerland

A. M. Abbas
National Research Institute of Astronomy and Geophysics, Helen, Cairo, Egypt

Keywords

Ultra-shallow seismic reflection · Refraction tomography · Mastaba

© Springer International Publishing AG, part of Springer Nature 2019
G. El-Qady, M. Metwaly (eds.), *Archaeogeophysics*, Natural Science in Archaeology,
https://doi.org/10.1007/978-3-319-78861-6_4

4.1 Introduction

Egypt contains many valuable monuments distributed throughout the country. They comprise approximately 33% of all archaeological findings worldwide. Nevertheless, the known monuments may only represent a small portion of all Egyptian archaeological treasures. A large number remain hidden beneath shallow veneers of unconsolidated sediments. Since standard archaeological excavations require large expenditures of effort and funds, archaeological prospection using advanced geophysical techniques has grown in importance in Egypt over the past few years. These techniques offer the possibility of providing fast and reliable information on the distribution of archaeological features to depths up to 50 m in a non-invasive manner. To function well, most geophysical techniques require large physical property contrasts between the archaeological features and the host natural ground or between the disturbed and natural ground.

In 2001, a team of Egyptian and Swiss geophysicists tested the ground-penetrating radar, seismic tomographic and ultra-shallow seismic reflection techniques at a site within the famous Saqqara necropolis, just south of Cairo (Fig. 4.1a). Near-surface sedimentary rocks characterised by very high electrical conductivities proved fatal for the ground-penetrating radar technique. In contrast, the seismic tomographic and ultra-shallow seismic reflection techniques yielded potentially valuable subsurface information. After presenting a brief review of the study site, we describe our seismic data acquisition strategy. We then show sequentially the results of tomographically inverting the direct and refracted arrivals and the results of processing the ultra-shallow reflections. An image that combines a seismic tomogram with the ultra-shallow seismic reflection section reveals the presence of subhorizontal geological layers interrupted by prominent low-velocity blocks that may outline the locations of sand-filled shafts, tunnels and/or subsurface chambers.

4.2 Saqqara

The Saqqara necropolis is situated on a plateau at the eastern edge of the Libyan Desert, opposite the ancient Egyptian capital of Memphis. This enormous cemetery, which extended ~6 km from north to south and had a maximum width of ~1.5 km (Fig. 4.1a), was an important burial site for royalty and high-level government officials from the Early Dynastic Period until the end of the Greco-Roman Period (Baines and Malek 1992; Black and Norton 1993; Vendel 2002; Kinnaer 2003; Raffaele 2003; Sitek 2003). It has often been referred to as the city of the dead.

By studying structures at Saqqara it is possible to trace the evolution of Egyptian tombs from the earliest mastabas to the famous pyramids (Baines and Malek 1992; Black and Norton 1993; Kinnaer 2003; Vendel 2002; Raffaele 2003; Sitek 2003). Mastabas are large rectangular flat-topped structures built of sun-dried mud bricks with rooms for offerings, funerary chapels, shrines and one or more shafts leading to tombs at depth. A number of Early Dynastic mastabas are found along the northeastern edge of the Saqqara plateau (Fig. 4.1a).

Constructions of the first ever pyramid and associated funerary complex was the next major step in the evolution of Egyptian tombs (Baines and Malek 1992; Black and Norton 1993). The Step Pyramid at Saqqara was built more than 4600 years ago by the vizier Imhotep for the third dynasty Pharaoh Netjeryjhet, better known as Djoser or Zozer. To create the Step Pyramid, Imhotep increased the horizontal dimensions of an original 63 × 63 m mastaba and then progressively stacked mastaba-like structures of decreasing size on the enlarged lowest mastaba. The final product was a pyramid 140 × 118 m at its base with six tiers reaching a height of ~62 m (Black and Norton 1993). It overlay a 28 m deep shaft leading to an oblong tomb. The Step Pyramid was the first massive structure to be constructed entirely of stone (i.e. locally quarried clayey

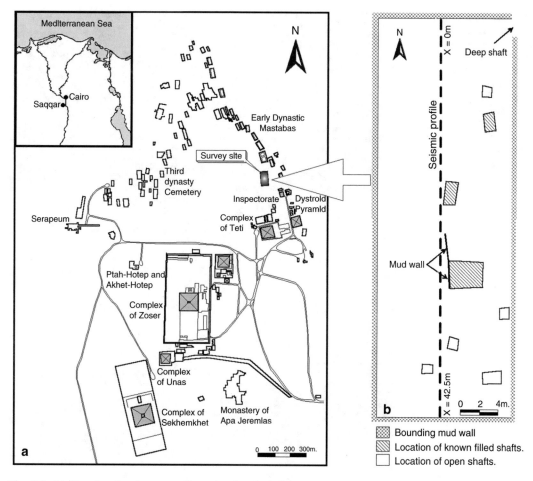

Fig. 4.1 (**a**) Map showing the survey site and archaeological features in the northern part of the Saqqara necropolis (modified after Black and Norton 1993). (**b**) Sketch of the investigated mastaba showing the seismic profile location

sandstone and limestone). At the time, it was probably the largest building on Earth.

Numerous Pharaohs had their pyramids built at Saqqara. For example, the last Pharaoh of the 5th dynasty, Unas, decorated his tomb with the first of the famous Pyramid Texts, spells written to help the Pharaoh ascend to heaven and return again. The last royal pyramid at Saqqara was erected during the 8th dynasty. Subsequently, Saqqara was a preferred burial site for powerful government functionaries and an unusual cemetery for revered animals. The Serapeum, a huge catacomb containing the corpses of the sacred Apis Bulls, was in use through to Greco-Roman times (Black and Norton 1993; Kinnaer 2003).

4.3 Survey Site

Most of the Saqqara necropolis is covered by a thin layer of sand overlying horizontal to sub-horizontal layers of limestone with intercalations of marl, cross-bedded sandstone, conglomerate, shale and claystone (Said 1990). The current investigation is focussed on the interior of a large Early Dynastic mastaba situated along the northeastern edge of the Saqqara necropolis (Figs. 4.1a, b, and 4.2a). A lot of the sand has been removed from this location, such that flat-lying bedrock is quite close to the surface. The interior of the mastaba covers an area of 42.5 × 15 m and is surrounded by a sun-dried

Fig. 4.2 Photographs showing features around and within the surveyed mastaba and location of the seismic profile. (**a**) View of mastaba with surrounding mud walls, stored pottery and mud-brick piles (west side of mastaba) and open shafts. (**b**) Acquisition of seismic data using a sledgehammer source

mud-brick wall that is up to ~2 m thick and now varies in height from a few centimetres to ~2 m. Within the mastaba are a number of shafts and a small remnant of an inner mud-brick wall (Figs. 4.1b and 4.2a, b). Several of the shafts are open at the surface. Some have a facing of mud bricks or limestone blocks, whereas others are simple engineered holes. At the base of several open shafts, which have dimensions of 1–2 m and depths of 1–5 m, there are small rooms and blocked-off tunnels. Other shafts open at the surface are blocked off within a meter or two of the surface. There is also evidence for several shafts that are completely filled. Filled shafts may also

be hidden beneath the sand. Finally, at the northeastern corner of the mastaba, barricaded horizontal entrances are observed on vertical walls extending to ~6 m depth.

4.4 Seismic Surveying in Archaeology

Although P-wave seismic methods are only rarely employed in archaeology, a few successful applications of these techniques in archaeological investigations have been reported in the literature. Seismic refraction methods have been used to locate Roman-age ditches in Britain (Goulty et al. 1990; Goulty and Hudson 1994; Ovenden 1994), prehistoric pits inside an Israeli cave (Beck and Weinstein-Evron 1997), a Persian-built canal in Greece (Karastathis and Papamarinopoulos 1997; Karastathis et al. 2001) and tombs buried inside Greek tumuli (Tsokas et al. 1995). Except for the work of Tsokas et al. (1995) and Karastathis et al. (2001), these studies involved acquiring profile data using 0.5–2.0 m spaced geophones and sparse source points. Interpretations were based on the relatively simple plus-minus, time-term or generalised reciprocal methods. Tsokas et al. (1995) employed a primitive fan-shooting approach, whereas Karastathis et al. (2001) used a combination of the generalised reciprocal method and the Zelt and Smith (1992) inversion scheme. Shallow P-wave seismic reflection methods have been used to map faults and cavities within a pyramid at Giza in Egypt (Dolphin 1981), a crypt at Champaign in Illinois (Hildebrand et al. 2002) and the previously mentioned Persian-built canal in Greece (Karastathis and Papamarinopoulos 1997; Karastathis et al. 2001).

4.5 Ultra-Shallow Seismic Reflection Surveying

Although shallow P-wave seismic reflection methods have been used to detect cavities within homogenous rock masses and tunnels (Steeples et al. 1986; Branham and Steeples 1998; Miller and Steeples 1991; Miller et al. 1995; Kourkafas and Goulty 1996), such techniques are usually inappropriate for investigating features at depths shallower than 5 or 10 m. For this depth range, the recently introduced ultra-shallow seismic reflection methods (Bachrach and Nur 1998; Bachrach et al. 1998; Cardimona, et al. 1998; Baker et al. 1999, 2000, 2001; Deidda and Balia 2001; Diedda and Ranieri 2001; Bachrach and Mukerji 2001; De Iaco et al. 2003; Schmelzbach et al. 2005) are suitable. The principal difference between shallow and ultra-shallow seismic reflection methods is the spacing between the receivers (2.0–5.0 m for the former versus 0.05–1.25 m for

the latter). To observe reflections returning from the upper ~5 m of the subsurface, it is necessary to transmit and receive broad-band waves rich in high frequencies.

4.6 Seismic Data Acquisition at Saqqara and Traveltime Picking

Each receiver spread included forty-eight single 14-Hz geophones deployed at 0.2 m intervals (Fig. 4.3; Table 4.1). To cover the entire length of the surveyed mastaba, it was necessary to lay-out the receiver spread five times (four are shown in Fig. 4.3). The source comprised three to five blows of a 11.25 kg sledgehammer on a 20 × 20 cm steel baseplate fixed to the ground via a 20 cm spike (Fig. 4.2b). Source points were distributed at 0.2 m intervals within the receiver spread, 0.4 m intervals for a distance of 4.8 m on either side of the spread, 0.6 m intervals for the next 7.2 m, and 0.8 m for longer distances (Fig. 4.3). A total of 460 forty-eight-trace source gathers were recorded with a temporal sampling interval of 0.125 ms.

Figure 4.4a shows a typical recording with the source located at one end of the receiver spread. The offsets in this figure are relative to the geophone nearest to the source baseplate, which was usually placed a very short distance on either side of the receiver spreads to avoid damaging the geophones. For the first ~4.0 m, the first arrivals are high-frequency air waves travelling at about 340–360 m/s. Such high air-wave velocities are typical for the 35–45 °C air temperatures experienced during the seismic survey. Closely following the air wave at offsets ≤0.8 m are waves that travel through the very near-surface unconsolidated dry sands with velocities of ~150 m/s (see Bachrach and Nur (1998) and Bachrach et al. (1998) for a discussion of the very low velocities observed in shallow dry sands); the actual velocities are difficult to determine because of interference with the air wave. In the 0.8–4.0 m offset range, velocities of the first arriving waves that travel through the ground vary from 400–650 m/s. At greater distances, the first

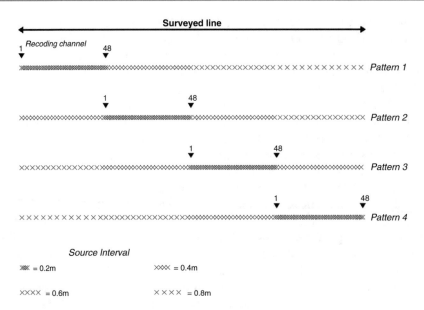

Fig. 4.3 Distribution of sources and receivers for four of the five deployments of the 48-geophone receiver spread. Except for very short lateral offsets to avoid damaging the geophones, receivers were coincident with the sources in the regions marked "recording channel 1–48"

Table 4.1 Acquisition parameters employed for the ultra-shallow seismic survey at Saqqara

Source	11.25 kg sledgehammer striking a baseplate with a spike
Vertical stacks	3–5
Number of 48-trace source gathers	460
Source spacing	0.2–0.8 m
Receivers	14 Hz geophones
Number of receiver stations	205
Receiver spacing	0.2 m
Offset range	0.2–42.5 m
Number of channels	48
Seismograph	StrataView (24 bit A/D conversion)
Sampling interval	0.125 ms
Record length	250 ms
CMP bin size	0.1 m
Subsurface fold	50–80

arrivals probably travel through the bedrock with velocities of 1000–2500 m/s. A noticeable feature of this arrival is its irregular character, with traveltime undulations of up to 3 ms at offsets of 6.4–7.4 m.

After converting the seismic data to SEG-Y format, an automatic picker (PROMAX 1997) was used to identify the times of direct and refracted arrivals from the raw seismic sections (e.g. Fig. 4.4). This technique worked well for traces distinguished by high signal-to-noise ratios (S/N) and first arrivals that had travelled through consolidated bedrock (e.g. at offsets 4.4–9.0 m in Fig. 4.4a). Manual picking was required for traces distinguished by moderate S/N or interfering air and ground waves (e.g. at offsets 0.6–4.2 m in Fig. 4.4b). From the 22,080 recorded traces, the times of approximately 21,000 direct and refracted arrivals could be estimated with an accuracy of 1.0–2.0 ms.

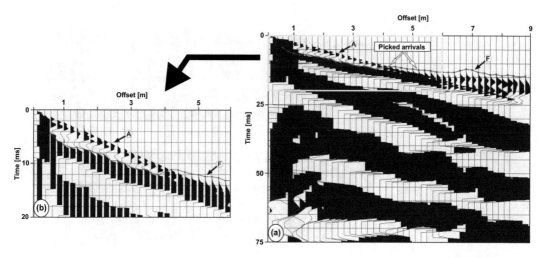

Fig. 4.4 Typical trace-equalised source gather showing the picks of the direct and refracted P-wave arrivals. Gain and plot parameters have been chosen to emphasise the direct and refracted arrivals (see Fig. 4.9 for a common-midpoint gather displayed using more conventional plot parameters). Note the variable nature of traveltimes at offsets of 6.4–7.4 m. (**b**) Zoom of region marked by yellow box in (**a**) highlighting the distinctly different arrival times of the high-frequency air waves and the later arriving low-frequency direct and refracted waves. The choice of the plotting parameters in (a) and (b) results in clipping. A—airwave; red line marked F—direct and refracted arrivals

4.7   Tomographic Inversions with Different Input Models

Traveltime inversion is a non-linear problem that usually requires a starting model close to the true subsurface model (Lanz et al. 1998). Since ray-paths through this starting model need to sample the entire depth of interest, velocities should increase gradually with depth. Any low-velocity regions with limited lateral extent would be detected during the inversion process, whereas laterally extensive low-velocity zones bounded by higher velocity layers are unlikely to be resolved on the basis of direct and refracted arrival times alone (Lanz et al. 1998; De Iaco et al. 2003).

To determine the robustness of the inversion process and the capability of the picked arrivals to constrain the subsurface velocity models, a large number of 1-D starting velocity models were tested. Representative P-wave tomograms that result from three of these models are shown in Figs. 4.5, 4.6, and 4.7. A series of tests to determine optimum smoothing and damping values was also conducted. Tomograms shown in Figs. 4.5, 4.6, and 4.7 were

obtained after 20 iterations of the inversion scheme that was controlled 5% by the damping, 47% by the smoothing and 48% by the data.

All three initial 1-D P-wave velocity models begin with a velocity of 150 m/s at the surface. They have gradients of 500, 200 and 100 m/s/m (Figs. 4.5a, 4.6, and 4.7a, respectively). Below 12 m depth, the initial models have a constant velocity of 3000 m/s. As shown in Figs. 4.5c, 4.6, and 4.7c, the P-wave tomograms that result from inverting the traveltime data using the three different initial models are very similar. To highlight features that may delineate the location of cavities (i.e. regions of anomalously low velocity), the full colour scale is used to represent only a limited range of velocities (0–1500 m/s), with all velocities >1500 m/s appearing as the same reddish brown colour. Calculated arrival times for the tomograms in Figs. 4.5c, 4.6, and 4.7c match the observed times equally well (e.g. Figs. 4.5b, 4.6, and 4.7b), with the root-mean-square differences varying between 1.6 and 1.7 ms (recall, that the estimated picking errors are 1.0–2.0 ms). The ray diagrams shown in Figs. 4.5d, 4.6, and 4.7d demonstrate that the

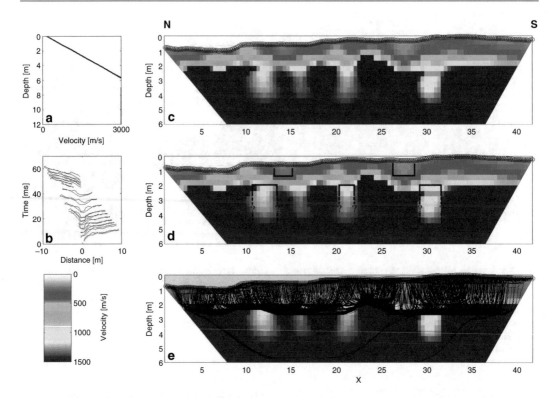

Fig. 4.5 Various diagrams showing the results of tomographically inverting all picked arrival times from the seismic data recorded at the Saqqara survey site (for location see Fig. 4.1b). (**a**) Initial 1-D model with 150 m/s velocity at the surface and a velocity gradient of 500 m/s/m. At depths greater than 12.0 m, the velocity is a constant 3000 m/s. (**b**) Comparisons of observed P-wave traveltimes from selected source records (blue lines) with traveltimes calculated from the final model (red lines). Except for source 1, all traveltime curves are shifted by 4 ms with respect to the previous one for display purposes. (**c**) Final tomogram that results from inverting all picked arrival times. Black circles at the surface identify source and geophone positions. To emphasise shallow parts of the tomogram (<4 m depth; all depths are relative to the highest topographic point of the survey line), the colour scale is limited to 0–1500 m/s, such that velocities >1500 m/s appear with the same reddish brown colour. (**d**) As for (c), with the thick black lines delineating regions of anomalously low seismic velocity that may represent hidden shafts, tunnels and/or tombs; continuous lines outline areas well sampled by rays, whereas dashed lines outline areas that are either poorly sampled or not sampled at all by rays in the final model. (**e**) As for (c), with the corresponding ray paths. (c), (d) and (e) are plotted with a vertical: horizontal exaggeration of 1.65:1

subsurface is well sampled in the upper 2.0–3.0 m, with the maximum depth of dense sampling dependent on the location. It can be concluded that velocities at shallow depths are well resolved. In contrast, the rather different distributions of rays at greater depths in the three tomograms suggest that lateral variations of velocity below ~3.0 m are not well resolved, despite the >30 m length of the recording spread. The similarity of the velocity tomograms at depths greater than ~3.0 m is a result of the smoothing imposed on the inversion process; as the inversion progresses, the effects of the low-velocity blocks are gradually pulled down into the poorly sampled regions of the tomograms. It is only possible to state with confidence that velocities below ~3.0 m depth are mostly >1500 m/s.

4.8 Final P-wave Velocity Tomogram

The similarity of the P-wave velocity tomograms in Figs. 4.5c, 4.6, and 4.7c demonstrates that inversions of the direct and refracted arrival times are only weakly dependent on the initial velocity model. Since these tomograms are so alike, our

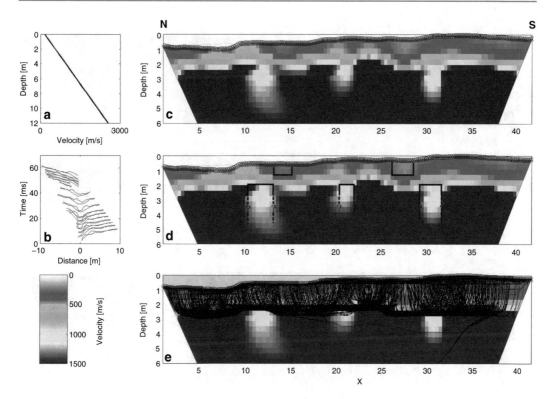

Fig. 4.6 (**a**) Initial 1-D model with 150 m/s velocity at the surface and a velocity gradient of 200 m/s/m. At depths greater than 5.7 m, the velocity is a constant 3000 m/s. (**b**), (**c**), (**d**) and (**e**) are as for (b), (c), (d) and (e) in Fig. 4.5

discussion and interpretation will be limited to that shown in Fig. 4.6. Moreover, because velocities at the two ends of this tomogram are close to the initial input model, they are not included in the interpretation; blank triangular zones in the tomograms of Figs. 4.5, 4.6, and 4.7 delineate regions that are not sampled by rays. In the following, all depths are given relative to the topographically highest point of the seismic profile near its southern end.

To a first-order approximation, the P-wave velocity structure beneath the mastaba is horizontally layered (note the vertical exaggeration of Fig. 4.6), in accord with the flat-lying geology of this region. The top layers with velocities in the range 150–650 m/s extend from the surface to ~2.0 m depth. They comprises variable thicknesses of loose sand overlying friable white limestone. There is one broad region of anomalously high velocity (>500 m/s) that reaches the surface between x = 5 and 11 m and two regions of anomalously low-velocity (150–350 m/s), one centred near x = 13 m and the other centred near

x = 27 m (see shallow regions marked by solid lines in Fig. 4.6d).

At ~2 m depth, velocities increase abruptly from uniformly less than 650 m/s to generally more than 1000 m/s, with most values >1500 m/s. Below this level, layers of consolidated limestone and shale are observed in numerous shafts within and near to the mastaba (Fig. 4.1b). Particularly prominent at depths >2 m are three blocks of unusually low velocity centred at x = ~12.0, ~21.5 and ~31.0 m. The depths and lateral positions of the upper boundaries of these low-velocity blocks appear to be quite well determined, whereas their thicknesses are not resolved (note the number of rays that sample the relevant regions in Fig. 4.6e).

4.9 Reflection Data Processing

A relatively standard processing scheme (Büker et al. 1998; De Iaco et al. 2003; Schmelzbach

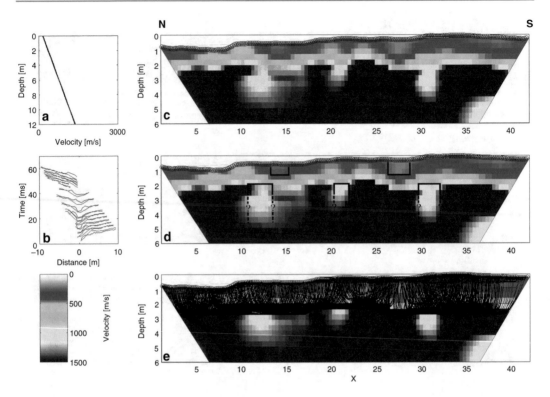

Fig. 4.7 (a) Initial 1-D model with 150 m/s velocity at the surface and a velocity gradient of 100 m/s/m. At depths greater than 12.0 m, the velocity is a constant 3000 m/s. (b), (c), (d) and (e) are as for (b), (c), (d) and (e) in Fig. 4.5

et al. 2005) was used to enhance the ultra-shallow P-wave reflections at the expense of various types of source-generated noise (i.e. direct, refracted, guided, surface and air waves; Robertsson et al. 1996a,b; Roth et al. 1998). In particular, various editing, filtering, static correction and display techniques were employed to improve the quality of reflections at traveltimes <50 ms (Fig. 4.8). The data processing benefited from the high fold of the data (Table 4.1); it is common for image quality to improve as the subsurface coverage increases (Lanz et al. 1996).

To demonstrate the effectiveness of the processing scheme, a typical CMP gather is used as an example (Fig. 4.9). Once the line geometry was established and the data edited (i.e. dead or poor quality traces eliminated), the trace-normalised raw shot and common-midpoint (CMP) gathers were inspected. Shot and CMP gathers in this form were dominated by high-amplitude low-frequency direct, refracted and guided phases and high-frequency air-waves

(Figs. 4.4 and 4.9a). At this early stage of the processing, there was little evidence for the presence of shallow reflections.

After application of a 40 ms automatic gain control (AGC) function to equalise amplitudes along the lengths of the traces and elevation static corrections to account for very small variations in topography, the data were subjected to spiking deconvolution with a filter length of 40 ms (Sheriff and Geldart 1995; Yilmaz 2001). This latter process was successful in reducing the effects of source-generated noise. Following these operations, a prominent hyperbolic reflection R could be discerned at very early times (<25 ms) on most CMP gathers (e.g. Fig. 4.9b).

Application of a 90-200-500-700 Hz bandpass frequency filter was then used to reduce random noise, with the upper limit being chosen to minimise the effects of high-frequency noise and the lower limit being selected to provide sufficient bandwidth to avoid ringing (Fig. 4.9c). Since the direct, refracted and air waves had frequencies

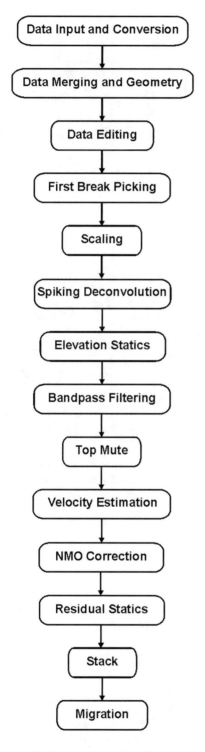

Data processing flow

Fig. 4.8 Flow chart of ultra-shallow seismic reflection data processing at Saqqara archaeological site

and move-outs that overlapped with those of the prominent shallow reflection, they could not be removed completely using standard frequency or multichannel filters (Büker et al. 1998; Spitzer et al. 2001). Consequently, it was necessary to apply carefully selected hand-picked top-mute functions to eliminate these unwanted arrivals (Fig. 4.9c).

To determine normal-moveout (NMO) corrections, apparent velocities of the prominent ultra-shallow reflection R were determined via a series of semblance analyses. Moveout velocities were found to vary from 230–380 m/s along the length of the profile (e.g. see moveout velocity marked on Fig. 4.9b and c). The effects of very near-surface velocity heterogeneities on reflection coherency were somewhat reduced after estimating and applying surface-consistent residual static corrections (maximum allowed time shifts were ± 1 ms). Application of NMO corrections with maximum stretching of 30% (wavelets with greater stretch were automatically muted) and residual static corrections resulted in CMP gathers on which reflection R appeared as a coherent flat event (e.g. Fig. 4.9d) ready for stacking.

4.10 Ultra-Shallow Seismic Reflection Section

The P-wave seismic reflection section in Fig. 4.10a was the result of stacking the CMP gathers using the laterally varying velocities determined for the ultra-shallow reflection R. Using a smoothed version of these same velocities, the time section of Fig. 4.10a was then converted to the depth section of Fig. 4.10b.

Although migration is rarely applied to ultra-shallow seismic reflection data (see discussion by Black et al. 1994), the moderate dip of reflection R between x = 9 m and x = 14 m suggested that such an operation may be useful. To test this, Fig. 4.10c shows a time-migrated version of the northern half of the seismic section; the rather noisy character of the ultra-shallow seismic reflection beneath the southern half of the seismic profile resulted in a poor migrated image.

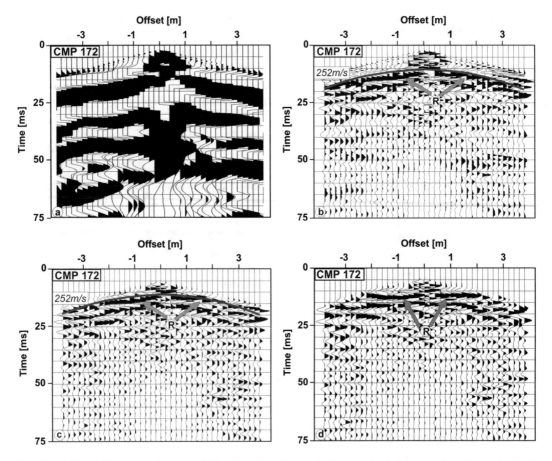

Fig. 4.9 (**a**) Typical trace-equalised raw CMP gather. (**b**) As for (a) after amplitude scaling using a 40 ms automatic gain control (AGC) function and application of spiking deconvolution with an operator length of 40 ms and elevation static corrections. Red line delineates reflection hyperbola R (velocity is shown in the adjacent box). (**c**) As for (b) after application of a top mute function to remove direct, refracted and air waves and bandpass filtering (90-200-500-700). (**d**) As for (c) after application of NMO and residual-static corrections

Figure 4.10c suggests that the reflection R may be discontinuous over a distance of ~2 m centred about x = 13 m (but see discussion in the next section).

Figure 4.10, which is plotted with a vertical: horizontal exaggeration of roughly 3:1, emphasises fine details of the P-wave velocity tomogram and seismic reflection section. To demonstrate the dominant horizontal nature of the reflecting boundary and related velocity structure, the same information is plotted with no vertical exaggeration in Fig. 4.11. Examination of the seismic reflection sections in Figs. 4.10 and 4.11 demonstrates that the single ultra-shallow reflection R distinguished by two peaks and an inter-

mediate trough has been imaged along most of the mastaba's long axis.

4.11 Interpretation

Before presenting our interpretation of the ultra-shallow seismic data, it is worthwhile to consider the wavelengths of the seismic waves that have travelled through the ground. Wavelength is a function of the frequency and the velocity of the waves. The dominant frequency of reflection R is ~250 Hz, whereas the seismic P-wave velocities are ~150 m/s for the loose sands, 400–650 m/s for the friable limestone and >1500 m/s for the

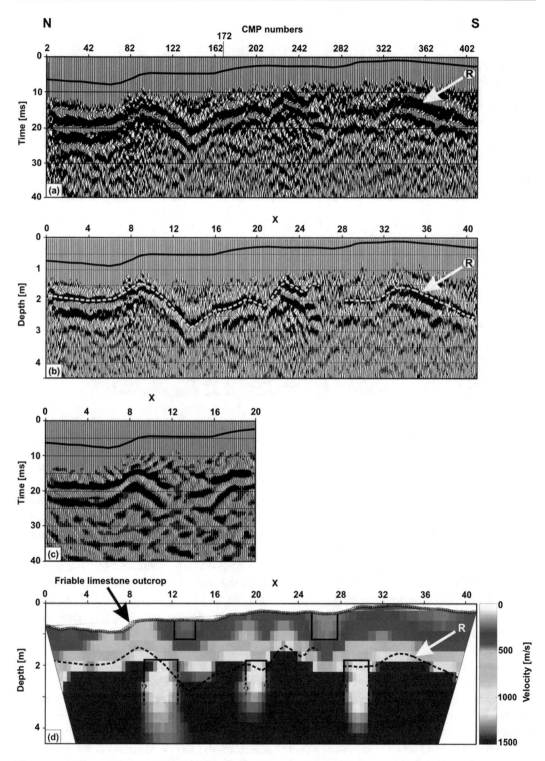

Fig. 4.10 (**a**) Stacked P-wave section after application of the processing scheme outlined in Fig. 4.8. (**b**) As for (a) after simple time-to-depth conversion using a smoothed version of the stacking velocities. (**c**) Time-migrated version of the stacked seismic section between CMP's 2 and 202. Second part of the section did not migrate well due to the presence of incoherent and presumed out-of-plane events. (**d**) Comparison of the depth to the picked reflector (dashed line in (b)) and the final tomographic refraction model of Fig. 4.6d. (b) and (d) are plotted with a vertical:horizontal exaggeration of 2.9:1

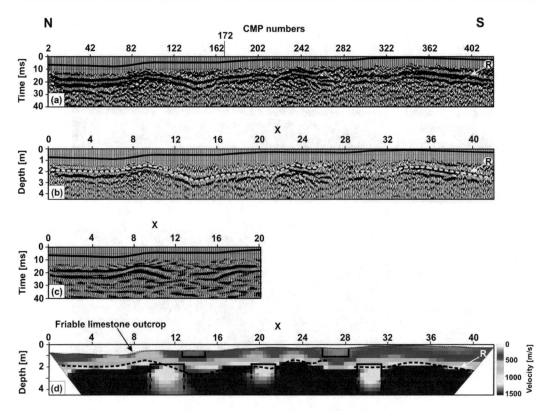

Fig. 4.11 As for Fig. 4.10, but with no vertical exaggeration

consolidated limestone and shale. Combining this information demonstrates that the seismic wavelengths vary from ~0.6 m in the loose sands through 1.6–2.6 m in the friable limestone to >6 m in the consolidated limestone and shale. Dividing these wavelengths by a factor of 4 yields conservative estimates for the vertical resolution of reflections in the different media (Sheriff and Geldhart 1995).

Based on visual observations and the seismic velocities shown in Figs. 4.10d and 4.11d, we conclude that the thickness of the loose sand varies along the length of the mastaba from zero to a few tens of centimetres. Friable limestone directly underlies the sand to a uniform depth of ~2 m below the highest topographic point of the survey line. The relatively high velocities (>500 m/s) that reach the surface between x = 5 m and x = 11 m are associated with an outcrop of the friable limestone. We interpret the two anomalously low velocity blocks that approach the surface at x = 13 m and x = 27 m

as shallow shafts filled with sand. Over time, this fill has been compacted, such that the velocities attain values up to 350 m/s.

Figures 4.10d and 4.11d suggest that the high-amplitude reflection R is generated at the interface across which the velocity increases abruptly from less than 650 m/s to more than 1000 m/s. As a consequence, it is interpreted as a reflection from the contact between the friable limestone and the underlying consolidated limestone and shale. Undulations and discontinuities in this reflection may be artefacts caused by abrupt lateral velocity variations. Where it shallows near x = 9 m, anomalously high velocities of the friable limestone reach the surface (Figs. 4.10d and 4.11d). In contrast, the adjacent deepening or disruption of reflection R near x = 13 m and its disruption near x = 27 m occur beneath the anomalously low velocities of the interpreted sand-filled shafts. It is well known that velocities based on semblance analyses or tomographic inversions may be inaccurate in regions characterised by abrupt lateral velocity changes. Moreover, when standard processing procedures are employed (e.g. Büker et al. 1998), strong lateral velocity variations may result in a deterioration of reflection quality, even if the reflecting horizon is continuous.

The distinct low velocity blocks that occur within the consolidated limestone and shale are the most interesting features revealed by our study (Figs. 4.10d and 4.11d). It is notable that the tops of these blocks lie at a uniform depth of 2 m, coincident with the upper boundary of the consolidated limestone and shale. Considering the number and extent of voids that are known to exist beneath the mastaba, it is likely that the low velocity blocks represent tunnels and/or subsurface chambers. That the estimated 1000–1200 m/s velocities are higher than values expected for air-filled cavities is not unexpected, because various synthetic studies (e.g. Musil et al. 2003) have shown that estimated velocities within cavities based on standard least-squares tomographic inversions are invariably too high. Taking into account the long wavelength nature of the seismic waves at 2 m depth and the variability in the tomograms of Figs. 4.5a, 4.6a and 4.7a, the times of the direct and refracted arrivals in the seismic data are consistent with the existence of three cavities of nearly uniform width (2–4 m) separated by regular distances of ~9.5 m.

4.12 Conclusions

An ultra-shallow seismic data set has been recorded along the length of an Early Dynastic mastaba within the ancient Saqqara necropolis. The short receiver and source spacings and high fold of the data (Table 4.1) resulted in very high-resolution subsurface information. The most important details were provided by velocity tomograms determined from inversions of the direct and refracted P-wave traveltimes (Figs. 4.5, 4.6, and 4.7). These tomograms revealed dominantly horizontally layered structures beneath the mastaba, consistent with the flat-lying geological units observed in numerous shafts within and near the study site. The

upper layer of loose sand and friable white limestone had velocities of 150–650 m/s. It extended from the gently north-dipping surface (elevation change of less than 1 m over a distance of ~32 m) to a nearly horizontal boundary at ~2.0 m depth below the highest point of the survey line. This well-defined boundary was the likely source of a strong reflection that extended along the length of the mastaba (Figs. 4.10 and 4.11). Underlying this boundary were layers of consolidated limestone and shale with velocities uniformly greater than 1000 m/s, with most values >1500 m/s.

Two anomalously low velocity blocks found within the upper geological layers and three discovered within the lower layers of consolidated limestone and shale could represent structures of archaeological significance (Figs. 4.10d and 4.11d). Those within the upper layers may delineate sand-filled shafts, whereas those within the lower layers may outline the upper boundaries of tunnels and/or subsurface chambers. Abrupt velocity variations may have affected the processing of the seismic reflection data, such that depressions or discontinuities in the principal ultra-shallow reflection R were observed below the two shallow low-velocity blocks.

Acknowledgements We thank various staff at National Research Institute of Astronomy and Geophysics for their contributions to the fieldwork, the National Research Institute of Astronomy and Geophysics and the Swiss National Science Foundation for financially supporting the project and SASKS for supporting the first author's study period in Switzerland.

References

Bachrach R, Mukerji T (2001) Fast 3D ultra-shallow seismic reflection imaging using a portable geophone mount. Geophys Res Lett 28:45–48

Bachrach R, Nur AM (1998) High-resolution seismic experiments in sand, Part 1: Water table, fluid flow, and saturation. Geophysics 63:1225–1233

Bachrach R, Nur AM, Dvorkin J (1998) High-resolution seismic experiments in sand, Part 2: velocities in shallow unconsolidated sands. Geophysics 63:1234–1240

Baines J, Malek J (1992) Atlas of Ancient Egypt. Andromeda, Oxford

Baker GS, Schmeissner C, Steeples D (1999) Seismic reflections from depths of less than two meters. Geophys Res Lett 26:279–282

Baker GS, Steeples DW, Schmeissner C, Spikes KT (2000) Ultra-shallow seismic reflection monitoring of seasonal fluctuations in the water table. Environ Eng Geosci VI:271–277

Baker GS, Steeples DW, Schmeissner C, Pavlovic M, Plumb R (2001) Near-surface imaging using coincident seismic and GPR data. Geophys Res Lett 28:627–630

Beck A, Weinstein-Evron M (1997) A geophysical survey in the el-Wad Cave, Mount Carmel, Israel. Archaeol Prospect 4:85–91

Black AC, Norton WW (1993) Blue Guide Egypt. Bedford, London

Black R, Steeples D, Miller R (1994) Migration of shallow seismic reflection data. Geophysics 59:402–410

Branham K, Steeples D (1998) Cavity detection using high-resolution seismic reflection methods. Min Eng 40:115–119

Büker F, Green AG, Horstmeyer H (1998) Shallow seismic reflection study of a glaciated valley. Geophysics 63:1395–1407

Cardimona SJ, Clement WP, Kadinsky-Cade K (1998) Seismic reflection and ground penetrating radar imaging of a shallow aquifer. Geophysics 63:1310–1317

De Iaco R, Green AG, Maurer HR, Horstmeyer H (2003) Seismic reflection and refraction study of a landfill. J Appl Geophys 52:139–156

Deidda GP, Balia R (2001) An ultra-shallow SH-wave seismic reflection experiment on a subsurface ground model. Geophysics 66:1097–1104

Deidda GP, Ranieri G (2001) Some SH-wave seismic reflections from depths of less than 3 metres. Geophys Prospect 49:499–508

Dolphin LT (1981) Geophysical methods for archaeological surveys in Israel. Menlo Park, CA, Stanford Research International

Goulty NR, Hudson AL (1994) Completion of the seismic refraction survey to locate the Vallum at Vindobala, Hadrian's Wall. Archaeometry 36:372–335

Goulty NR, Gibson JPC, Moore JG, Welfare H (1990) Delineation of the Vallum at Vindobala, Hadrian's Wall, by shear-wave seismic refraction survey. Archaeometry 32:71–82

Hildebrand JA, Wiggins SM, Henkart PC, Conyers LB (2002) Comparison of seismic reflection and ground-penetrating radar imaging at the controlled archaeological test site, Champaign, Illinois. Archaeol Prospect 9:9–21

Karastathis VK, Papamarinopoulos S (1997) The detection of King Xerxes' canal by the use of shallow seismic reflection and refraction seismics-preliminary results. Geophys Prospect 45:389–401

Karastathis VK, Papamarinopoulos S, Jones RE (2001) 2-D velocity structure of the buried ancient canal of Xerxes: an application of seismic methods in archaeology. J Appl Geophys 47:29–43

Kinnaer J (2003) The ancient Egyptian site. http://www.ancient-egypt.org/index.html

Kourkafas P, Goulty NR (1996) Seismic reflection imaging of gypsum mine working at Sherburn-in-Elmet, Yorkshire, England. Eur J Environ Eng Geophys 1:53–63

Lanz E, Maurer HR, Green AG (1998) Refraction tomography over a buried waste disposal site. Geophysics 63:1414–1433

Lanz E, Pugin A, Green AG, Horstmeyer H (1996) Results of 2- and 3-D high resolution seismic reflection surveying of surficial sediments. Geophys Res Lett 23:491–494

Miller R, Steeples D (1991) Detecting voids in a 0.6 m coal seam, 7 m deep, using seismic reflection. Geoexploration 28:109–119

Miller R, Xia J, Harding R, Neal J, Fairborn J, Steeples D (1995) Seismic investigation of a surface collapse feature at Weeks Island Salt Dome, Luisiana. Am Assoc Petrol Geol Div Environ Geosci 2:104–112

Musil M, Maurer HR, Green AG (2003) Discrete tomography and joint inversion for loosely connected or unconnected physical properties: application to crosshole seismic and georadar data sets. Geophys J Int 153:389–402

Ovenden SM (1994) Application of seismic refraction to archaeological prospecting. Archaeol Prospect 1:53–63

PROMAX (1997) A reference guide for the ProMAX Geophysical Processing Software (2-D,7.0), 2 volumes. Advanced Geophysical Corporation, Englewood, CO

Raffaele F (2003) Early dynastic Egypt. http://members.xoom.virgilio.it/francescoraf/index.html

Robertsson JOA, Holliger K, Green AG (1996a) Source-generated noise in shallow seismic data. Eur J Environ Eng Geophys 1:107–124

Robertsson JOA, Holliger K, Green AG, Pugin A, De Iaco R (1996b) Effects of near-surface waveguides on shallow high-resolution seismic refraction and reflection data. Geophys Res Lett 23:495–498

Roth M, Holliger K, Green AG (1998) Guided waves in near-surface seismic surveys. Geophys Res Lett 25:1071–1074

Said R (1990) The geology of Egypt. Balkema, Rotterdam

Schmelzbach C, Green AG, Horstmeyer H (2005) Ultra-shallow seismic reflection imaging in a region characterised by high source-generated noise. Near Surf Geophys 3:33–46

Sheriff RE, Geldart LP (1995) Exploration seismology, 2nd edn. Cambridge University Press, Cambridge

Sitek D (2003) Ancient Egypt: history and chronology. http://www.narmer.pl/indexen.html

Spitzer R, Nitsche FO, Green AG (2001) Reducing source-generated noise in shallow seismic data using linear and hyperbolic t-p transformations. Geophysics 66:1612–1621

Steeples DW, Knapp RW, McElwee CD (1986) Seismic reflection investigations of sinkholes beneath interstate highway 70 in Kansas. Geophysics 51:295–301

Tsokas GN, Papazachos CB, Vafidis A, Loukoyiannakis MZ, Vargemezis G, Tzimeas K (1995) The detection of monumental tombs buried in Tumuli by seismic refraction. Geophysics 60:1735–1742

Vendel O (2002) Absolute egyptology. http://nemo.nu/ibisportal/0egyptintro/index.html

Yilamz O (2001) Seismic data analysis. Processing, inversion and interpretation of seismic data, volumes 1 and 2. Society of Exploration Geophysics, Tulsa, OK

Zelt CA, Smith RB (1992) Seismic traveltime inversion for 2-D crystal velocity structure. Geophys J Int 108:16–34

Archaeological Geophysics in Portugal: Some Survey Examples

5

António Correia

Abstract

The first attempts to apply geophysical methods to archaeological sites in Portugal date from the mid-sixties of the last century. Since then, geophysical methods have been used more and more frequently to help with archaeological site recognition, delineating buried structures, and help with excavating strategies. The first geophysical methods used in Portugal were geoelectrical methods followed by magnetic methods. Today these two methods are still used but the georadar and the electrical resistivity tomography methods have also been used on a routine basis whenever local conditions permit.

Four archaeological sites will be described as examples on the use of geophysical methods in Archaeology. Two of them are from roman times (the Roman Villa of Tourega in central Portugal and the Roman town of Troia in the west coast of Portugal), one is from Neolithic times (a burial mound in central Portugal) and the last one is a recent archaeological site (eighteenth century) and has to do with the location of a crypt known to exist in the garden of the Portuguese Legislature in Lisbon.

Only electrical resistivity tomography and georadar were used. The sites were chosen because in all of them there were already previously excavated areas or there were plans for future excavation. When choosing these sites the idea was to be able to compare the interpretations of the geophysical data with the results of future excavations.

Keywords

Portugal · Geophysics · Neolithic site · Roman site · Nineteenth century crypt

5.1 Introduction

To the author's knowledge, the first geophysical methods used in archaeological prospection in Portugal date from the early sixties of the last century (dos Santos and Esteves 1966; Tite and Alldred 1965–1966). Since then many other researchers have been using different geophysical methods for detecting, delineating, and studying areas where archaeological remains are suspected to exist underground.

Geoelectrical methods were the first ones to be used in archaeological prospecting in Portugal. In the beginning of the nineties of the last century, in addition to geoelectrical methods other methods such as magnetics and georadar began to be used.

Nowadays, and following the general trend around the world, almost all geophysical methods

A. Correia (✉)
Department of Physics and Institute of Earth Sciences, University of Évora, Évora, Portugal
e-mail: correia@uevora.pt

© Springer International Publishing AG, part of Springer Nature 2019
G. El-Qady, M. Metwaly (eds.), *Archaeogeophysics*, Natural Science in Archaeology,
https://doi.org/10.1007/978-3-319-78861-6_5

are used to study archaeological sites. Georadar (with several antennae), electromagnetic methods, electrical resistivity tomography in two and three dimensions, magnetic gradiometer surveys and magnetic susceptibility surveys are routinely used in archaeology as a means of uncovering buried artefacts in sites with archaeological interest.

In this chapter four archaeological sites where geophysical methods were used are presented. The first is a Neolithic burial mound called Anta das Moitas; it is located in central Portugal near the town of Proença-a-Nova. The second site is an isolated roman villa located near the town of Évora in Central Portugal (the Roman villa of Tourega). The third site is a Roman town with a fish paste factory located near the sea and close to the town of Setubal (Roman ruins of Troia). The last site is a crypt which is located under the garden and parking lot of the Portuguese Legislature in Lisbon. All the sites that will be described in this chapter will be excavated sooner or later which means the geophysical interpretations will be compared with new information from excavation activities.

Figure 5.1 shows the locations of the four archaeological sites.

5.2   The Archaeological Sites

5.2.1   Neolithic Burial Mound of Anta das Moitas in Proença-a-Nova

5.2.1.1   Introduction
The municipality of Proença-a-Nova, in cooperation with Emerita Ltd., has been excavating several archaeological sites near the town of Proença-a-Nova. The region is well known for the abundance of archaeological sites from the Neolithic period. One of the sites (see Fig. 5.1 for location), near the village of Moitas, is a Neolithic burial mound, known as Anta das Moitas (Fig. 5.2). The excavation in the site started in the summer of 2013 and is still progressing (Fig. 5.3); however, before excavating the site a geophysical survey using ground penetrating radar (GPR) and electrical

resistivity tomography (ERT) was done as an attempt to find the location of the burial mound's chamber and its main entrance. Both ERT and GPR profiles were measured along the same directions shown in Fig. 5.2. In principle GPR would allow the identification of the slabs of schist which form the walls and the cover of the dolmen and the possible entrance, from the clay and silt that cover the structure. Since the moisture in the soil was relatively large, ERTs should also give information about depth and orientation of the schist slabs, which have higher electrical resistivity than the soil.

Since 2013 there has been an archaeology summer school funded by the municipality of Proença-a-Nova to excavate and prepare students in archaeological activities and, at the same time, improve and allow the access of the general public to the sites. All these activities are integrated in a wider study of pre-historical dolmen burial sites that is taking place in Portugal.

For those interested in seeing the area of the burial mound, the geographical coordinates in the Google Earth are: 39°43′28.50″N, 7°51′33.37″W. The average altitude of the site is 375 m a.s.l.

5.2.1.2   Method
For the Anta das Moitas archaeological site two geophysical methods were used; electrical resistivity tomography and ground penetrating radar (Fig. 5.4). Both were carried out along the profiles shown in Fig. 5.2; however, for the GPR method three parallel profiles, 0.5 m apart, were carried out along the two profiles. Both ERT and GPR profiles were 39 m long. The ERT profiles were done using a Wenner configuration with 40 stainless steel stacks 1 m apart. As can be seen in Fig. 5.2, profile 1 crosses the centre of the mound. In the figures where the ERT profiles are shown bluish colours represent low electrical resistivities and reddish colours represent high electrical resistivities.

5.2.1.3   Some Results
Figures 5.5 and 5.6 show the ERT obtained along the profiles 1 and 2. No figures are shown for the GPR profiles done; the results were inconclusive, as is explained later.

Fig. 5.1 Location of the four archaeological sites (red triangles) of this chapter. 1 refers to the burial mound in Proença-a-Nova, 2 the Roman villa of Tourega, 3 the Roman ruins of Troia, and 4 the Portuguese Legislature in Lisbon

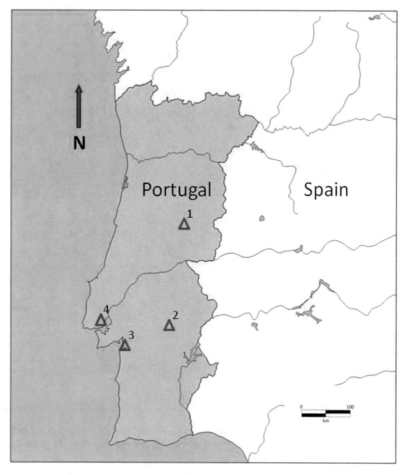

The ERT along profile 1 (Fig. 5.5) shows that there are basically three areas, from left to right: a shallow reddish area near the limits of the profile (between 0 and 13 m, and between 28 and 38 m) which show high electrical resistivity values; a central and shallow area also with high electrical resistivities but lower than in the first area (yellow and brown colours) (between 17 and 19 m); a deeper area with bluish colours (between 8 and 17 m, and between 21 and 27 m) with relatively low electrical resistivities.

The first area in the ERT was interpreted as a zone in the mound with blocks of superficial rocky material which were visible after cleaning the first layer of the soil covering it. The second area was interpreted as the possible entrance to the chamber of the dolmen which is assumed to

be full with soil and small rocks/pebbles fallen from the upper part of the ground. The third area was interpreted as finer soil (clay or silt) saturated with water.

As the excavation proceeded it was apparent that the geophysical interpretation was close to what was being uncovered (Fig. 5.3).

The ERT along profile 2 (Fig. 5.6) shows that, in geoelectrical terms, there are basically two areas: a shallow area (with reddish colours) located in the extremes of the profile (between 0 and 12 m and between 27 and 39 m) with high electrical resistivity values, which was interpreted, as in profile 1, as a zone covered with blocks of superficial rocky material; a central area (between 11 and 26 m, with bluish colours) with depths that vary between 1 and 5 m was

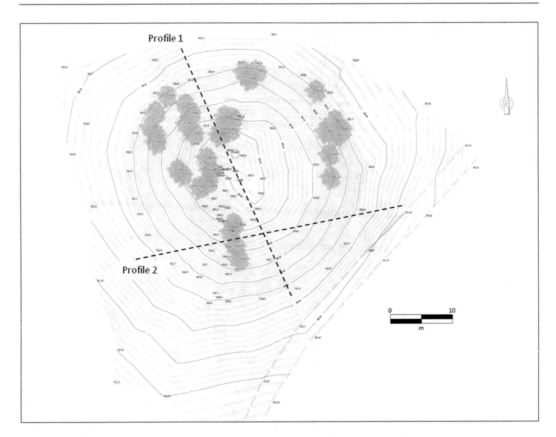

Fig. 5.2 Topographic map of the mound's area (Anta das Moitas). Profile 1 and 2 indicate the orientation of the two electrical resistivity tomography and ground penetrating radar profiles done. In green the location of trees

interpreted as clayey or silty material that was used to fill the area around the dolmen, which was confirmed during the excavation stages.

The results from the ERT profiles and from the excavation allowed understanding as to why the GPR did not give any good results in this particular archaeological site. As a matter of fact, after starting the excavation and cleaning the first layers of soil it was seen that they covered blocks of rocks that are used to protect the filling material (clay and silt) that was used to cover the dolmen. These rocks behaved as intense diffractors of electromagnetic energy making the obtained radargrams not very useful for a geophysical interpretation of the buried structures in the ground.

5.2.1.4   Conclusions

From the two geophysical methods used up to now in the Anta das Moitas archaeological site,

only the electrical resistivity tomography has shown good potential to detect and delineate the structure of the dolmen buried in the site. This contrasts with ground penetrating radar which was not very useful to detect those same structures.

With the ERT profiles it was possible to infer that the mound was covered by blocks of rocks which were placed on top of clay and silt, possibly to protect them from erosion. The slabs of schist that compose the walls and the cover of the dolmen chamber were also identified by the ERT profile 1.

Future geophysical surveys will concentrate on trying to discover the main entrance of the dolmen which in Iberia is normally oriented to the east. It is also expected to use magnetic methods (magnetic and gradiometer surveys) in the summer school to take place in 2017.

Fig. 5.3 Excavated area at the end of the summer of 2014. The red arrow indicates the geographical north. The blocks that constitute a protection cover for the clay and silt underneath can be seen as well as the slabs of schist that make the walls of the dolmen

Fig. 5.4 Using the GPR during the summer of 2013 along profile 2 of Fig. 5.2

This archaeological site appears to be a very interesting place to test several geophysical techniques, even more so because it will be completely excavated in the near future; this will allow comparing geophysical interpretations from several methods with the results of the excavation.

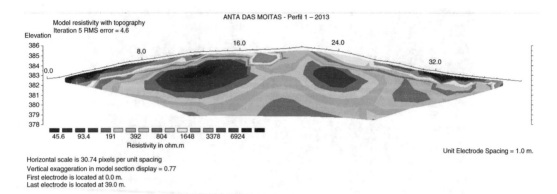

Fig. 5.5 Electrical resistivity tomography along profile 1. Bluish colours correspond to low electrical resistivities while reddish colours correspond to high electrical resistivities. See text for interpretation

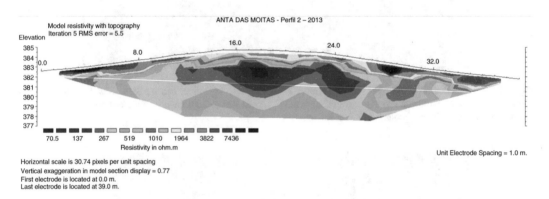

Fig. 5.6 Electrical resistivity tomography along profile 2. Bluish colours correspond to low electrical resistivities while reddish colours correspond to high electrical resistivities. See text for interpretation

5.2.2   Roman Villa of Tourega (I–IV a.D.)

5.2.2.1   Introduction

In the process of locating and mapping the most appropriate archaeological site for testing new ground penetrating radar (GPR) acquisition techniques, a subsurface survey of the surroundings of exposed structures was conducted in the Roman villa of Tourega. The villa is located about 15 km southwest of the town of Évora, in the Alentejo region in central Portugal. A bathhouse structure as well as a large water tank reservoir have been previously excavated (Fig. 5.7). At this particular site only GPR methods were used with the goal of finding and delineating possible extensions of the villa

complex; the main GPR target was then to identify linear archaeological structures, basically building walls.

From archaeological artefacts it was possible to infer that the villa was occupied from the first to the fourth century a.D. A funerary inscription for a roman senator, dating from the early third century a.D., suggests the villa belonged to a senatorial family for some period and the pottery found indicates a connection of Tourega to roman trade routes of the time (Vaz Pinto et al. 2004). Figure 5.8 shows an interpretation of the excavated structures.

This site was chosen for a geophysical survey using the GPR method because of the expected linear structures associated with roman buildings. Furthermore, the already excavated area could

Fig. 5.7 Picture of the remains of the Roman villa of Tourega, about 15 km southwest of the town of Évora in central Portugal

Fig. 5.8 Cartoon of the excavated portion of the site. There were three main phases of construction which are represented by a different colour

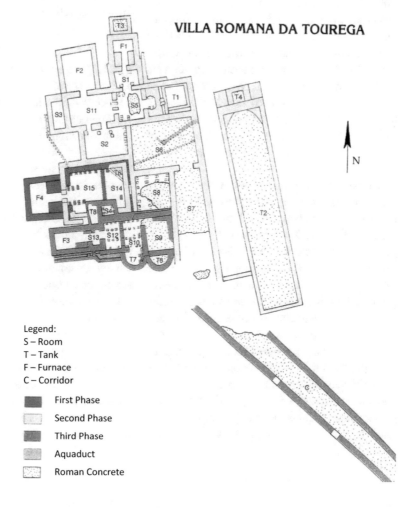

serve as guidance for type, orientation and depth of the expected structures.

For this archaeological site a summary of the results of these surveys is presented here as they contain useful information on the location and possible extension of still buried structures.

The location and depth information is of sufficient quality to be used in the planning of future excavations and in the planning of more extensive geophysical surveys with the sole objective of mapping archaeological remains.

For those interested in seeing the area of the roman villa, the geographical coordinates in the Google Earth are: 38°30′6.95″N, 8°01′41.38″W. The average altitude of the site is 199 m a.s.l.

5.2.2.2   Method

A Sensors & Software Inc. Noggin 500 MHz GPR system and a Sensors & Software Ink PulseEKKO system with bistatic 200 MHz antennae were used for the main subsurface mapping survey; however, the results obtained with the 200 MHz antennae will not be shown here. The survey was constrained to an area delimited to the North and West by the fence enclosing the archaeological site, to the East by the excavated site itself and areas of high grass and thick shrubbery, and to the South by another fence and zones of slightly more abrupt topography.

To make the acquisitions more convenient, a main grid was laid out and subdivided into several square or rectangular sub-grids. The most common line spacing used was 1 m, which is generally too coarse for 500 MHz data but was sufficient in our case for locating test areas. Three sub-grids were re-acquired using a more appropriate 0.50 m line spacing to assess the reliability and resolution degradation of the main data set. GPR lines were collected in both orthogonal directions (X and Y); the X axis approximately corresponds to the N–S direction and the Y axis approximately corresponds to the E–W direction. The radar antennae were dragged directly on the ground; data acquisition was generally complicated by the overgrown grass. A straight-line

progression was difficult to achieve, and a consistent and even pacing of the profiles was hard to maintain throughout the survey. This inevitably resulted in a degraded positioning accuracy which is difficult to quantify. Overall, the 500 MHz data consist of a total line length of 2180 m in the N–S direction and 1470 m in the E–W direction, plus an additional 220 m for the slanted grid. This represents a total of 3870 m.

A maximum time window of 75 nanoseconds (ns) was used, which, based on average wave velocity, corresponds to a maximum depth of investigation of approximately 3.5 m. 200 MHz and 500 MHz common-offset and 200 rapid multi-offset data were collected for processing experiments. Processing was standard and consisted of dewowing, time-zero shift, spherical and exponential gain, bandpass filtering, and fk migration.

Figures 5.9 and 5.10 show time slices for 8 and 10 ns, respectively. A velocity of 0.12 m/ns was used.

5.2.2.3   Some Results and Conclusions

The most obvious result is that GPR has proved to be successful in imaging buried stone structures at the Tourega site. GPR is used fairly routinely for the prospection and study of Roman period sites in areas with well developed soils and sedimentary bedrock, mostly in Northern Europe. The success in the case of structures built directly onto granitic bedrock with relatively little soil was not assured. The Tourega results are therefore important as they demonstrate that this technique can be used very effectively in a wide variety of conditions. An abundance of buried structures can be seen in direct connection with the end of the current excavation (Figs. 5.9, 5.10 and 5.11). The corridor does seem to end at the end of the excavation; it appears to be connected to another structure that makes an angle with it. It is clear that the south fence does not mark the end of the site in this direction. There is an obvious continuation of the structures S and W of the fence. The continuation of the structures to the E of the surveyed area is not so obvious but is

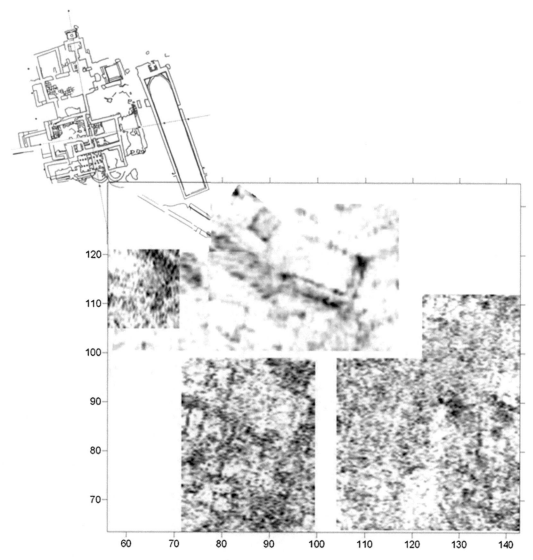

Fig. 5.9 Collage of the sketch of the excavated area and 8 ns (about 0.48 m depth) GPR slices for different areas surveyed. Distances in m. Sketch in the upper left corner indicates the location of the excavated area

likely. The results by themselves provide very clear evidence that significant structures will be found if the excavation is resumed.

5.2.3 Roman Town of Troia (I–V a.D.)

5.2.3.1 Introduction

The Roman Ruins of Troia are known since the sixteenth century. After several stages of excavation in the nineteenth and twentieth centuries it was finally established that the area near the tip of the Peninsula of Troia has been the place of a roman town with fish factories. The 25 factories identified up to now had a total of 160 tanks where fish was salted or transformed into fish paté or fish sauces (of which the *garum* was the most famous one). These products were appreciated by wealthy romans around the Roman Empire and so the town flourished from the first to the fifth century

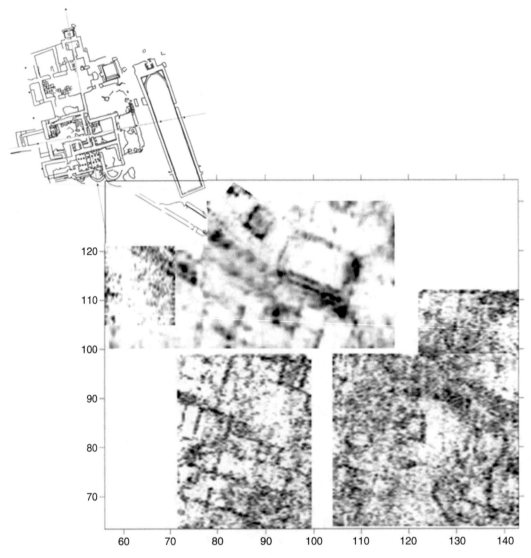

Fig. 5.10 Collage of the sketch of the excavated area and 10 ns (about 0.69 m depth) GPR slices for different areas as in Fig. 5.3. Distances in m. Sketch in the upper left corner indicates the location of the excavated area

a.D. Archaeological information indicates that fish factories finished their activity in the first half of the fifth century, and the town was completely abandoned in the sixth century. Even though a large area has been already excavated, uncovering dwellings and several necropolis, many other areas have not been excavated.

Since 2006 the roman ruins of Troia belong to the Troia Resort which, following up on a suggestion by the Geophysical Centre of the University of Évora, has allowed the surveying of areas

that were not excavated yet but would be in the future. Again, the idea was to compare the results from the geophysical surveys with the structures expected to be found after excavation. An interesting aspect of this site is that all the buried structures are covered with sand, which is soaked in rain water at the surface and sea water in the deeper layers. So, in principle, there is a measurable contrast of the physical properties associated with different geophysical methods. Up to now only electrical resistivity tomography and ground

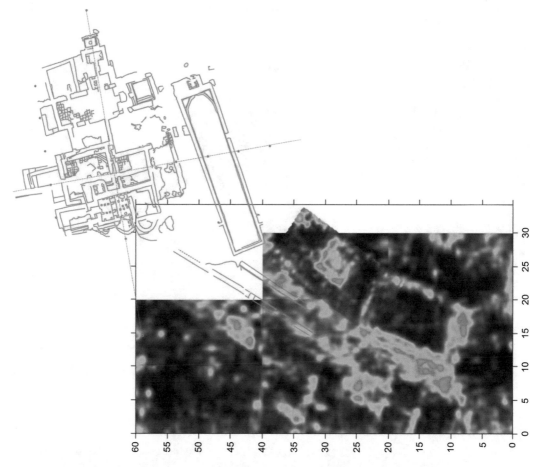

Fig. 5.11 Colour detail of Fig. 5.10. The continuation of the structures already excavated is obvious. Distances in m

penetrating radar have been used to find and delineate buried stone structures such as walls and floors.

A preliminary survey was done in June 2013 and it is expected to continue the geophysical work in the future in areas that are planned to be excavated. Figure 5.12 is a map of what is excavated in the Roman Ruins of Troia and shows three areas that were initially chosen to carry out the geophysical surveys. In the end only areas 1 and 2 were chosen to do the ERT and GPR surveys. Area 3 was not considered because of the existence of metal structures for use of pedestrians visiting the ruins.

The location of the Roman Ruins of Troia can be seen in Fig. 5.1. For those interested in seeing the area of the roman site, the geographical coordinates in the Google Earth are: 38°29′9.82″N,

8°53′5.32″W. The average altitude of the site is 3 m a.s.l.

5.2.3.2 Method

The geophysical methods used in the area of the Roman Ruins of Troia were ERT and GPR. In each of the two areas (1 and 2) chosen for the surveys (see Fig. 5.12) one electrical resistivity tomography profile and three parallel ground penetrating radar profiles were measured; the three GPR profiles were done so that the central GPR profile was coincident with the ERT profile and the other two GPR profiles were 0.5 m away from the central one to each side.

The ERT profiles were 39 m long and were carried out using a Wenner configuration with 40 stainless steel stacks 1 m apart. The GPR profiles were done using a 400 MHz antenna

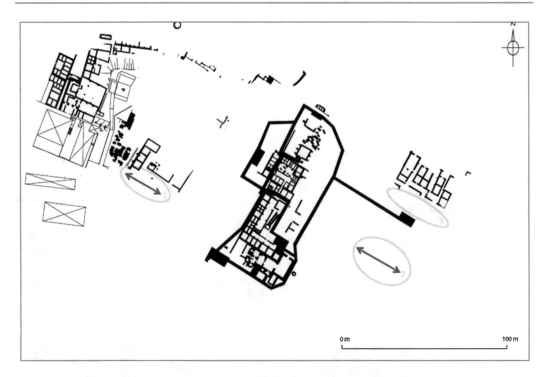

Fig. 5.12 Location of the two areas where ERT and GPR profiles were done (brown ellipses with double red arrows inside); double red arrows indicate the orientation of the ERT and GPR profiles

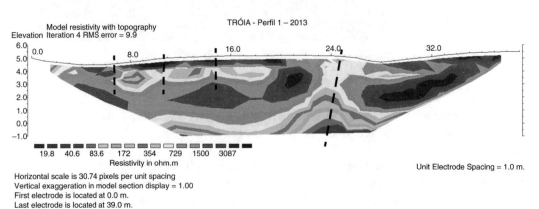

Fig. 5.13 Electrical resistivity tomography (profile 1) done in area 1 of Fig. 5.12. Reddish colours correspond to high electrical resistivity values; bluish colours correspond to low electrical resistivity values. Black dashed lines are interpretations of possible contacts between structures with different electrical resistivities

5.2.3.3   Some Results

Figures 5.13 and 5.14 show the coincident ERT and central GPR profiles done in area 1 of Fig. 5.12. Figures 5.15 and 5.16 show the coincident ERT and central GPR profiles done in area 2 of Fig. 5.12. and were 40 m long in area 1 and 42 m long in area 2.

The ERT profile in area 1 (Fig. 5.13) indicates that ground in the area presents a compartment structure: up to 24 m there are zones with intermediate electrical resistivities (green and yellow colours) imbedded in zones of high electrical resistivity (red and orange colours). At 24/25 m there is a vertical (or sub-vertical) contact which separates two high electrical resistivity zones.

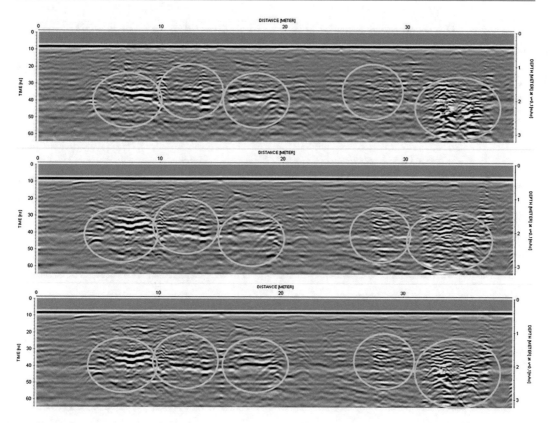

Fig. 5.14 Three radargrams done with the orientation of ERT along profile 1 of Fig. 5.12. Only the central radargram coincides with the ERT shown in Fig. 5.13. The other radargrams are located 0.5 m to the NE (upper radargram) and to the SW (lower radargram) of the central one. Yellow ellipses indicate the most prominent reflections

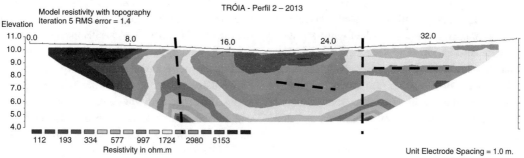

Fig. 5.15 Electrical resistivity tomography (profile 2) done in area 2 of Fig. 5.12. Reddish colours correspond to high electrical resistivity values; bluish colours correspond to low electrical resistivity values. Black dashed lines are interpretations of possible contacts between structures with different electrical resistivities

The bluish zones are probably sand with sea water because the electrical resistivities are very low. As a preliminary interpretation, the compartments (which show lower electrical resistivity values) are separated by zones of intermediate electrical resistivity values (black dashed lines in Fig. 5.13)

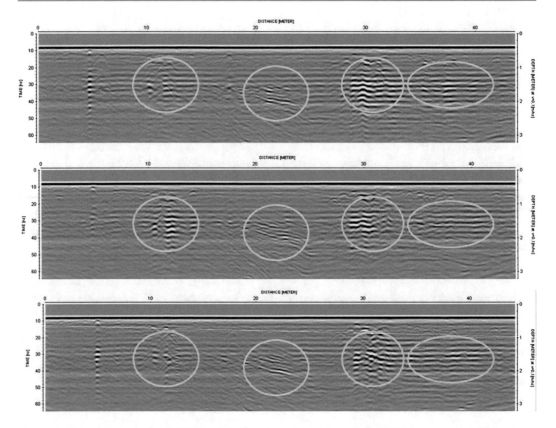

Fig. 5.16 Three radargrams done with the orientation of ERT along profile 2 of Fig. 5.12. Only the central radargram coincides with the ERT shown in Fig. 5.15. The other radargrams are located 0.5 m to the NE (upper radargram) and to the SW (lower radargram) of the central one. Yellow ellipses indicate the most prominent reflections

which are probably associated with rock walls observed in areas already excavated. The referred compartments are probably filled with rain water soaked sand and have lower electrical resistivity than the interpreted walls. For distances larger than 24/25 m, electrical resistivities are very high and show a horizontal pattern, probably indicating a large rock concentration or stone floors.

It is interesting to note that the central GPR profile (Fig. 5.14), coincident with the ERT profile in area 1, corroborates the above interpretation. In the radargram of Fig. 5.14 the most important reflections are shown inside yellow ellipses. As a matter of fact, there are many superficial electro-magnetic reflections which correspond to shallow rocks from crumbled walls. However, deeper reflections also indicate the existence of compartments in the same zones as the ones interpreted in the ERT profile. It is also interesting to note that the most intense reflections of electro-magnetic energy are horizontal or nearly so. Finally, in the right portion of Fig. 5.14 (after 32 m) there are strong reflections which indicate strong dielectric constant contrasts and so important buried structures are expected to be found there; this same conclusion can be inferred from the ERT profile (Fig. 5.13).

The compartment structure observed in the ERT profile (Fig. 5.13) can also be inferred in the left portion of the radargram shown in Fig. 5.14.

The ERT profile in area 2 (Fig. 5.15) is less complex than the ERT profile of area 1. The geoelectrical structure indicates three large compartments which are separated by the vertical black dashed lines. There are, however, two zones separated by two horizontal black dashed lines which probably correspond to rock/sand contacts; however, these two lines were drawn after interpreting the radargram coincident with the ERT. There is also an important vertical contact at about 11/13 m that separates two high electrical resistivity media, the left medium being less resistive than the right one.

There is another important vertical contact at about 26/27 m which also separates two geoelectrically different media; the left medium is less resistive than the right one. In this profile the possible compartments appear to be wider, as if the buried structures were wider and better defined in horizontal terms.

As happened for the ERTs done in areas 1 and 2, the central radargram obtained in area 2 (Fig. 5.16) is also less complex that the radargram obtained in area 1 (Fig. 5.14). In the former the number and intensity of electromagnetic reflections is less numerous (yellow ellipses in Fig. 5.15); there are also shallow hyperbolae probably from shallow rocks. In this radargram there are reflections at 4 m, 10/12 m, between 18 and 24 m, between 28 and 34 m, and between 35 and 42 m. Except in the case of the reflection at 4 m, (which may correspond to a buried shallow unknown object or the top of a stone wall) all other reflections are coincident with the electrical resistivity contrasts observed in the ERT profile.

In general terms it can also be said that there is a good coincidence between the electromagnetic reflections observed in the radargram in area 2 and the geoelectrical structure of the ERT for the same area; it appears that there are buried structures that were detected and delineated by both ERT and GPR. However, the number of buried structures is less in area 2 than in area 1.

5.2.3.4 Conclusions

From the two electrical resistivity tomographies and ground penetrating radar done in the two areas in the Roman Ruins of Troia, it appears to be possible to conclude that there are buried structures that have a clear geophysical signature. Their interpretation in archaeological terms is, however, more complex; one thing, though, is

evident: the geophysical data obtained by ERT and GPR in areas 1 and 2 of the Roman Ruins of Troia are consistent with each other. Area 1 appears to have more buried structures and is more complex than in area 2. In this case, the utility of both ERT and GPR is evident and produced interpretable data. Excavation is thought to start again in 2017.

5.2.4 Crypt of the Marquises of Castelo Rodrigo

5.2.4.1 Introduction

The museum of the Portuguese Legislature, knowing, from newspapers dating from the twenties of the last century, that there was a crypt buried in the gardens of the Portuguese Legislature decided to try to locate it. The idea was to excavate the site and prepare it for public visits. However, before initiate excavation of the site it was decided to do a geophysical prospection in the garden and in the parking lot of the legislature building to locate the crypt. In logistics terms, the area where the geophysical surveys should be done is complex: during the working days the deputies leave the cars in the parking lot and so the space could not be used for any geophysical surveys. So, all geophysical surveys were done on Saturdays and Sundays when there were no cars in the parking lot. Besides trying to locate the crypt, it was thought as a good idea trying to locate underground tunnels, which are known to exist, that may connect the crypt, the main building and other buildings in its vicinity.

The Portuguese Legislature has is sessions in a building that is known as Sao Bento's Palace. It is located well inside the town of Lisbon and is the building of the Portuguese Legislature since 1834. It was built as a Benedictine monastery at the end of the sixteenth century. In the seventeenth century the crypt of the marquises of Castelo Rodrigo was built in what is now the garden and parking lot of the building; its exact location was lost after several construction works inside and outside the original monastery which covered the crypt.

For those interested in seeing the area of the Portuguese Legislature, the geographical coordinates in the Google Earth are: 38°42′45.58″N, 9°9′15.95″W. The average altitude of the site is 32 m a.s.l.

5.2.4.2 Method

The geophysical methods chosen to try to locate the crypt and tunnels associated with it were electrical resistivity tomography (ERT) and ground penetrating radar (GPR) with two different antennae (400 and 200 MHz). However, the results with 200 MHz were not good and will not be shown here. The GPR surveys were done using a grid which was subdivided into several rectangular sub-grids. The line spacing used was 0.5 m. GPR lines were collected in both orthogonal directions (X and Y), the direction X coinciding with the orientation of the main outside wall of the building. The radar antennas were dragged directly on the ground. Figure 5.17 shows the area outside the building where the geophysical surveys were done as well as the six sub-areas.

Three ERT profiles were done (Fig. 5.18): two of them had 39 m long crossing near the building's façade and one 59 m long in the garden parallel to the building's façade. The smallest ERTs were done using a Wenner configuration with 40 stainless steel stacks 1 m apart while the longest was done using a hybrid roll along technique.

5.2.4.3 ERT Results

Figure 5.18 shows the rear of the building of the Portuguese Legislature where the three ERTs were done on June 25 and 26, and August 6, 2011. For the three electrical resistivity tomographies (a, b, and c in Fig. 5.18) bluish colours correspond to low electrical resistivity values while reddish colours correspond to high electrical resistivity values.

ERTa in Fig. 5.18 is 39 m long and indicates that, except for a shallow thin layer (which is cobblestone), electrical resistivities are generally low; the blue spots between 1.3 and 2.9 m might correspond to pockets of clay and water from watering the garden. The orange spots between 4.0 and 6.0 m might correspond to structures associated with the crypt (such as undiscovered tunnels).

ERTb in Fig. 5.18 is also 39 m long and, in general terms, presents the same characteristics of ERTa. Electrical resistivities are, though, slightly higher than in ERTa. The blue spots are interpreted again as water from watering the building's garden. However, at 13/14 m there is high electrical resistivity pocket which might correspond to a continuation of buried structures associated with the crypt.

ERTc in Fig. 5.18 is 59 m long, was done in an area that has no cobblestone and is about 1 m higher than the other two ERTs. In general terms, ERTc presents the lowest electrical resistivities of the three electrical resistivity tomographies which have to do with the water that percolates from the soil in the garden towards the underground of the parking lot. Since the ERTs were carried out during the summer, deeper layers present lower electrical resistivity values. The stairs in marble in the middle of the ERT are well identified by high electrical resistivity values in the middle section.

So, as a conclusion, the three electrical resistivity tomography profiles were not able to show unequivocally the presence of the crypt or extensions of it.

5.2.4.4 GPR Results

Even though a few GPR surveys were done inside the building of the Portuguese Legislature, only the results obtained for the areas outside the building are shown; the GPR results inside the building were very poor because the level of noise was too high to do any processing. Figure 5.17 shows the relative positions of the six outside areas where GPR surveys was done. The GPR coverage for each area was done in such a way that a three-dimensional picture of the ground could be obtained; for that the GPR Slice software was used. For all six areas of Fig. 5.17, each GPR run was separated from the next by 0.5 m. As already said, 400 and 200 MHz antennas were

Palace's garden

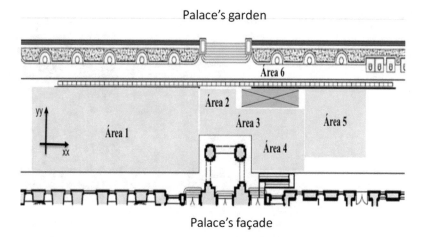

Palace's façade

Fig. 5.17 Sketch of the area where GPR surveys were done in the parking lot of the Portuguese Legislature. The X and Y directions are defined in area 1 and were used for all other areas also shown. The lower part of the figure corresponds to the building's façade

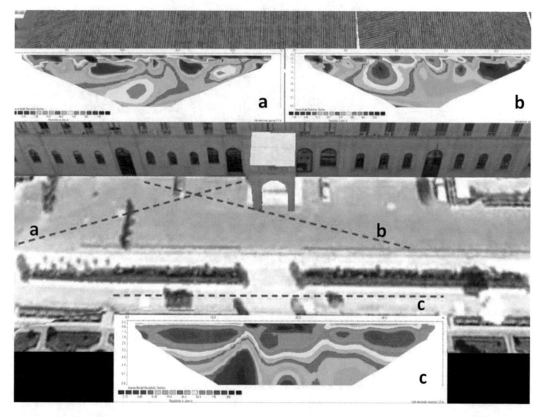

Fig. 5.18 Sketch of the area where ERTs were done (red dashed lines for orientation of ERTs a, b, and c) and ERT results in the parking lot of the Portuguese Legislature. The middle part of the figure corresponds to the building's façade and the upper part, in brown, corresponds to the building's roof

Fig. 5.19 GPR slices for
Area 1 in Fig. 5.17. Slices
correspond (from top to
bottom) to depths of
0.22–0.39 m, 0.44–0.60 m,
0.87–1.04 m, 1.53–1.70 m,
and 2.07–2.18 m

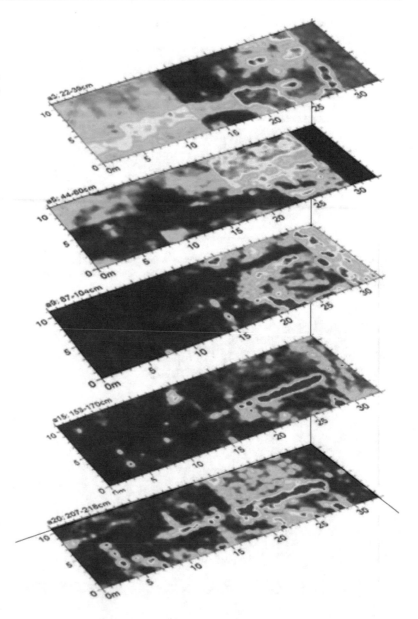

used; however, the data from 200 MHz antenna had poor quality and was not used for further processing. In Figs. 5.19, 5.20, 5.21, 5.22 and 5.23 the results obtained are shown for all areas except area 6 which was not wide enough for proper processing and interpretation. Figure 5.24 is a collage of the GPR results for the surveyed area (Areas 1–5) for 1.4–1.6 m depth.

5.2.4.5 Conclusions

To try to find the location of the crypt of the marquises of Castelo Rodrigo two methods were

Fig. 5.20 GPR slices for Area 2 in Fig. 5.17. Slices correspond (from top to bottom) to depths of 0.22–0.39 m, 0.44–0.60 m, 0.87–1.04 m, 1.53–1.70 m, and 2.07–2.18 m

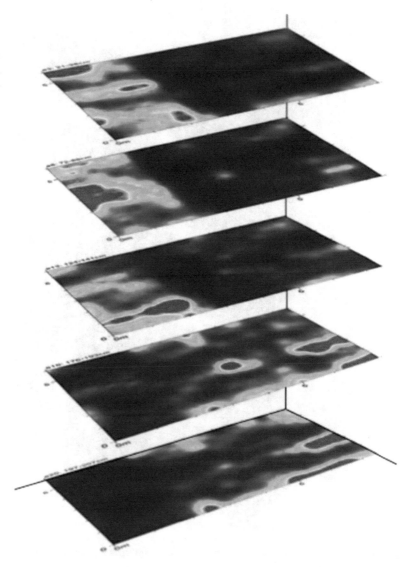

used: ground penetrating radar and electrical resistivity tomography. The former was effective in finding structures associated with the crypt; the same cannot be said about electrical resistivity tomography. ERT profiles were only able to detect strong contrasts of moisture in the ground related with watering activities in the palace's garden. During the field work a dome filled with soil and rocks was found (Fig. 5.25) near the area where the geophysical surveys were being done (Fig. 5.26).

Acknowledgments The short examples of application of geophysical methods to archaeology in Portugal, presented in this chapter, could not have been done without the cooperation of several colleagues and researchers from different institutions. The author would like to acknowledge and thank the participation in the field work, the discussions, and the authorisations to use the collected data in this publication to Inês Vaz Pinto (Troia Resort), Teresa Parra da Silva and Joaquim Soares (Museum of Portuguese Legislature), João Caninas (Emerita Ltd.), Isabel Gaspar (Municipality of Proença-a-Nova), and Brooke Berard and Jean-Michel Maillol (University of Calgary). All the equipment used during the field work belongs to the Geophysical Centre of the University of Évora (now Institute of Earth Sciences, Portugal) and the Department of Geology and Geophysics of the University of Calgary (Canada).

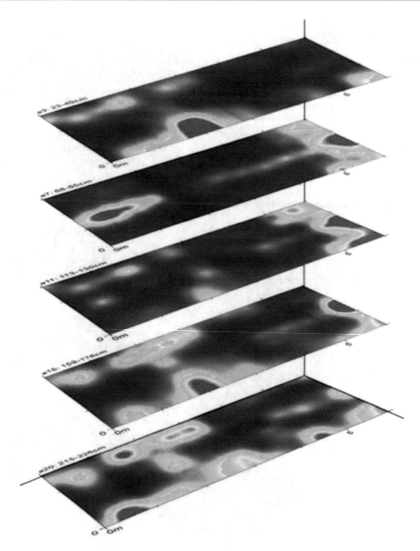

Fig. 5.21 GPR slices for Area 3 in Fig. 5.17. Slices correspond (from top to bottom) to depths of 0.22–0.39 m, 0.44–0.60 m, 0.87–1.04 m, 1.53–1.70 m, and 2.07–2.18 m

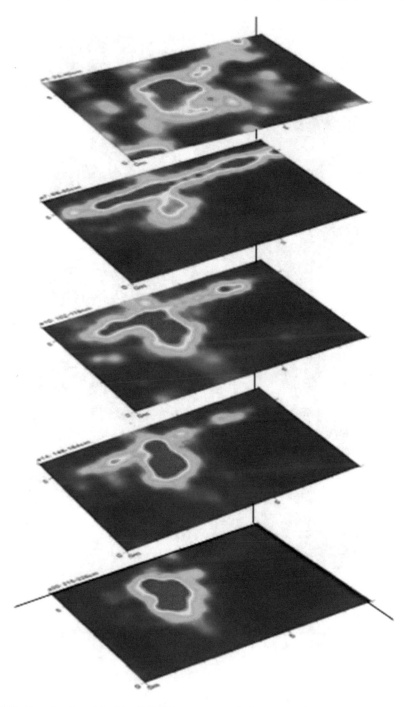

Fig. 5.22 GPR slices for Area 4 in Fig. 5.17. Slices correspond (from top to bottom) to depths of 0.22–0.39 m, 0.44–0.60 m, 0.87–1.04 m, 1.53–1.70 m, and 2.07–2.18 m

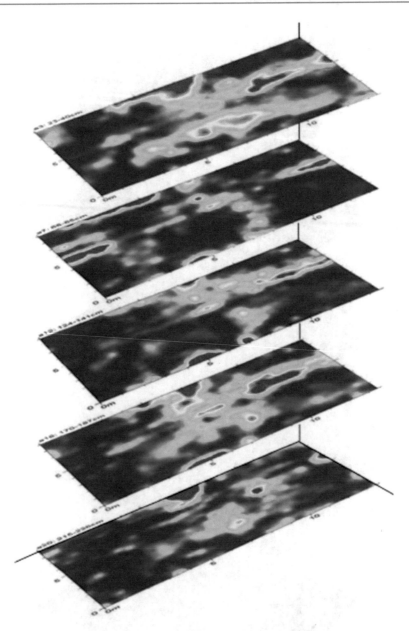

Fig. 5.23 GPR slices for Area 5 in Fig. 5.17. Slices correspond (from top to bottom) to depths of 0.22–0.39 m, 0.44–0.60 m, 0.87–1.04 m, 1.53–1.70 m, and 2.07–2.18 m

Fig. 5.24 Collage of the GPR results for the surveyed area (Areas 1–5) for 1.4–1.6 m depth

Fig. 5.25 Picture of the structure (ceiling/tunnel?) found during construction work in the area of the geophysical survey. The structure is filled with sand, rocks, and dirt. The dimension of the structure can be appreciated by comparing its size with the upper part of the ladder in the right portion of the picture

Fig. 5.26 Collage of the GPR results for the surveyed area (Areas 1–5), similar to Fig. 5.24 but for 0.5–0.7 m depth. The black ellipse indicates the location of the structure shown in Fig. 5.25

References

dos Santos MF, Esteves JM (1966) Possibilidade de aplicação do método da resistividade eléctrica na prospecção arqueológica. Ethnos V:313–335

Tite SM, Alldred JC (1965–1966) Aplicação de métodos científicos de prospecção em estações arqueológicas portuguesas. Trabalhos de Antropologia e Etnologia XX (1–2):147–160

Vaz Pinto I, Viegas C, Dias F (2004) Terra sigillata and amphorae from the Roman villa at Tourega (Evora, Portugal). In: Pasquinucci M, Weski T (eds) Close encounters: sea- and riverborne trasde, ports and hinterlands, ship construction and navigation in antiquity, the middle ages and modern times, British Reports, vol S1283, p 11

Integrated Geophysical Methods for Detecting Archaeological Han Dynasty Tombs

6

Man Li, Zhiyong Zhang, David C. Nobes, and Jun Yang

Abstract

A group of Han Dynasty tombs were found in Nanchang, Jiangxi Province, China, close to Poyang Lake. To investigate the coffin chambers, tunnels, burial pits, and remains of the tomb, a reconnaissance survey was carried out using integrated geophysical methods, including a ground magnetic survey, a self-potential (SP) survey, electrical resistivity tomography imaging, and ground penetrating radar (GPR).

A survey area measuring 85 m east to west and 120 m south to north completely contained the grave mounds. The survey grid was 5 m spacing in the east-west direction and 2 m spacing in the south-north direction. Positive anomalies in the magnetic field readings corresponded to the burial pits and the rammed foundation. A high SP value corresponded to the surface projection of the coffin chambers and tunnels, and also to the collapsed mausoleum building. Inversion of the multi-electrode resistivity data showed the positions of the coffin chambers, funeral pits, and tunnels very well. 2D and 3D resistivity inversion showed that tombs and burial pits have relatively low resistivity, because they lie below the groundwater table and relative to the surrounding soil contain more water.

Furthermore, the entry ramps have relatively high resistivity. GPR signals did not have good penetration because of the low-resistivity moist surficial soil, and thus could not detect the chamber and tunnels, but revealed clearly the modern tombs and building foundations that lie near the ground surface.

Keywords

China · Han dynasty · Tombs · Magnetic · Self potential · Resistivity · Ground penetrating radar

6.1 Introduction

A group of Han Dynasty (202BC–8AD, 25AD–220AD) tombs are located at Guodun Mountain (Fig. 6.1), which is 500 m northeast from the Guanxi District, Tangping Village, Xinjian Town, Nanchang City, Jiangxi Province, Peoples Republic of China, close to Poyang Lake. The Guodun grave is located in the fiefdom which belonged to the Haihun Marquises in the Han Dynasty (Li 1985), and near two other historic

M. Li (✉) · Z. Zhang · D. C. Nobes
School of Geophysics and Measurement-Control Technology, East China University of Technology, Nanchang, Jiangxi, China

J. Yang
Jiangxi Administration Bureau of Cultural Relics, Nanchang, Jiangxi, China

sites of an ancient city which also belonged to the Han dynasty. The cemetery is shaped in the form of a trapezoid; the width from south to north varies from 141 to 186 m, the length from east to west varies from 233 to 248 m. The total area of the ruins is about 46,000 m². There are no obvious surface traces except the two grave mounds which look like inverted hoppers. The burial mounds face south and should belong to the couples of the marquises. They are typically the same style of burial but have different coffin-type tombs. Unfortunately, there are two vertical holes dug by archaeological thieves into each tomb mound.

We concentrated on a research area approximately 85 m east to west and 120 m south to north in the middle of the cemetery (Fig. 6.1b), focusing on local archaeological studies. We applied integrated multi-geophysical methods, including a ground magnetic survey, a self- potential (SP) survey, multi-electrode resistivity imaging, and ground penetrating radar (GPR) imaging. The magnetic and self-potential surveys were used first, in order to quickly investigate the distribution of archaeological traces. Magnetic and self-potential data can indicate lots of anomalous features, but it is difficult to calculate the depths of the anomalous bodies and determine what those anomalous bodies might be.

GPR and multi-electrode resistivity data were then acquired. We can easily identify abnormal features at depths no more than 4 m from the GPR data, so GPR was used to detect shallow tombs and buried remains of old buildings. Multi-electrode resistivity was applied to investigate much deeper anomalous bodies such as the chambers and tunnels of the two big tombs, accompanying pits, etc.

The potential of multi-method surveying strategies has been well recognized in archaeological prospecting (e.g., Dabas et al. 2000; Chavez et al. 2001; Drahor 2006; Kvamme 2006; Keay et al. 2009; Böniger and Tronicke 2010; Sambuelli et al. 2011; Landry et al. 2015). It is an effective way to reduce uncertainties and ambiguities by analyzing the multiple geophysical data sets which are collected at the same archaeological site.

6.2   Geophysical Characteristics and Method Validity Analysis

When the tombs, passages used to enter into tombs, grave mounds, the pits, etc., were built the primary soil structure should be destroyed so the physical property underground should be changed. These are the physical basics of archaeological geophysics. During the field

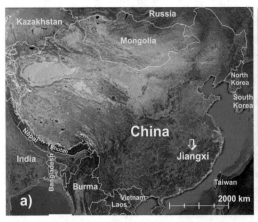

Fig. 6.1 (**a**) Location of Jiangxi Province in China, and the location of the site within northern Jiangxi (indicated by the arrow). (Adapted from GoogleEarth image.) (**b**) Aerial image of the Guodun Grave site after excavation.

The features in the center of the airphoto are the excavated tombs and associated features (Adapted from GoogleEarth image)

investigation, the main background medium was clayey soil and reviewing the relevant information about Han Dynasty's cemeteries, we speculate that there are not only two tombs include passages used to enter the tombs and at least one chamber of each tomb, but also accompanying pits underground. There should also be some buildings for worship and for the living people who guarded the cemetery in ancient times. So the abnormal media in the study area included: the grave mound, the building materials for the tombs, the buried remains of old buildings, modern tombs, a variety of funerary goods, and so on.

For magnetic surveying, remanent magnetism and magnetic susceptibility are the basic properties of the medium. During the process of burial building, the original soil column was disturbed and the distribution of the former magnetic characteristics was changed. Under the effect of the Earth's magnetic field, induced magnetization occurred in the surrounding soil and the external contour wall of the tomb. The ancient graves have the anomalous features which are usually negative in the north and positive in the south in Northern Hemisphere. Because of the ferromagnetic and thermal remanent effects, metallic buried pits and the burned remains of building materials have a strong magnetic response.

The medium's spontaneous polarization produces the self-potential field, which has a relationship with groundwater and its flow. The electronic conduction is transected by the water table surface; above the water table, the environment is oxidizing, and below the water table, the environment is reducing. Thus an oxidation-reduction electric field is formed. Groundwater flow causes the differences of migration velocity and absorption between positive and negative ions, creating an electric field; similarly a spring causes an ascending electric field, and so on. The different water content and ability for migration of water, and thus ions, in the rammed area, chamber, and passages, are the basics for the application of the self-potential method in this area (Vichabian and Morgan 2002; Drahor 2004).

The resistivity of a medium is related to the porosity and moisture content (e.g., Milsom and Eriksen 2011; Reynolds 1997). Within the depth of penetration, the loose accumulation is made up of sticky soil and cultivated soil containing weathering byproducts, so the background resistivity is not high, about 60–80 ohm-m. The resistivity is measured using an array of four small symmetric poles. If the chambers and passages are filled with water or mud, they should have much lower resistivities compared with the background; in contrast, if they are air-filled, the resistivities will be much higher. After a long period of weathering and compaction, the resistivity of the buried remains of these collapsed building is similar to the surrounding medium. Rammed earth construction areas in the cemetery usually have higher resistivity due to the high rammed density.

The material dielectrical permittivity is mainly influenced by the water content. Based on the previous analysis, the chambers and passages filled with water will have higher dielectric coefficients, and the modern tombs above the water table will have lower dielectric coefficients.

In general, the physical properties in this area are suitable for applying high-precision ground magnetic surveys, self-potential method surveys, multi-electrode resistivity imaging and ground penetrating radar (GPR) to detect the underground archaeological features.

6.3 Data Acquisition and Results

The 85×120 m^2 survey area completely contained two buried tombs and surface tumuli; the survey grid was 5 m spacing in the east-west direction and 2 m spacing in the south-north direction. A local coordinate system was established, and the coordinate origin was set at the southwest corner of the survey area, with the positive directions pointing in the due north and due east directions, respectively. To ensure good data continuity, we insert a line between two adjacent lines when necessary. We numbered the lines from west to east as Line 1, Line 3, Line 5 . . . to Line 35. For the magnetic survey, there are 18 lines and 60 points along each line.

6.3.1 Magnetic Survey

About 1110 magnetic data points were collected using a Geometrics G856 proton precession magnetometer. During the measurements the sensor was used at a constant height (1.6 m) above the ground surface. The measurement points were taken at the sampling grids shown in Fig. 6.2. The proton precession magnetometer is used to detect magnetic anomalies in the Earth's magnetic field (e.g., Reynolds 1997). It uses the precession of spinning hydrogen atoms in a hydrocarbon liquid to measure the magnetic field. When a current is injected in the coil around the detector which is filled with a hydrogen-rich liquid, the hydrogen protons align to the magnetic field created by coil because the hydrogen protons act as small magnets. When the current is turned off, the protons precess at a frequency proportional to the Earth's total magnetic field. This causes a small signal in the coil due to the polarization of the protons.

The residual magnetic field response is shown in Fig. 6.3a. There is a sheet-metal shed in the

south of the survey area; the area has been blanked where the data were adversely affected.

Most of the magnetic anomalies are positive; they reflect the features of coffins, rammed soil, funerary pits, and building remains. Figure 6.3b shows the magnetic field "reduced to the pole" (RTP). "Reduction to the pole" (Sheriff 2002) is a mathematical process whereby the magnetic field over a survey area is transformed to appear as if it is at the pole, that is, the Earth's inducing field appears to be vertical. This removes any residual bipolar anomalies associated with the mid-latitude location of the site, and repositions the anomalous readings directly over top of the features giving rise to the anomalous readings. This allows for easier interpretation.

We can infer the rammed earth area from the distribution of positive anomalies; the left higher positive anomaly may be produced by companion pits. However notice that there are two positive anomalies which are produced by iron tubes used to support canvas and provided cover while the thieves dug their access hole (iron interference areas shown in Fig. 6.3b). There are five negative

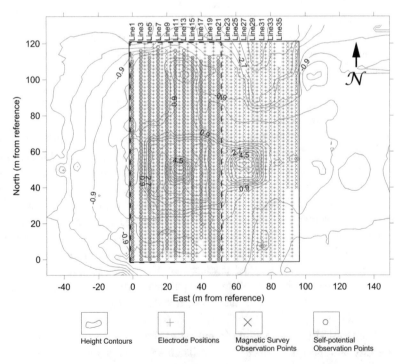

Fig. 6.2 Layout of geophysical prospecting work at the study site, superimposed on the undisturbed topography prior to any excavations. The areas of the magnetic survey (solid rectangle, Fig. 6.3) and SP survey (dashed rectangle, Fig. 6.4) are outlined. The line numbers at the top correspond to the GPR survey lines

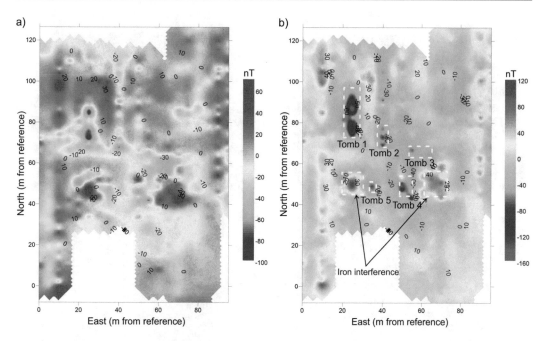

Fig. 6.3 (a) The raw original anomalous magnetic field. (b) The magnetic field after reduction to the pole (RTP) processing. The main anomalies are labeled

anomalies; we infer they are produced by voids associated with the tombs. The biggest negative anomaly is inferred as the companion tomb; the others will be modern tombs.

6.3.2 Self-potential Method

The self-potential (SP) survey is one of the oldest methods in geophysical exploration, and it is still used for solving many problems in applied geophysics. In the self-potential method, there are two non-polarizable electrodes; one electrode is fixed as the zero potential reference, and the other is moved along the measure points (e.g., Milsom and Eriksen 2011; Reynolds 1997).

The potential difference between the two electrodes is measured. Eleven lines of SP measurements were completed in this area. The original self-potential anomaly is showed in Fig. 6.4a. A 2D low pass filter was used to deal with the self-potential data as the field can be easily disturbed by small surficial features; the result is shown in Fig. 6.4b. The main anomaly was highlighted after filtering; overall, the

positive high-value anomaly looks like a cross glyph. The negative anomaly which adjoins the positive anomaly (A1) is affected by steep ridges in the terrain surface and the rammed soil underground. The opposite anomaly, A2, is mainly caused by the rammed soil. The high-value A3 and A6 anomalies correspond with the ancient building ruins; A3 appears as a negative anomaly in the magnetic data. Because there is an iron house near the south edge of researching area, we edited the data affected by the house including A6. The long strip positive-value A4 and A5 anomalies will be caused by the grave mound and the buried tunnels. When we compare Figs. 6.3 and 6.4, we see that A4 overlaps with and is partly correlative with the biggest negative magnetic anomaly.

6.3.3 Ground Penetrating Radar

Ground penetrating radar (GPR) is a high-frequency electromagnetic imaging method (e.g., Milsom and Eriksen 2011; Reynolds 1997). The GPR transmitting antenna generates

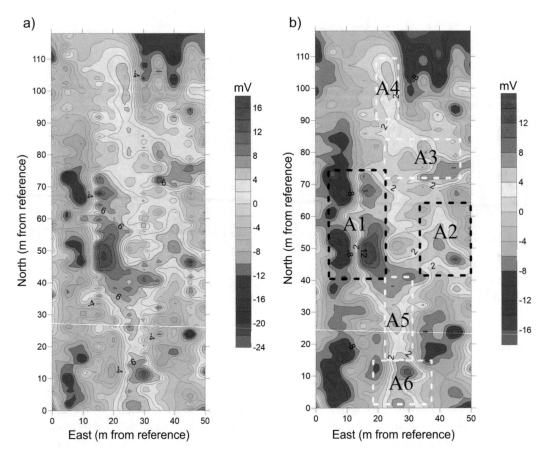

Fig. 6.4 (**a**) Raw self-potential anomaly map. (**b**) Self-potential anomaly map after processing with a 2D low-pass filter. The negative anomaly adjacent to the positive anomaly (A1) is caused by steep ridges in the terrain and the rammed soil underground. The other positive anomaly, A2, is mainly caused by rammed soil. The A3

and A6 anomalies correspond with the ancient building ruins. The long strip A4 and A5 anomalies are due to the grave mound and the buried tunnels. The A4 anomaly correlates with the biggest negative magnetic anomaly (Fig. 6.3)

a high frequency electromagnetic signal; this signal passes through the earth, much like a seismic wave propagates, and will encounter subsurface materials of varying dielectric coefficients, electric conductivities, and magnetic permeabilities. At these boundaries, the signal can be reflected.

When detected by the receiving antenna at the surface within the vicinity of the transmitter, the reflected signal will provide information about the subsurface interfaces. The signal can be attenuated in the material, by scattering, and by electromagnetic induction losses.

The SIR system 3000 made by GSSI was used to survey the site. Two sets of antennae, 100 and

500 MHz, were used for the investigation. GPR was used in a continuous profiling mode, operating with 16 bits dynamic range, 512 samples per scan, and 16 "stacks" per scan. The time window was set to either 60 nanoseconds or 100 nanoseconds. Most of the survey was along the S-N profiles; some cross profiles were also acquired.

In order to suppress random noise and low frequency interference which commonly occur, the ground penetrating radar data were band pass filtered. The 500 MHz data were processed with a 100–2000 MHz band pass filter. The 100 MHz data were processed using a 20–400 MHz band pass filter. Then automatic gain amplification was applied to balance the energy of the signals across the profiles.

As the soil moisture content is higher, the resistivity is lower; GPR had a good correspondence with the shallow burial tombs and historical sites. There were 16 GPR anomalous areas; most of them had strong reflections shaped like umbrellas (Fig. 6.5), which are diffractions caused by scattering of the GPR signals from discrete features (cf. Field et al. 2001) and are deduced as the modern tomb and the building foundations. The depth of burial is shallow; to get the true depth of the reflection, we analysed the diffraction curves, and obtained a GPR velocity of 0.076 m/ns. When we compare the results, we find that the anomalies are correlated with the SP and magnetic anomalies.

6.3.4   Multi-Electrode Resistivity

The multi-electrode resistivity survey is a technique which can quickly and accurately collect direct current (DC) resistivity data from the simultaneous arrangement of all of the electrodes in a profile (e.g., Milsom and Eriksen 2011). Data are collected through an electronic switching box which controls the selected pairs of direct current injected electrodes and measured voltage electrodes by pre-selected survey array design. The known magnitude of the transmitted current magnitude, the measured voltage, and the active electrode geometry are used to calculate an apparent resistivity value for each measurement. The true resistivity section can be obtained by inverting the entire observed apparent resistivity (e.g., Loke and Barker 1996). There are many

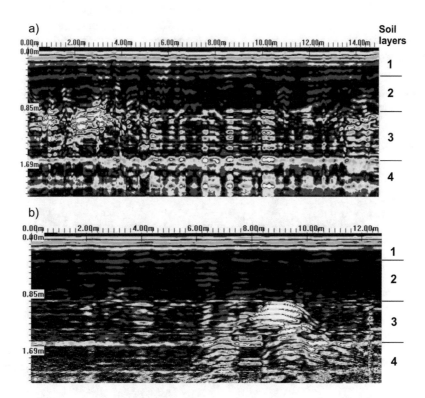

Fig. 6.5 Sample GPR cross sections from (**a**) the latter half of Line 11 and (**b**) Line 29, showing the characteristic GPR response. Line 29 (b), in particular, shows a clear example of an "umbrella- shaped" response used to estimate the near-surface GPR velocity

types of array geometries for the four active electrodes used in each measurement; the dipole-dipole array was adopted in order to detect the local anomalous body, such as tombs.

The DUK-2A system made by ChunQing Geo-instrument Company was used to survey the site. A total of 60 electrodes were used per profile and a total of 18 south-north cross section data were collected. The profiles were first inverted using Res2DINV, as illustrated in Fig. 6.6, and we then further processed the data from the 18 multi-electrode resistivity sections by 3D inversion using Res3DInv. We obtained the 18 two-dimensional subsurface models (Fig. 6.7) by inverting the apparent resistivity using the smoothness-constrained least-squares method (Loke and Barker 1996). The companion pits were found beneath the northern profiles and the passages for the two tombs were found toward the south.

6.4   Discussion

The main burial chambers were characterised by low resistivities, between 80 and 100 ohm-m. The multi-electrode resistivity results are consistent with the self-potential observations. In the north of the site, the magnetic field anomalies were negative after being reduced to the pole. The west tomb (Tomb 1) is much bigger than the east one (Tomb 2), and there appears to be a tunnel directed south beneath the water table. GPR has no clear signal from the deep burials.

The funeral pits appear as moderately high-resistivity features, from 160 to 180 ohm-m. They also had positive magnetic anomalies. The SP responses were positive above them, and there were clear GPR reflection signals.

The buried remains and the ancient relics of the burials had high resistivities, between 220 and 240 ohm-m, and had obvious anomalous GPR responses. Most of them had strong reflections, like the umbrella shapes noted in Fig. 6.5. In the raw original magnetic data, there were positive-field anomalies above these deeper features, but after RTP, the anomalies were reduced or disappeared.

The RTP positive magnetic anomalies represent a normal geomagnetic field in this location, but the RTP negative anomalies represent the cavities of the tombs. The high SP values

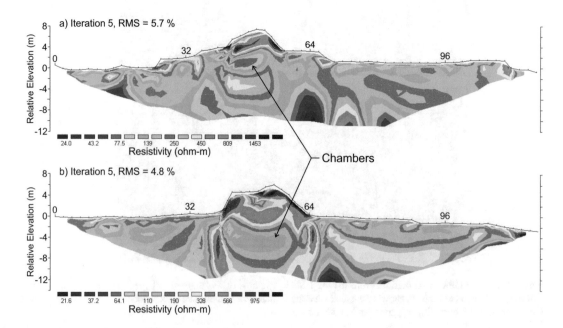

Fig. 6.6 Sample 2D tomographic best-fitting models for multi-electrode electrical profiles crossing (**a**) Tomb 1 (Line 11) and (**b**) Tomb 2 (Line 25), showing typical low-resistivity chamber responses

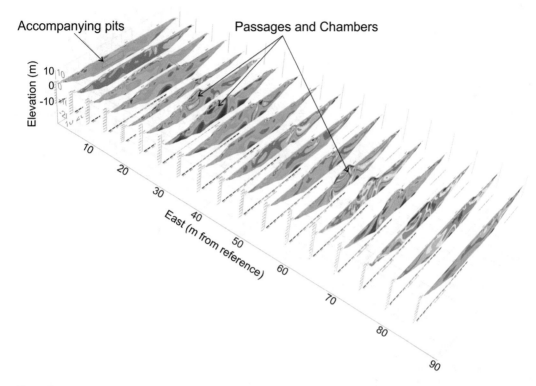

Fig. 6.7 Perspective view of multi-electrode resistivity south-north cross section results, viewed as if looking from the southwest. The main features are labeled and highlighted

corresponded to the surface projection of the coffin chambers and tunnels; and the collapsed mausoleum building has a relatively high-value SP anomaly. Multi-electrode resistivity inversion showed the positions of the coffin chamber, funeral pits, and tunnels. The GPR had weak penetration because of the low-resistivity moist soil, and could not detect the chamber and tunnels, but showed the modern tombs and building foundations well. Given the correspondence between the results from the different methods, the synthesis of the 3D resistivity inversion results is thus generally representative of the composite results (Fig. 6.8). The water table was at a depth of about 6.3 m in the survey area.

The layout of the burial sites highlighted by the geophysical surveys and later confirmed by archaeologists is summarised in Fig. 6.9.

6.5   Conclusion

The combination of magnetic prospecting, self-potential method, multi-electrode resistivity and ground penetrating radar detected the distribution of archaeological sites, coffin chambers, tunnels, and funeral pits in an area containing Han dynasty tumuli. We deduced two typical tombs in this area, and identified the orientations of the burials. Two building ruins can also be recognized, as well as the funeral pits distributed around the tombs. Site excavations began in April 2011, and many historical remains have been found. Tens of thousands of precious cultural relics were unearthed at the site, e.g. gold vessels, bronze ware, iron ware, jade objects, porcelain, bamboo weaving, straw braid, textiles, and wooden slips (http://mt.sohu.com/20151105/n425319729.shtml, in Chinese)

Fig. 6.8 3D diagram of the natural logarithm (ln) of the resistivity, viewed from the southeast. Tomb 1 and Tomb 2 are the main burial chambers, including buried remains and ancient relics. Tunnels 1 and 2 are labelled. Note the anomalous response from the accompanying pit at depth

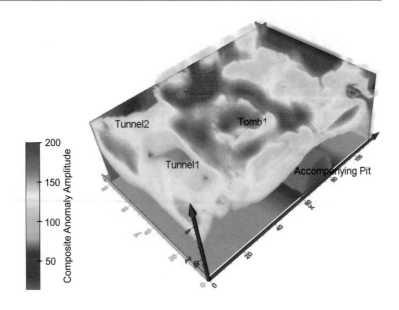

Fig. 6.9 Interpreted layout of the Han Dynasty Guodun grave site, superimposed on the original undisturbed topography

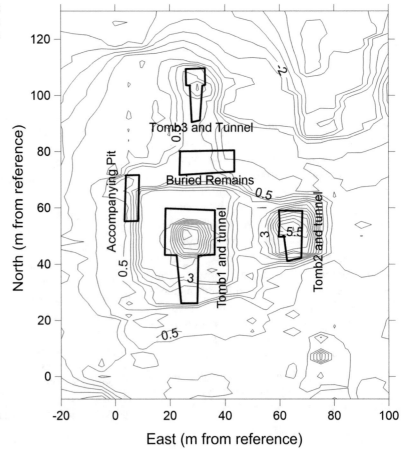

Acknowledgements The field exploration was supported by Jiangxi Administration Bureau of Cultural Relics. Data processing was supported by the Natural Science Foundation of China, grants No. 41304055 and No. 41304056. Thank you to all the people who supported this work and who provided useful comments and suggestions for this chapter.

References

Böniger U, Tronicke J (2010) Integrated data analysis at an archaeological site: a case study using 3D GPR, magnetic, and high-resolution topographic data. Geophysics 75(4):B169–B176

Chavez RE, Cámara ME, Tejero A, Barba L, Manzanilla L (2001) Site characterization by geophysical methods in the archaeological zone of Teotihuacan, Mexico. J Archaeol Sci 28:1265–1276

Dabas M, Camerlynck C et al (2000) Simultaneous use of electrostatic quadrupole and GPR in urban context: investigation of the basement of the Cathedral of Girona (Catalunya, Spain). Geophysics 65(2):526–532

Drahor MG (2004) Application of the self-potential method to archaeological prospection: some case histories. Archaeol Prospect 11:77–105

Drahor MG (2006) Integrated geophysical studies in the upper part of Sardis archaeological site, Turkey. J Appl Geophys 59(3):205–223

Field G, Leonard G, Nobes DC (2001) Where is Percy Rutherford's grave? In: Jones M, Sheppard P (eds) Australasian connections and new directions: proceedings of the 7th Australasian Archæometry Conference, vol 5. Research in Anthropology and Linguistics, University of Auckland, Auckland, pp 123–140

Keay S, Earl G, Hay S, Kay S, Ogden J, Strutt KD (2009) The role of integrated geophysical survey methods in the assessment of archaeological landscapes: the case of Portus. Archaeol Prospect 16:154–166

Kvamme KL (2006) Integrating multidimensional geophysical data. Archaeol Prospect 13:57–72

Landry DB, Ferguson IJ et al (2015) Combined geophysical approach in a complex Arctic archaeological environment: a case study from the LdFa-1 Site, Southern Baffin Island, Nunavut. Archaeol Prospect 22:157–170

Li KY (1985) Jiangxi city census summary. Jiangxi Hist Relics 2:19–24

Loke MH, Barker RD (1996) Rapid least squares inversion of apparent resistivity pseudosections by a quasi-Newton method. Geophys Prospect 44:131–152

Milsom J, Eriksen A (2011) Field geophysics, 4th edn. Wiley, Chichester, UK, p 287

Reynolds J (1997) An introduction to applied and environmental geophysics. Wiley, Chichester, UK, p 806

Sambuelli L, Comina C et al (2011) Magnetic, electrical, and GPR waterborne surveys of moraine deposits beneath a lake: a case history from Turin, Italy. Geophysics 76(6):B213–B224

Sheriff RE (2002) Encyclopedic dictionary of exploration geophysics, 4th edn. Society of Exploration Geophysicists, Tulsa, OK, p 429

Vichabian Y, Morgan FD (2002) Self potentials in cave detection. Lead Edge 21(9):866–871

Geomagnetism Exploration of the Egyptian Archaeology: Thirty-Years of Success and Challenges

7

T. Abdallatif, H. H. Odah, A. E. El Emam, and A. Mohsen

Abstract

The very long history of Egypt is a matter of attraction for many scientists, researchers from different countries all over the world. Egypt has many unexplored archeological sites. The archeologists know where to search, but they like to know where to dig.

The problem is: Traditional and random archaeological excavations consume more time, effort and Money. The best way is to explore the archeological remains using passive geophysical tools (e.g. magnetic and EM methods) without destroy or dig in the area.

The ancient Egyptian archaeological constructions are composed of different materials (e.g. Stone, mud, mortar, wood, Alabaster, granite, sandstone...etc.). Limestone and mud bricks are, however, basic materials in most of these constructions. Mud bricks are mainly composed of clay and mud extracted from the Nile River, and in some cases, they are mixed with a binding materials such as rice husks or straw. They were used to some extent in pre-Roman Egypt, and their use increased at the time of Roman influence.

Mud brick features within archaeological sites possess rather high magnetic properties than the surrounding soil. They are enriched with magnetic minerals (i.e. magnetite) that respond effectively to magnetic measurements. This indicates highly remanent and induced magnetization of the mud bricks, resulting in rather strong magnetic field, which can be measured by high-resolution instruments.

The application of geomagnetic in archaeological prospection has been begun in Egypt has started by pioneers (Hussein, 1983) after around 30 years of worldwide application. In the middle of Nineties (1990s) of the last century in Egypt, a new technology (i.e. magnetic gradiometry) with high speed and sampling rate, has proven to be ideal for near-surface mud bricks investigation, leading to real archaeological discoveries at different places in Egypt.

The present study aimed to analyze the development of archaeomagnetic prospection in Egypt during the last 30 years in order to assess the future directions and possible potential in this important aspect for an historical country like Egypt.

Seven examples from different districts in Egypt have been introduced to show the usage of magnetic gradiometry to explore the shallow structures made of mud bricks.

T. Abdallatif · H. H. Odah · A. E. El Emam (✉) ·
A. Mohsen
National Research Institute of Astronomy and
Geophysics, Cairo, Egypt
e-mail: akotb@nriag.sci.eg

© Springer International Publishing AG, part of Springer Nature 2019
G. El-Qady, M. Metwaly (eds.), *Archaeogeophysics*, Natural Science in Archaeology,
https://doi.org/10.1007/978-3-319-78861-6_7

Keywords

Magnetic gradimetry · Processing sequence ·
Statistical detection · Mud bricks · Cemetery ·
Hawara · Deahshour · Egypt

7.1   Introduction

The application of geophysical methods in the archaeological investigations is steadily increasing, particularly during the past 30 years, and also in line with the introduction of technologically advanced instruments, which yield high-quality data in less time, effort, and cost. Geophysical techniques have made a significant contribution to archaeology (David 1995) since they are nondestructive techniques and capable to reduce the need for traditional excavation. Several geophysical methods can be used together or separately to outline archaeological structures. Magnetic methods are from such techniques that have proved a remarkable success in delineating shallow archaeological features (Abdallatif et al. 2003). They have been used to detect features that contain reasonable amount of magnetic minerals inherited in different objects such as buried walls and structures, kilns, bricks, roof tiles, fire pits, buried pathways, tombs, buried entrances, monuments, and inhabited sites. Most of these objects are detected because they are more magnetic than the surrounding or covering material. Other features, such as certain types of walls and tombs, are not magnetic themselves, but displace a uniformly magnetic soil (Abdallatif 1998).

Two main magnetic components are controlling the behavior of natural/archaeological features; these are the remanent magnetization and the magnetic susceptibility. Several attempts have been made to measure the remanent magnetization and the magnetic susceptibility of the fired objects made from mud (i.e., bricks, ceramics, kilns, etc.). These attempts have concluded that the remnant magnetization and the magnetic susceptibility are usually enhanced by firing. This concluded approach is very relevant to magnetic archaeoprospection because fired materials have a strong magnetic field that can be measured by a magnetometer. Recently, several magnetic-gradiometer surveys were conducted in Egypt, and successfully yielded very expressive results on the ancient Egyptian relics buried in different kinds of soils (Abdallatif 1998; El-Bassiony 2001; Kamei et al. 2002; Ghazala et al. 2003).

Generally, magnetic survey techniques can be used for a variety of purposes since they are fast, inexpensive, and accurate. Their application is non-destructive and most geophysicists consider magnetic surveys one of the best geophysical reconnaissance tools. In archaeoprospection, the objective of magnetic methods is to detect the inherent magnetic minerals (e.g., magnetite and maghemite) incorporated in the ancient archaeological features like pyramids, temples, and houses.

The ancient Egyptians have used mud bricks and limestone in the construction of their buildings/structures. The mud bricks are mainly composed of clay and mud materials extracted from the Nile River, which, at that time, was the artery of life for Egyptian people. These clay and mud materials are enriched with magnetic minerals that respond effectively to magnetic survey systems.

Historically, magnetic methods were first used in the 1950s (Aitken et al. 1958) and have since become the backbone of archaeological prospecting, successfully detecting many ancient relics, particularly those made of mud bricks (Abdallatif 1998; Kamei et al. 2002; Ghazala et al. 2003; Herbich 2003; Abdallatif et al. 2003, 2005, 2010; Odah et al. 2005; El Emam et al. 2014).

Previous applications of magnetic prospection in Egypt have played a prominent role in uncovering the country's cultural heritage. They have also provided important information about the composition and dimensions of various buried archaeological features (Hussain 1983; Abdallatif 1998; Odah et al. 1998; El-Bassiony 2001; Mousa et al. 2001; Kamei et al. 2002; Abdallatif et al. 2003; Ghazala et al. 2003; Khozyam 2003).

Archaeologically, traditional methods of archaeological investigation have been varied all over the world and in Egypt in particular because

Egypt contains a remarkable percentage of the unearthed relics in the world.

The diversity of these methods aims to reach to all things left by the ancient humans that can take advantage of our present and future time, including knowledge, treasures and civilization. Although different modern geophysical methods have appeared in the past 50 years, however, Egyptians were still—to a short time ago—using traditional methods to find the buried relics based on the historical information contained and recorded in some books, and in some cases the Egyptian archeological features were incidentally discovered (Shaw 2000).

There are many reasons, which put the geophysical methods as a great value for archaeologists. Some of these reasons are:

Archaeology is destructive. As a site is dug up it is systematically destroyed, hence each step of the dig must be painstakingly slow with careful documentation at each level. Geophysical probing on the other hand is rapid, non-destructive, and does not disturb the site.

Not all archaeological sites can be excavating. Examples would be historic buildings, churches, mosques, the pyramids, parks, and areas which underlie modern urban development. Again, geophysical methods are non-destructive and very rapidly employed, hence often cost effective in the end. In some cases, these methods may be all that the archaeologist is able to use at some sites.

Archaeologists can be greatly helping in setting his digging priorities if geophysical methods can be used ahead of time. Geophysical surveying can in many cases reveal artifact-laden vs. barren ground, and disclose important underground features: buried walls, voids, tunnels, ancient streets, etc.

Many decades may be required to explore a given site, such as a Tell. In fact, total excavation of a site may be impractical. Geophysical survey work at a given sites can usually be done in a few days or weeks of effort, the results of which are useful for many years of subsequent excavation work.

Salvage archaeology has become important as urban sites encroach on archaeological sites in many parts of the world. Thanks to modern legislation, substantial funding for archaeological research prior to the clearing of an area and construction of new buildings may be available. In many such cases, however, the time available for the archaeological effort may be very limited. Geophysical methods may be of great value, as the site will be often destroyed by the new construction.

7.2 Egyptian Archaeology

Egyptian archaeology is concerning with all features left by ancient Egyptians that simply reflect their culture, customs and civilization at the time they lived, in particular, legacy of science, knowledge, industry, elements of sovereignty, art treasures…etc.

Some archaeologists refer to the meaning of "Egyptian archaeology" as all remains of ancient Egyptians that help reading their history whether they are material or intellectual, standing on ground or transported, superficial or buried, stone or manuscripts…etc. In our opinion, one can say that Egyptian archeology is the study of all legacies of the ancient Egyptians which having a remarkable impact on the environment in which they lived, whether these legacies are valuable material or having a moral value, or both.

Non-Egyptian archaeologists and scientists (Rogers 1992; Williams and Stocks 1993; Black and Norton 1993) have been recorded their observations and experiences about Egypt in different literatures. However, Bard's book on Egyptian archaeology (2007) describes the archaeology of Egypt in a simple interesting and wonderful manner. Egyptian archeology indicates outstanding long history reflecting the past civilization of ancient Egyptian, including architecture, sculpture, embalming and burial customs and other more and more.

Civilization is a complex form of culture, the learned means by which human groups adapt to

Fig. 7.1 Different forms of Egyptian archaeology. (**a**) Giza pyramids, (**b**) Mortuary temple of Hatshepsut, (**c**) Sacred Lake, and (**d**) Drawings of ancient Egyptian life

their physical and social environments. Ancient Egyptian Civilization, with its unique monuments and works of art, has left very impressive remains (Bard 2007). It emerged between 3200–3000 BC, when a large region stretching along the lower Nile River and Delta was unified and then controlled by a centralized kingship. Its distinctive characteristics—the important institutions of kingship and state religion, monumental tombs and temples, the art which decorated these monuments, and hieroglyphic writing—emerged at this time and continued for over thousands of years.

For today Egyptians and foreign visitors to Egypt, Egyptian antiquities are visible everywhere in the country. They are very diverse in their structure with amazing shapes, attracting the mind and eyes of all people as long as they stay in Egypt, for example, one can see the pyramids—one of the seven ancient wonders of the world—the temples, tombs, statues, papyrus, obelisks. . .etc. (Fig. 7.1).

Besides, the Egyptian antiquities after the Islamic conquest of Egypt has been varied to take the form of palaces and mosques characterized by beautiful Islamic architecture, manuscripts collections, wood art, metal, ceramics, glass, different weapons featured every age, and the texture of beautiful rugs. . .etc. (Fig. 7.2).

A closer look at the archaeological evidence provides information about how the Egyptians built their monuments. Using systematic methodology, not fantasy, archaeologists who study ancient Egypt interpret archaeological evidence, providing a more rational, down to earth—and much more interesting—understanding of the past, including interpretations of "why they did it." There is also a large corpus of preserved texts, which adds to our understanding of the cultural meanings of these works, and how this civilization functioned (Bard 2007).

About the time from when there is evidence of early agriculture in Egypt, there is also evidence

Fig. 7.2 Some forms of Islamic heritage after the Islamic Conquer of Egypt. (**a**) Sultan Hasan Mosque in Old Cairo area, and (**b**) Rostrum of Mosque

of the use of pottery, and increasing numbers of potsherds are found at archaeological sites. Potsherds (broken pieces of pots) are important sources of information because pottery styles tend to change rapidly through time and are generally culture specific. Potsherds are useful for classifying late prehistoric as well as Dynastic sites by period and/or culture; sometimes imported, foreign pots are also identified at Egyptian sites.

Archaeological fieldwork in Egypt has been conducted according to the research problems and priorities of particular expeditions. For Egyptologists an important focus of fieldwork is often epigraphic studies, and philologists study ancient texts. Traditional archaeological methods were not sufficient to cover large areas in order to explore the hidden archaeology, and thus, new fast methods to help and support archaeologists and archaeological work are required.

7.3 Geophysics and Egyptian Archaeology

Worldwide, the application of geophysical methods in archaeology has begun almost half century ago in the European continent and then spread subsequently in the whole world. In 1946 and 1958, geoelectrical and magnetic methods were preliminary applied in England. Recently, the use of modern technology—from sophisticated devices and high-speed and precision software—has improved the application of geophysical methods in the field of archeology and led to many significant and fruitful results. Geophysicists worldwide have been worked in Egypt and obtained good results over the past decades. Meanwhile, several Egyptian geophysicists have contributed in the application of geophysics in archaeology; their results in collaboration with other foreign scientists are recorded and published (Hussain 1983; O'Connor 1997; Abdallatif 1998; El-Bassiony 2001; Mousa et al. 2001; Fassbinder et al. 2001; Abdallatif et al. 2003, 2005, 2010; Ghazala et al. 2003; Herbich 2003; Khozyam 2003; Odah et al. 2005; El Emam et al. 2014).

In the past few years an increasingly application of geophysics in the near-surface section of the earth has led to realistic and tangible results in different areas of the human life. The characterization and contribution of the near-surface geophysics in delineating different problems of today's requirements are well addressed by Butler (2005). Among the geophysical tools, magnetic methods effectively play an important role in enhancing the investigational practices. They have been extensively used in various areas, mostly in archaeology, to image the magnetized findings without digging.

Contribution of geophysics in all aspects of life including archaeoprospection is performed through different applicable methods such as, gravity, magnetic, seismic, geoelectric, electromagnetic, radioactive, thermal, induced polarization, spontaneous potential, ground penetrating radar...etc. Near-surface geophysical investigation in a country like Egypt has a prime consideration, in particular in the field of archaeology and prior to the commencement of any excavation work to help protecting the monuments and other valuable features from damage or destruction.

Historically, not all geophysical methods can be/were applied at one time, and also, not all methods were efficient in their first use. Later, after a lot of different experiments/applications, and in line with the continuing evolution in the manufacture of high technological geophysical equipments, the archaeological investigations have proved excellent efficiency and precise results.

Therefore, the emergence and application of geophysical methods in this field has had a significant impact in overcoming problems of traditional archaeological work, in particular the time factor, which plays an important role in the implementation and the establishment of economic and environmental projects. Geophysical methods have shown superiority and success over the past five decades on counterparts of other methods in locating archaeological features accurately and at different depths, without any damage or destruction despite their characteristic high speed.

Reaching to the excavation work in Egypt requires the contribution and input of many specialists from different disciplines including surveyors, geophysicists, Philologists, Egyptologists...etc. On-ground remote sensing (i.e. geophysical prospecting), different geophysical equipments are used to help archeologists to safely reach to the excavation stage and to accordingly locate buried remains such as magnetometers, resistivity-meter, ground-penetrating radar...etc.

The importance of applying geophysics in archaeology is basically to delineate the buried Egyptian relics in less time, efforts and cost. Geophysics aims to monitor and measure the physical properties (e.g. magnetic susceptibility, electrical resistivity, electrical conductivity, dielectric constant, elastic properties...etc.) of archaeological features buried in the ground or in the deep sea, and therefore know, their exact location for further excavation work.

The growing interest of different countries to find and maintain their ancient heritage, whether archaeological or cultural or otherwise, has pushed the governments of these countries to support all possible non-destructive means, like geophysics, to deal with potential large discoveries, and also to properly evaluate the situation of pristine and undiscovered archaeological areas. The reported use of geophysics benefits archaeologists in this aspect due to the advantages that the geophysical method own (e.g. high speed, cost effective, non-destructive...etc.). These advantages were also witnessed in similar geotechnical investigation pertaining to engineering, environmental and others.

Since, the excavation work is very important in the field of archaeology, it requires careful and accurate treatment. However, due to the expedited action of some archaeologists to carry out the excavation in a short time, unintentional damage may occurred for some unearthed relics, particularly those exist below the surface at about only a half meter, and thus, the inevitable result may lack deliberation and precision. The role of geophysics is required in such case as it is very capable to direct the archaeologists to see clearly at which and where to dig safely without any damage.

The inevitable result of this success achieved by the geophysical methods in archaeoprospection has led to change concepts of archaeologists concerning excavation and drilling plans. Majority of foreign expeditions who are working in Egypt are commonly dependent on geophysical surveys prior to any archaeological excavations, which prove the importance of applying geophysics in archaeology.

7.4 Approach of Magnetic Method in Archaeoprospection

Magnetic methods are mainly used to monitor and record an important natural universal phenomenon, namely geomagnetism or earth's magnetic field, originated from the interior and exterior of the earth. Theoretical and experimental work of geomagnetic scientists assumes that the earth represents as a huge magnet inducting a magnetic field with variable intensities. This field owes to two main sources, the first, is Earth's interior (from magma rotation) which indicates the largest part of the magnetic field (~95%), and the second source is Earth's exterior (from solar winds) which indicates the smallest part of the magnetic field (~5%) (Fig. 7.3).

To measure the geomagnetic field and its components, two main procedures are applicable:

1. Periodic measurements of the total magnetic intensity and other related components (Fig. 7.4) at constant stations named as geomagnetic observatories.

2. Field survey measurements whether on the land, or on the sea or from the air. This survey could be carried out at regular points along specified-spacing lines or traverses. In certain cases, the strategy of survey imposes the surveyor to follow a certain design of equidimensional grid patterns (Fig. 7.5) with small or large intervals depending on the scope of work. For example, if the scope of work is to find small-size objects (i.e. archaeological features, mineral...etc.), the readings must be taken over a very small intervals (i.e. 0.25 or 0.5 m), but if the scope of work is to find large sources (i.e. -geological

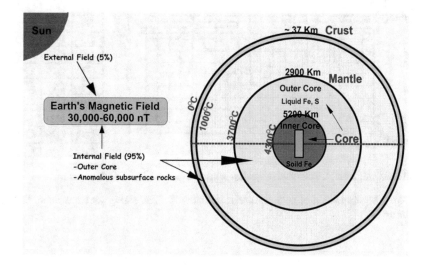

Fig. 7.3 Geomagnetic field and its components

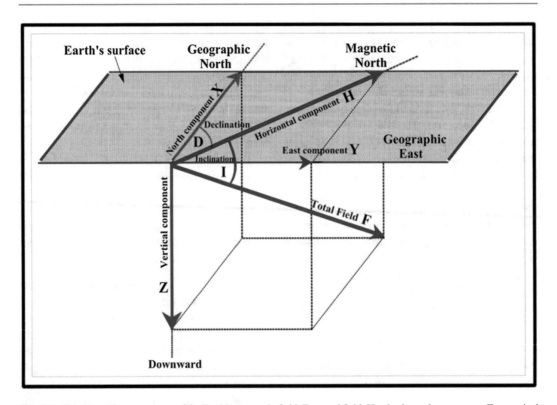

Fig. 7.4 The magnetic components of the Earth's magnetic field: F = total field, H = horizontal component, Z = vertical component, X = north component, Y = east component, D = declination, and I = inclination

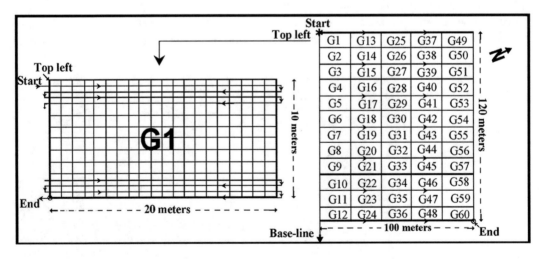

Fig. 7.5 Distribution of field grids showing the grid sizes, zigzag pattern method, base line, traverses and sample intervals

Fig. 7.6 Examples of magnetic instruments. (**a**) Fluxgate gradiometer (FM256 of Geoscan Research 2005), (**b**) proton magnetometer, (**c**) and magnetic susceptibility meter (MS2 meter and MS2B sensor of Bartington Instruments 1999)

structures), the readings are to taken over large intervals of up to several hundred meters.

In both of the previously mentioned procedures, different equipments (Fig. 7.6) are used to measure the total magnetic intensity (**F**) of the Earth's magnetic field and other associated components (horizontal component (H), vertical component (Z), declination (D) and inclination (I) (Fig. 7.4)). Also, through these measurements, the horizontal and vertical magnetic gradient can be estimated, mathematically and experimentally (Pedersen 1989).

We refer here also that the layers-forming rocks are gaining this magnetic field based on their remnant magnetization and magnetic susceptibility, which rely mainly on the magnetic mineral they possess, or simply the ratios of inherited iron characterized by high magnetization.

Earth's magnetic field is distributed with different strengths in various kinds of rocks constituting Earth's layers according to the variation of natural iron content inherited in them. The magnetic method has the capability to measure the change in the geomagnetic field caused by small objects buried in the ground depending on the contrast between these objects and the surroundings. Thus, it is highly possible to conduct magnetic measurements by the use of different kinds of magnetometers in order to pick up possible magnetic anomalies (magnetic contrast with the surrounding rocks), which can be interpreted and construed archaeologically or geologically in accordance with the scope of work. For precise results, it is strongly advised to keep all magnetic measurements far away from any external influences (i.e. sources of noise) made of iron such as surface facilities, architectural buildings, metallic fences, power lines...etc.

Magnetic method is used to determine the variation in the geomagnetic field and hence detect the locations of subsurface sources and estimate their depths, particularly those emerging from basement rocks (suitable places for oil accumulation) that characterized by very high magnetization. Also, and based on the capability of the method to detect magnetic contrast of the near surface objects, it can be used in determining the occurrences of precious metals, ancient monuments, and modern tools manufactured by humans, such as water pipes, electrical cables, metallic tanks and utilities networks and others, generally made of iron. In addition, few other uses are also applicable such as identifying sites suitable for the establishment of runways and location of landmines...etc.

The application of magnetic method in the field of archaeology has been started in 1958 in England, where it proved a great success in locating archaeological finds, consisting of mud and fire bricks, which contain high concentrations of magnetite minerals.

Different examples of Egyptian archaeological features/structures are considered strategic targets of the magnetic survey such as tombs, pyramids, ditches, kilns, solar boats, ruins of ancient cities along with some of old metallic tools (e.g. Artefacts). In addition, the measurements of magnetic susceptibility of the topsoil were used in mapping areas that witnesses life activities of ancient human.

In Egypt, magnetic method was applied in the early of eighteenth using less-sensitive proton magnetometer (Hussain 1983), later on, in the mid of Nineteenth, a modernized technique of the method was applied by Abdallatif (1998) under a joint project between the *National Research Institute of Astronomy and Geophysics* in Egypt and *Bayerisches State Office for Heritage* in Germany using state of the art instruments (e.g. Fluxgate gradiometer FM36 and Caesium vapour magnetometers). These instruments have given excellent results in the detection of many

archaeological findings made of mud brick (e.g. tombs, wall remains of the ancient city, solar boats, kilns and archaeological ditches) (Fig. 7.7).

Technique of measuring the vertical magnetic gradient (VMG) (Fig. 7.8) is the most effective approach of the magnetic techniques to precisely detect magnetic features few meters beneath the soil. It has proven worth in detecting small man-made sources and archaeological features (Gibson 1986; Clark 1990; Scollar et al. 1990; Abdallatif et al. 2005, 2007). The VMG has various advantages over the total field data (e.g. Hood 1965; Slack et al. 1967; Hood et al. 1979; Ackerman 1971; Barongo 1985; and Keating and Pilkington 2004; Abdallatif et al. 2005). The reason is that it resolves better the small sources with high resolution, and removes all magnetic effects originating at a great distance from the sensor such as diurnal variations and regional fields accompanying to the geomagnetic field.

Magnetic gradiometers are extensively used in today's environmental and geotechnical investigations. Results are optimized when the sensors of these gradiometers set at a suitable height from the ground. Poor interpretation will be made if this height kept undetermined. An ongoing research on the vertical magnetic gradient, the authors of this article examine the impact of the sensor height (SH) on the anomaly detectability (AD) in the near surface applications using the vertical magnetic gradient (VMG) techniques. Calculated detectability and upward continued data to various higher distances of an Egyptian archaeological site emphasize good detectability of the near-surface features at a vertical distance up to 0.3 m. Results after this height will be foggy and image visibility will be faint, leading to poor interpretation.

Isolation of near surface sources with high detectability is the main goal of today's magnetic survey. This goal gets much easier with the use of high sampling gradiometers, which almost remove the regional field generated by the deep sources (Breiner 1973) where the VMG from distant sources is so small that it is negligible compared to the VMG from nearby sources. Accordingly, the difficulty of source isolation

Fig. 7.7 Examples of Egyptian archaeological findings. (**a**) Ancient Egyptian pottery kiln from Al-Lahun, (**b**) Mud brick walls from Abydos, and (**c**) Solar boats from Abydos (Abdallatif 1998)

which most geophysical methods suffer from is almost unconsidered here, and the high detectability is rising up instead, particularly with subtle sources. This detectability ultimately depends on the SH of the magnetic instruments. Variability of this SH due to any reason will surely affect the detectability and thus the data quality. For example, the slight distortion of the geomagnetic field caused by subtle sources will be less detectable with the upward increase of the SH, particularly with the hand-carry instruments (e.g. FM256) and also in historical areas where the archaeological features show a small magnetic variation. Note that in most regional magnetics, the SH is negligible comparing to the depth to the subsurface sources.

As a general rule, the closer the sensor to the ground, the greater the spatial resolution and detectability can be achieved. However, the proximity of the sensor to the ground may cause a contamination of the data from the noises of the topsoil layer. Thus, determining specific and allowable height can save the data from such risk of surface noises and at the same time achieve good detectability.

As a rule of thumb, the detectability of the magnetic anomaly changes if the plane of magnetic observations changes (Fig. 7.9). In the near-surface investigation, three variables are directly affecting the detectability of the anomalous features (Fig. 7.8): the operator length, the SH, and the depth. The operator length looks

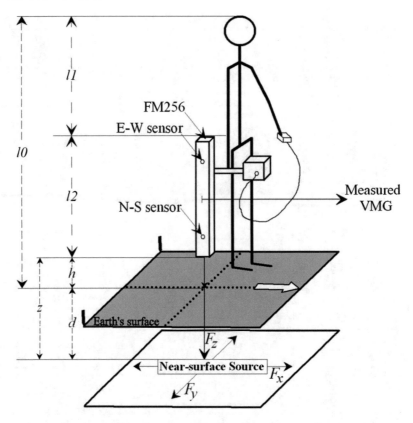

Fig. 7.8 A sketch of the fluxgate gradiometer (FM256), showing the operator length l_0, the distance from the operator vertex to the instrument head l_1, the FM256 length l_2, the SH h and the depth d

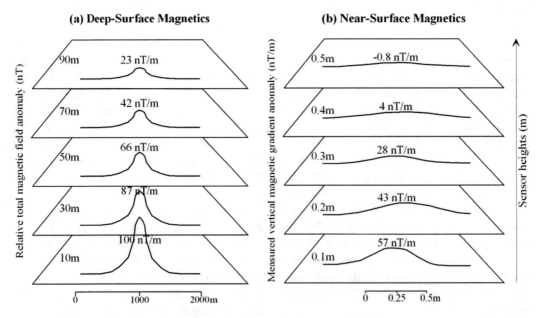

Fig. 7.9 Observations at different sensor heights of deep and near-surface magnetics

wondering, but it is very possible when more than one operator of different lengths do survey for a large site with variable carrying styles. Although the SH is variable, it can be determined and fixed, while the depth is immeasurable and usually unchangeable. Thus the SH is the main controller of the AD and is directly impacted by the operator length (Fig. 7.8).

The vertical magnetic gradient anomaly produced at two different heights is mathematically expressed as

$$\frac{\partial \boldsymbol{F}}{\partial z} = \frac{F_B - F_T}{\Delta z} \qquad (7.1)$$

Where F_B & F_T are the total magnetic field at the bottom and the top sensors, Δz is the interval between the top and the bottom sensors, and $\partial F/\partial z$ is the derivative of the Earth's magnetic field in Z direction (i.e. vertical gradient). Equation (7.1) indicates that the gradiometer anomaly therefore varies at a higher fall-off than its corresponding total field anomaly.

For any generalized magnetic sources, regardless of their sizes, Breiner (1973) referred to three main factors affecting the detectability of magnetic anomaly, the first is the distance between the magnetometer and the object, the second is the amount of ferromagnetic material associated with the object in contrast with the surrounding material, and the third is the expected background magnetic noise arising from such sources as geology or man-made materials and electric current.

For subtle magnetic sources, however, the detectability perhaps depend on many other factors such as the source parameters (e.g. depth, size, nature), the site conditions (e.g. geological, archaeological), the method used (e.g. magnetic), the instrument (e.g. spectral resolution, SH, sampling, memory), the field work procedures (pattern, sampling interval), the processing techniques (isolating, filtering), and the operator (e.g. skill and experience). The AD is, accordingly, a function of the measurable criteria of the previous factors. For example, the AD is a function of the depth and size of the sources being investigated. More

deeply sources are more difficult to detect than shallowly ones. Also, structures of small size (e.g. pits) have lower detection probabilities than large structures (e.g. tomb, ditch).

Majority of geomagnetic scientists believe that using the technique of the VMG in the near surface applications is very productive and effective in providing more spatial details and enhancing detectability of the small magnetic anomalies, which in turn help interpreting the data comprehensively.

7.5   Applications, Examples, Case Studies

7.5.1   Example 1: Saqqara, Giza, Egypt

A detailed gradiometer survey west of Zoser pyramid, Saqqara (Fig. 7.10 a) was conducted in the mid of 1997 using the FM36 to explore for the near-surface archeological features as a result of a cooperation of the BSCOM, IGAG, NRIAG, and Polish Institute of Archaeology and Ethnology (Abdallatif 1998). Saqqara was selected because it is archeologically important for being contains more than ten pyramids including the globally known one (Zoser), and also because it contains mud brick cemeteries of different Egyptian dynasties. These cemeteries are located at shallow depth, surrounded by smoothed sand, and overlain by sandy soil which possesses low magnetic susceptibility.

The processed magnetic plot by Abdallatif (1998) revealed various significant magnetic anomalies (Fig. 7.10c) west of Zoser Pyramid. The most important one is a square-shaped feature located at grid B1 (Fig. 7.10c), which later confirmed by excavation as enclosure of a burial shaft (i.e. tomb) composed of mud brick walls, and located at ~0.5 m depth (Herbich 2003). The measured VMG of this shaft indicates a low deformation of the geomagnetic field ranges from −1.3/11.9 nT/m.

Fig. 7.10 Field example from Egypt. (**a**) Location map of the Saqqara archaeological site, including the study area. (**b**) Raw VMG data. (**c**) Low pass filtered VMG image shows revealed archaeological and man-made features

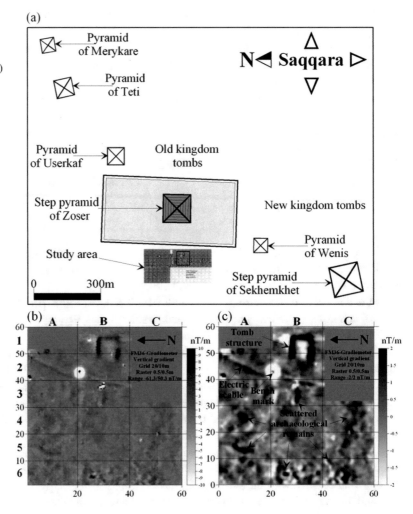

The estimated depth (z) of other revealed features is ~0.5 to ~0.65 m (Abdallatif 1998), thus by subtracting the h from z, we simply get the mean depth to the top of the near-surface features as: $d_{ave} = z_{ave} - h = 0.575 - 0.1 = 0.475$ m, or $= 0.405$ m if h reaches the maximum error (i.e. 0.07 m).

7.5.2 Example 2: Abu Sir, Giza, Egypt

Abdallatif et al. (2005) have carried out an integrated magnetic study at Abu Sir, a locality in northern Egypt also known as the "Land of the Forgotten Pyramids" (Fig. 7.11). Historical records of Abu Sir refer to the existence of a large necropolis or burial place for the kings of the 5th Dynasty who constructed their buildings with mud bricks.

Two magnetic tools were applied over an area of 25,600 m^2 (Fig. 7.11) in order to trace and detect hidden archaeological features near the Temple of the Sun. The acquisition of the magnetic data was initiated by measuring the magnetic susceptibility (Dearing 1994) of the topsoil samples collected within the entire study area (Fig. 7.11). This was followed by a gradiometer survey to measure the vertical gradient of the geomagnetic field over a restricted area of 14,400 m^2. The magnetic susceptibility results

Fig. 7.11 Location map of the land of forgotten pyramids of Abu Sir showing the study area

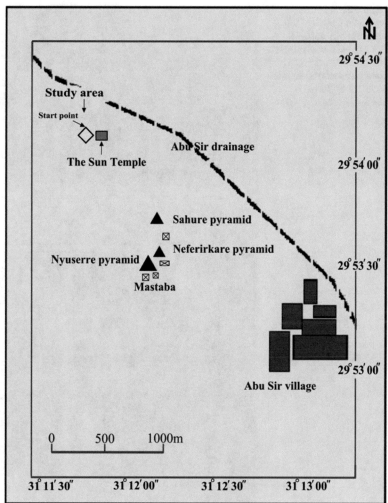

are characterized by high values in the middle of the study area and a small extension of high values to the southwest. The magnetic instruments applied in this study are presented in Fig. 7.12.

The magnetic susceptibility measurements identified regions of interest to be targeted during the gradiometer survey. The gradiometer results revealed the existence of numerous archaeological features of different shapes and sizes composed of mud bricks. These features may represent tombs, burial rooms, and dissected walls, and all of them probably belong to the 5th Dynasty of pharaohs. The depth of the expected buried archaeological features was

estimated from the gradiometer results and is about 1.2 m for deep features and 0.42 m for shallow features.

The resultant magnetic susceptibility map shows concentrations of low values at the southern, northern, northeastern, and northwestern parts of the study area with a minimum value of about 6×10^{-5} SI (Fig. 7.13). Relatively high values (maximum value of about 36×10^{-5} SI) were noticed in the middle of the study area with a small extension to the southwest. The parts of the study area characterized by high magnetic susceptibilities outline the main area of the magnetic features as confirmed by the gradiometer survey.

Fig. 7.12
(**a**) A photograph of the
Magnetic Susceptibility
meter (MS2B of Bartington
1999), and (**b**) A
photograph of the fluxgate
gradiometer (FM36 of
Geoscan Research 1987)

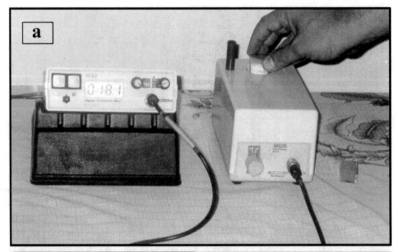

In general, the high magnetic susceptibility values indicate the proximity of buried archaeological features rather than industrial activities. Specifically, the regions of high values suggest cultural activities near the Temple of the Sun, and possibly ancient remains associated with the Temple.

The gradiometer survey, which was carried out over an area of 120 × 120 m (Fig. 7.11) targeted by the topsoil susceptibility survey, produced the most conclusive results. The gradiometer results (Fig. 7.14) confirmed the presence of hidden archaeological features within the high susceptibility area.

Figure 7.14, prepared from the low-pass filtered magnetic image, shows the presence of numerous features (positive and negative anomalies). The prevailing black features refer

Fig. 7.13 Magnetic susceptibility map showing the results of the topsoil magnetic susceptibility survey

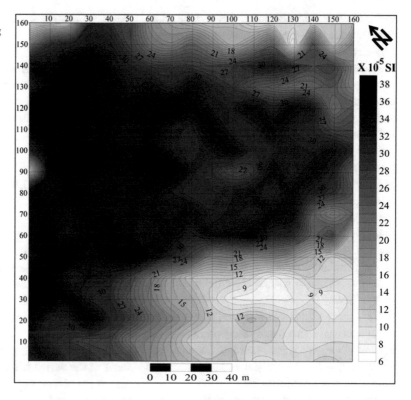

Fig. 7.14 Magnetic image of the study area after application of a low-pass filter (X = 2 and Y = 2)

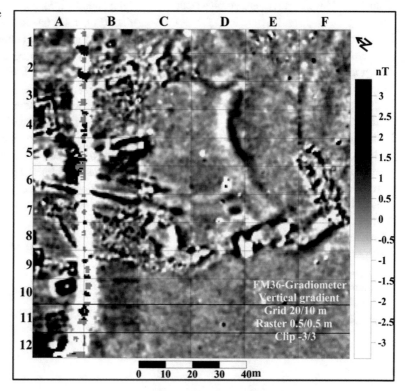

to the positive magnetic anomalies and the white features refer to the negative magnetic anomalies. The gradiometer results indicate the existence of big mud-brick structures (M1, M2, and M3) represented by positive magnetic anomalies and concentrated mainly at the right part of the image and extending to the left, upper left, and bottom left sides. These structures are located at the western side of the Sun Temple and represent the most important archaeological features revealed by the gradiometer survey. The authors believe that these structures are an extended ritual place of the temple. In addition, the results also show the presence of other archaeological features, including tombs, burial rooms, and dissected walls. All of them probably belong to the 5th Dynasty of pharaohs, who constructed their buildings with mud bricks.

This study shows that it can be advantageous to conduct topsoil magnetic susceptibility surveys prior to other geophysical surveys. The magnetic susceptibility results can guide subsequent and more intensive investigations.

The following advantages of the fluxgate gradiometer (FM36) in Egypt are emphasized:

1. It works well with shallowly buried features as it removes or minimizes the effect of the regional field that can be expected to originate from the deep sources. This favors the use of gradiometer survey for environmental and engineering purposes.
2. It takes less time and effort than the normal magnetometer survey, as well as resistivity and GPR surveys. This can facilitate the fast reporting of initial results that can then be used to guide any further survey work.
3. It accurately detects some important parameters of the expected sources and features (i.e., the vertical gradient of the geomagnetic field, the positions, the horizontal dimensions, and the polarity of the near-surface features). Most of these parameters are useful in geophysical interpretation.

7.5.3 Example 3: Karnak, Luxor, Egypt

In this study, Egyptologists have suggested there might be a hidden extension of the Karnak Temple, located in the Luxor governorate of Egypt (Fig. 7.15). They believe that the temple contains some buried sections on its northern and eastern sides. During January and September 2002, Abbas et al. (2005) have carried out ground-penetrating radar (GPR) and magnetic surveys on an 80 × 40 m area near the eastern gate (Fig. 7.16).

The magnetic gradiometer survey was conducted over a layout composed of 16 grids; each grid consists of successive zigzag lines. The magnetic data was acquired by the fluxgate gradiometer (FM36) at 0.5 intervals in order to measure the vertical magnetic gradient of the geomagnetic field.

The GPR data did not give a clear picture of some of the buried features. A clearer identification of the hidden features was accomplished by interpretation of the magnetic data using the analytic signal approach. Low pass filtered magnetic data (Fig. 7.16) has revealed the presence of coffin- and tomblike structures built of mud bricks. Limestone blocks were also detected, which may be attributable to destroyed pillars or statues (Fig. 7.17).

Conclusions concerning further extensions of the Karnak Temple will remain theoretical until some excavation has occurred. The geophysical investigation of this study helped identify the locations of interesting phenomena that should be considered targets for future archaeological excavation.

7.5.4 Example 5: Dahshour, Giza, Egypt

Abdallatif et al. (2010) have conducted near-surface magnetic investigations using gradiometer survey in the area east of Amenemhat II

Fig. 7.15 Map of Egypt showing the location of Karnak Temple

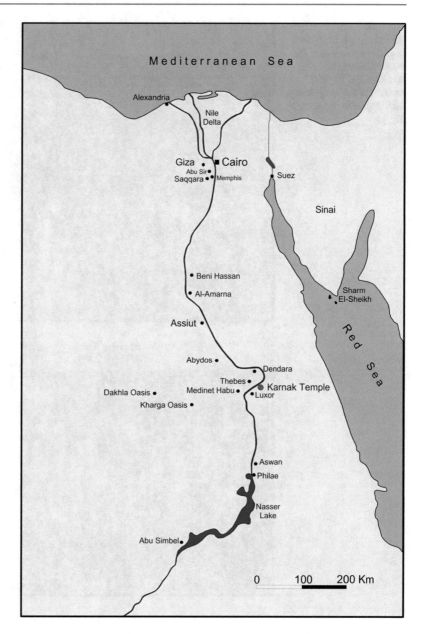

pyramid (Fig. 7.18) to measure the vertical magnetic gradient with a high resolution instrument at 0.5 m sampling interval. The application of the vertical magnetic gradient method over an area of 340 × 200 m has detected four archaeological features of different shapes and sizes that mostly consist of mud bricks. The data showed some undesirable field effects such as grid discontinuities, grid slope, traverse stripe effects, spikes and high frequencies originating from recent ferrous contamination. These undesirable effects were addressed to produce an enhanced display. The processing sequence of gradiometer data applied in this study was sufficient to enhance the appearance of the detected mud brick features.

Fig. 7.16
(**a**) Photograph of some
inspection activities at the
study area and its
surroundings. (**b**) The
location of the study area
with respect to the Karnak
Temple

The authors have successfully detected four main structures in the area east of the pyramid; the causeway that connected the mortuary temple with the valley temple during the Middle Kingdom of the 12th Dynasty, the mortuary temple and its associated rooms, ruins of an ancient working area and an Egyptian-style tomb structure called a Mastaba.

The processed data (Fig. 7.19) emphasized significant archaeological structures that possess appreciable magnetic content. This includes the causeway that connects the mortuary temple with the valley temple in the Middle Kingdom of the 12th Dynasty (Fig. 7.20), the mortuary temple and its associated rooms (Fig. 7.21), ruins of an ancient working area (Fig. 7.22) and an Egyptian-

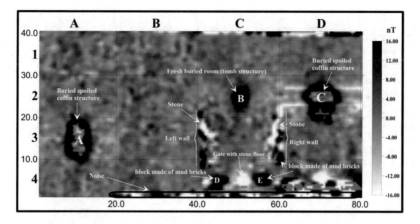

Fig. 7.17 Magnetic image showing the final results after application of a low-pass filter (X, Y = 2 Gaussian)

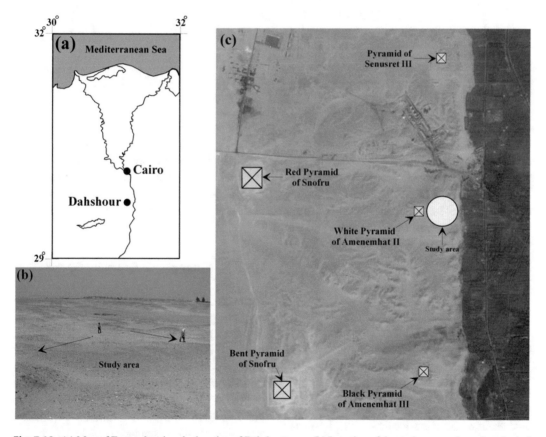

Fig. 7.18 (a) Map of Egypt showing the location of Dahshour area, (b) Location of the study area at the archaeological site of Dahshour, and (c) A satellite image of the location of Dahshour (reproduced from Google Earth)

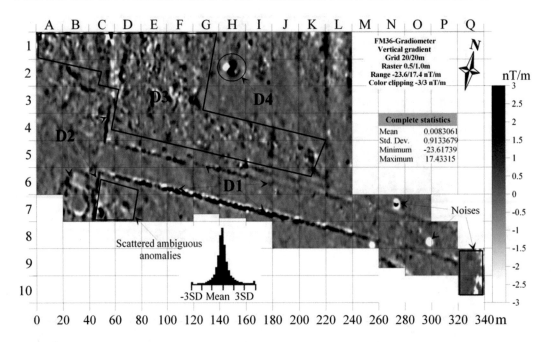

Fig. 7.19 Gradiometer data of Fig. 7.6b after applying low pass filter and interpolation functions. This plot shows the causeway and the mortuary temple of the pyramid of Amenemhat II in Dahshour. D1, D2, D3, and D4 delineate the main discoveries in this study

style tomb structure called a Mastaba (Fig. 7.23). The valley temple of the pyramid of Amenemhat II that should be connected to the recently revealed causeway is believed to be buried beneath urban construction on the farthest eastern side. Abdallatif et al. (2010) have carried out excavation of a small part within the study area in order to prove the reliability of magnetic discoveries. This excavated part indicated the shallowness of the detected features.

This example shows that it can be advantageous to conduct a gradiometer survey prior to other geophysical surveys. The results can guide subsequent and more intensive investigations. Furthermore, the fluxgate gradiometer accurately detects some important parameters of the discovered features (i.e., the locations, the horizontal dimensions and the polarity of the near-surface features).

A sketch diagram showing the causeway and the mortuary temple of the pyramid of Amenemhat II discovered by the magnetic method at Dahshour is presented in Fig. 7.24.

7.5.5   Example 6: Hawara, Egypt

El Emam et al. (2014) have applied a near-surface magnetic gradiometer survey on a site at Hawara, Fayoum city to delineate archaeological features made by mud bricks at shallow depths in the northeast of the pyramid of Amenemhat III (Fig. 7.25). Archaeology of Fayoum has been addressed by Hewison (1984), stating most of the important sites and locations of interest according to the discoveries made by foreign expeditions.

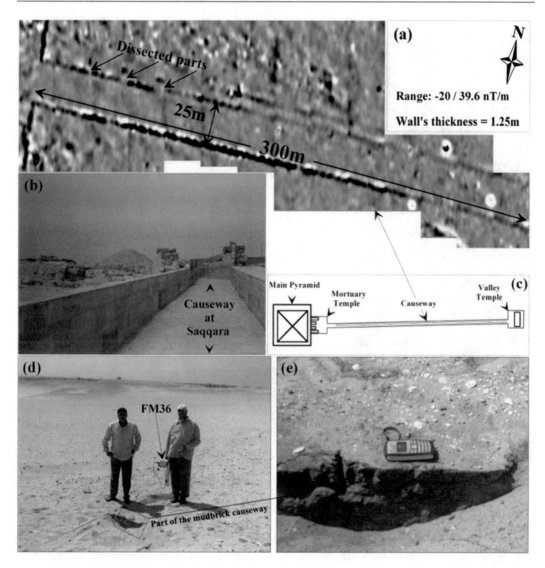

Fig. 7.20 (a) Causeway (D1), discovered by the magnetic method, indicates reasonably high magnetic amplitude, reflecting its composition of baked mud brick. It extends east-west for ~300 m. The causeway may connect the mortuary temple with the valley temple, (b) Example of a restored causeway at Saqqara, (c) A complete structure of the funeral style of the ancient Egyptians, (d) Excavation of a small part of the causeway discovered by the magnetic survey at very shallow depth, and (e) Enlargement of this part displaying its obvious mud brick composition

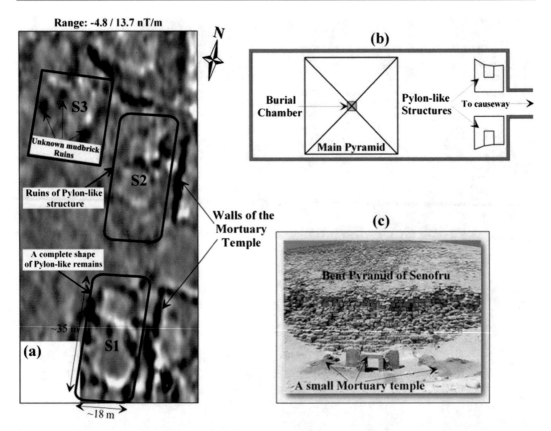

Fig. 7.21 (**a**) Mortuary temple (D2) of the pyramid of Amenemhat II discovered by the magnetic method. It shows three main mud brick features; the wall, a complete shape of pylon-like remains, and ruins of pylon-like structures, (**b**) A sketch ground plan of the pyramid's design showing the pylon-like structures, and (**c**) A surface example of a small mortuary temple of the bent pyramid of Senefro at Dahshour

Magnetic survey data were processed to a precise forward processing sequence, including clipping, de-sloping, removing stripes and edge discontinuities, de-spiking, filtering, interpolating, and finally detecting statistically significant results. The processed magnetic data indicate high amplitude anomalies in the northeastern part of the pyramid Amenemhat III, suggesting the presence of shallow cemeteries of mud brick walls.

Low pass filtering of the vertical gradient data helped reveal the relatively deep mud brick features in the generally shallow section. These features were isolated using a statistical detection technique, and verified by the correlation with the surrounding mud brick discoveries. This required the analysis of the magnetic behavior of exposed mud bricks and man-made objects for comparison with the magnetic signature of buried material. These were then matched with the vertical magnetic gradients of the exposed mud brick features from here and three other Egyptian localities. They are mostly the extension of the 12th Dynasty cemeteries, which exist in the western and southwestern parts of the study area.

The results of this study emphasize the significance and safe application of geophysics before archaeological excavation or engineering construction. The magnetic method, as a quick and cheaper reconnaissance geophysical method, can test for further surveyability of other geophysical tools, guide them to effectively contribute in the

Fig. 7.22
(**a**) Ruins of mud brick structures (D3) interpreted as an ancient working area that was totally demolished by unspecialized prospectors sometime in the past, (**b**) and (**c**) Example of an excavated site 300 m from the northern side of the study area. Remains of wall structures are composed of mud brick with an average thickness of 1–2 m

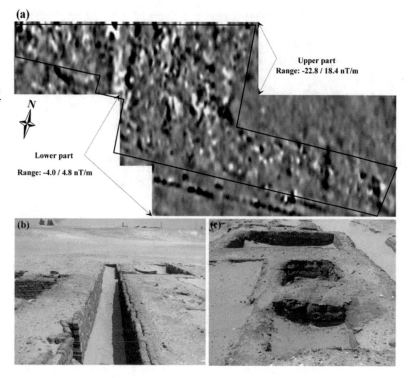

Fig. 7.23
(**a**) Dipole magnetic anomaly (D4) suggests a tomb-like structure—a "Mastaba", and (**b**) A surface example for a similar mud brick Mastaba at Dahshour, 2 km south of the study area

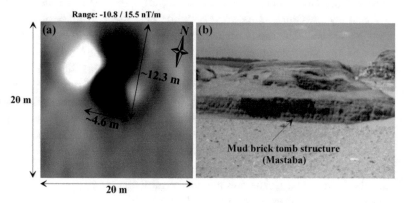

survey strategies, and assist accordingly the future plans of archaeologists.

The measurements of the vertical magnetic gradient using the fluxgate gradiometer (FM36) (Fig. 7.26) was successfully adapted in the archaeological site of Hawara, where it proved its high efficiency in detecting localized magnetic disturbance associated with mud brick features near the surface, and provided a better understanding of the history of ancient Egyptians.

In comparison with influence of man-made materials, mud brick anomalies have only a very

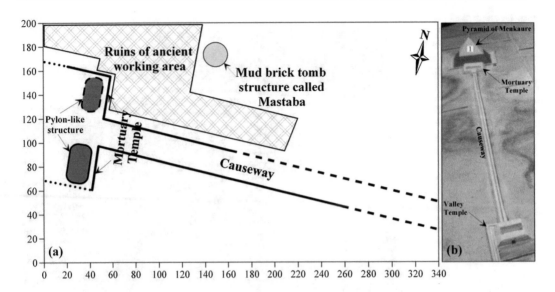

Fig. 7.24 (a) A sketch diagram showing the causeway and the mortuary temple of the pyramid of Amenemhat II discovered by the magnetic method at Dahshour, and (b) Results are expected to be similar to this construction of the pyramid of Menkaure and its associated structures at Giza (Reproduced from http://www.brekka.net/268S_egypt.html)

small effect on the geomagnetic field. This small effect was prominently depicted in the lower part of the study area due to its contrast with the surrounding sand. This study highlights the efficiency of low pass filtering application with gradiometer data to enhance the relatively deep archaeological features in the shallow section. It also recommends applying the statistical detection to low pass filtered data to isolate any possible remaining noise from sources of interest.

Following the processing sequence introduced in this study (Fig. 7.27), it is necessary to minimize the field errors, isolate and reduce the random noises, and enhance the displayed results. This sequence can involve more or less functions depending on the factors controlling the operator, instrument, geology, and the general conditions of fieldwork operation. This sequence can also be applied to other near- surface applications of environmental and engineering purposes.

Magnetic behavior of the VMG was a useful approach for describing and classifying mud brick features of different localities. The VMG of the mud bricks delineated in Hawara matches with the VMG of the mud bricks uncovered from the same site, and falls under the range of the VMG of Egyptian mud bricks (<30 nT/m). Mud brick features delineated northeast of Hawara pyramid are mostly the extension of the 12th Dynasty cemeteries, which exist in the western and southwestern parts of the study area (Fig. 7.28).

7.6   Conclusions

The application of magnetic in discovering Egyptian archaeological features through different researchers has proven excellent effectiveness and impressive results in detecting the past

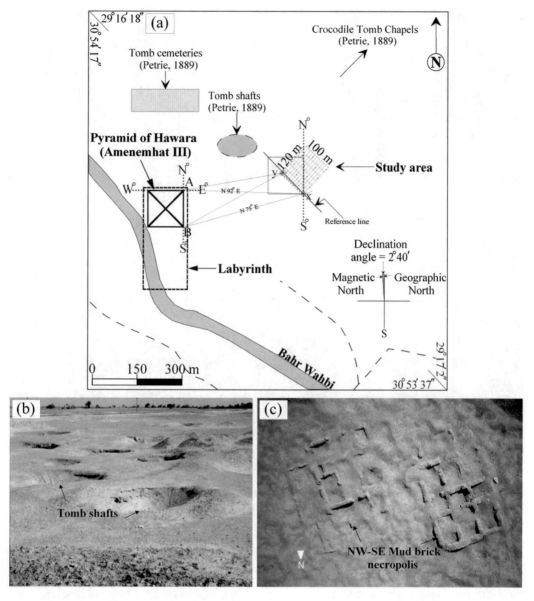

Fig. 7.25 Location of the study area in Hawara, northeast of Amenemhat III Pyramid. (**b**), (**c**) Photographs snapped by Uytterhoeven (2003) for the past excavation by Petrie (1889). Ancient cemeteries and mud brick structures are observed in the vicinity of the study area

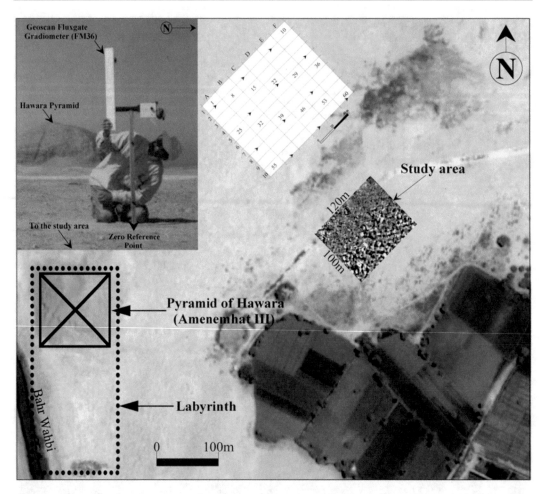

Fig. 7.26 Magnetic results of the study area northeast of Hawara Pyramid Site. Aligning and balancing the fluxgate gradiometer (FM36) at a selected zero reference point is shown in the top left corner

heritage and discovering the history. Some concluded points are emphasized here:

1. Application of the magnetic method or any other geophysical method is site specific.
2. Geophysical methods, including magnetic, are effectively used to supplement traditional subsurface investigations.

3. Methods which are successful at one site will not be necessarily successful at another.
4. Results depend on interpretation and should be field-verified by more definitive exploration/excavation techniques in order to provide the most integrated solution for the identification of subsurface hidden relics.

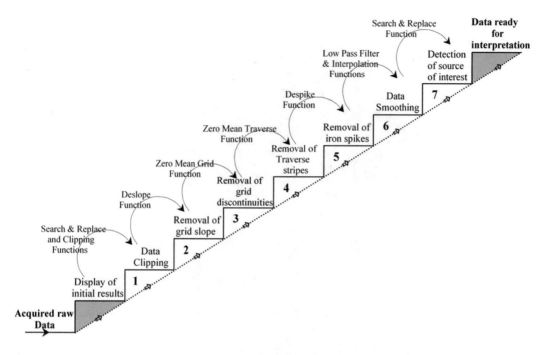

Fig. 7.27 Scheme of the forward processing sequence applied in this study. It is based mainly on the use of Geoplot functions (Geoscan Research 2005)

5. The gradiometer-magnetometer comparison had concluded the following:
   (a) The obtained gradiometer results are more clear and expressive for the expected hidden archaeological features due to the high resolution of the used instruments.
   (b) The expected interference of the subsurface shallow and deep geological sources may affect the obtained magnetometer data.
   (c) Archaeoprospection is not successful by using magnetometer instruments with low sensitivity, particularly in the absence of other sensitive instruments.
6. The application of magnetic method gives fast and accurate results than the traditional archaeological excavation. It saves time, effort and money.
7. The application of two magnetic tools at some sites like Abu Sir area have been

succeed to verify the scope of work in a relatively short time.
8. The magnetic susceptibility measurements are useful in large areas; it is capable to detect the zones to be focused by another magnetic tool, helping the general strategy of any further geophysical application.
9. The measurement of the vertical gradient of the geomagnetic field gives more accurate results than the measurement of the total magnetic field.
10. The fluxgate gradiometer (e.g. FM36) is preferable for shallow investigations (e.g. archaeoprospection) than the old fashion magnetometers.
11. The depth parameter has a second priority for the archeologists, since most of the investigated areas are usually contain very shallow objects (e.g. ~1.2 m at Abu Sir area).

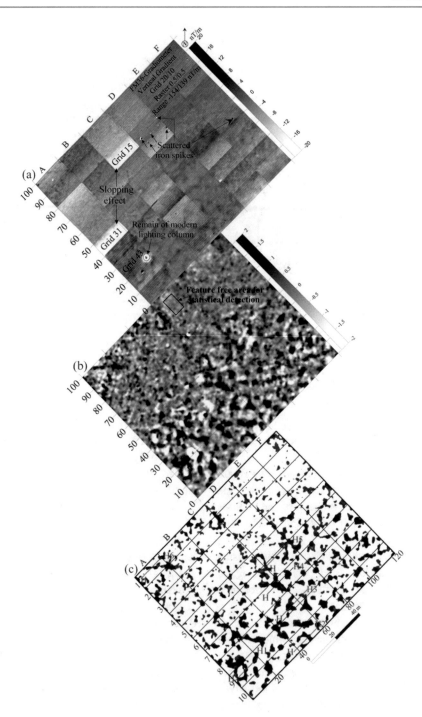

Fig. 7.28 (a) Raw gradiometer data show smeared anomalies due to the noisy effect of lighting column rubble at grid 43, (b) processed magnetic image followed the scheme of Fig. (47), and (c) archaeological features of interest isolated due to the application of statistical detection with threshold0 2.5 SD. The delineated magnetic anomalies represented by black color are interpreted in this study as mud bricks extension of the 12th Dynasty cemeteries in the archaeological site of Hawara. HH1, HH2, HH3, HH4, HH5, and HH6 are six selected profiles for the VMG analysis. The color scale refers to clipped data

References

Abbas AM, Abdallatif TF, Shaaban FA, Salem A, Suh A (2005) Archaeological investigation of the eastern extensions of the Karnak Temple using ground penetrating radar and magnetic tools. Geoarchaeology 20(5):537–554

Abdallatif TF (1998) Magnetic prospection for some archaeological sites in Egypt. Ph.D. thesis, Ain Shams University, Cairo, Egypt

Abdallatif TF, Abd-All EM, Suh M, Mohamed RM, El-Hemaly I (2005) Magnetic tracing at Abu Sir (land of forgotten pyramids), Northern Egypt. Geoarchaeology 20(5):483–503

Abdallatif TF, Mousa SE, Elbassiony A (2003) Geophysical investigation for mapping the archaeological features at Qantir, Sharqyia, Egypt. Archaeol Prospect 10:27–42

Abdallatif TF, Suh M, Oh J, Kim KH (2007) Impact of magnetic survey design on the imaging of small archeological objects: practicability in gradiometer surveying. Lead Edge 26(5):571–577

Abdallatif TF, El Emam AE, Suh M, El Hemaly IA, Ghazala HH, Ibrahim EH, Odah HH, Deebes HA (2010) Discovery of the causeway and the mortuary temple of the Pyramid of Amenemhat II using near-surface magnetic investigation, Dahshour, Giza, Egypt. Geophys Prospect 58:307–320

Ackerman J (1971) Downward continuation using the measured vertical gradient. Geophysics 36(3):609–612

Aitken MJ, Webster G, Rees A (1958) Magnetic prospecting. Antiquity 32:270–271

Barongo JO (1985) Method for depth estimation on aeromagnetic vertical gradient anomalies. Geophysics 50:963–968

Bartington Instruments Ltd (1999) Operation manual for MS2 magnetic susceptibility system. Bartington Instruments, Oxford

Bard KA (2007) An introduction to the archaeology of ancient Egypt, Wiley-Blackwell, 424p, 978-1-4051-1149-2

Black AC, Norton WW (1993) Blue guide Egypt. Bedford, London, 762p

Breiner S (1973) Application manual for portable magnetometers. Geometrics, Sunnyvale

Butler DK (2005) Near-surface geophysics. SEG, Tulsa

Clark JA (1990) Seeing beneath the soil: prospection methods in archaeology. B.T. Batsford, London

David A (1995) Geophysical survey in archaeological field evaluation. Research & Professional Services, Guideline No. (1), pp 1–35

Dearing JA (1994) Environmental magnetic susceptibility using the Bartington MS2 system. British library cataloguing in publication data. 0952340909

El-Bassiony AA (2001) Geophysical archaeoprospection in the Saqqara and Qantir areas, Egypt. Master thesis, Ain Shams University, Cairo, Egypt.

El Emam A, Abdallatif T, Suh M, Odah H (2014) Delineation of Egyptian mud bricks using magnetic gradiometer techniques. Ar J Geosci 7:489–503

Fassbinder J, Becker H, Herbich T (2001) Magnetometry in the desert Area west of Zoser's Pyramid, Saqqara, Egypt, w: Magnetic prospecting in archeological sites, Becker & Fassbinder (red.), Monuments and Sites VI, Munich, pp s.64–65

Geoscan Research (1987) Instruction manual version 1.0 (Fluxgate Gradiometer FM9, FM18, FM36). Geoscan Research, Bradford

Geoscan Research (2005) Instruction manual 1.01. (Geoplot 2.01). Geoscan Research, Bradford

Ghazala H, El-Mahmoudi AS, Abdallatif TF (2003) Archaeogeophysical study on the site of Tell Toukh El-Qaramous, Sharkia Governorate, East Nile Delta, Egypt. Archaeol Prospect 10:43–55

Gibson TH (1986) Magnetic prospection on prehistoric sites in Western Canada, University of Alberta. Geophysics 51:553–560

Herbich T (2003) Archaeological geophysics in Egypt: the polish contribution. Archaeol Polona 41:13–55

Hewison NR (1984) The Fayoum; A practical guide, The American University in Cairo Press, 85p

Hood P (1965) Gradient measurements in aeromagnetic surveying. Geophysics 3(5):891–902

Hood PJ, Holroyd MT, McGrath PH (1979) Magnetic methods applied to base metal exploration. In: Hood PJ (ed) Geophysics and geochemistry in the search for metallic ores: Geol Surv Can Econ Geol Rep, vol 31, pp 527–544

Hussain AG (1983) Magnetic prospecting for Archaeology in Kom Oshim and Kiman Faris, Fayoum, Egypt. ZES 110:36–51

Kamei H, Atya MA, Abdallatif TF, Mori M, Hemthavy P (2002) Ground-penetrating radar and magnetic survey to the west of Al-Zayyan Temple, Kharga Oasis, Al-Wadi Al-Jadeed (New Valley), Egypt. Archaeol Prospect 9:93–104

Keating P, Pilkington M (2004) Euler deconvolution of the analytic signal and its application to magnetic interpretation. Geophys Prospect 52:165–182

Khozyam A (2003) Geophysical prospection of some archaeological sites in Saqqara area, Giza, Egypt. Master thesis, Faculty of Science, Ain Shams University, Cairo, Egypt

Mousa SEA, Abdallatif TF, Hussain AG, AlBassiony AA (2001) Application of geophysical tools for mapping the archaeological features at Saqqara, Giza, Egypt. Proceedings of the Nineteenth Annual Meeting of the Egyptian Geophysical Society, pp 1–26

Odah H, Abdallatif TF, Hussain AG (1998) Micromagnetic prospecting to delineate the buried solar boats in Abydos area, Sohag, Egypt. J Environ Sci Mansoura University, Egypt 16:121–133

Odah H, Abdallatif TF, El-Hemaly IA, Abd El-All E (2005) Gradiometer survey to locate the ancient remains distributed to the northeast of the Zoser Pyramid, Saqqara, Giza, Egypt. Archaeol Prospect 12:61–68

O'Connor D (1997) Report on Abydos: the early dynastic project, New York University, pp 1–5

Pedersen LB (1989) Relations between horizontal and vertical gradients of potential fields. Geophysics 54:662–663

Petrie WMF (1889) Hawara, Biahmu and Arsinoe, London

Rogers A (1992) New complete guide of Egypt, Italy, 146p

Scollar I, Tabbagh A, Hesse A, Herzog I (1990) Archaeological prospecting and remote sensing, topics in remote
sensing. Cambridge University Press, Cambridge, pp
375–516

Shaw I (2000) History of ancient Egypt. Oxford University
Press, Oxford

Slack HA, Lynch VM, Langan L (1967) The geomagnetic
gradiometer. Geophysics 3(2):877–892

Uytterhoeven I (2003) Hawara in the Graeco-Roman
period. Life and death in a Fayum village. Doctoraat
in de Archeologie

Williams V, Stocks P (1993) Blue guide, Egypt, Atlas,
Street atlas of Cairo, maps, plans and illustrations.
Ernest Benn, London

The Utility of Geophysical Models in Archaeology: Illustrative Case Studies

8

M. Teresa Teixidó and José Antonio Peña

Abstract

This chapter uses case studies to illustrate the way geophysical techniques are integrated into archaeological approaches. The general idea is to show how the different geophysical models obtained using magnetic, electric, seismic and GPR data, are used in the archaeological praxis and the capacity they have to respond to particular questions.

To understand their area of application and how we can extract useful information from them, it is first necessary to establish the basic points on which the geophysical models are established. For this reason, the most general features of the geophysical exploration techniques are presented first, followed by, in successive sections, a description of their main areas of application illustrated with examples.

These areas of application have been grouped into three domains: (1) the use of geophysical models for working out the geoarchaeological context of a site; (2) for obtaining subsurface models containing the spatial distribution of buried structures; and (3) for making a detailed study of a particular structure.

We conclude this chapter by showing that geophysical information is a useful tool in archaeological research. Thanks to these methods, we are able determine the geological features of a site and assess their potential value in order to plan subsequent archaeological activities.

Keywords

Geophysical prospection · Geoarchaeology · Buried structures · Non-destructive exploration methods

M. T. Teixidó (✉)
Instituto Andaluz de Geofísica (IAG), Universidad de Granada, Campus Universitario de Cartuja, Granada, Spain
e-mail: tteixido@ugr.es

J. A. Peña
Instituto Andaluz de Geofísica (IAG), Universidad de Granada, Campus Universitario de Cartuja, Granada, Spain

Dpto. de Prehistoria y Arqueología, Universidad de Granada, Campus Universitario de Cartuja, Granada, Spain

8.1 General Features of Geophysical Models

The application of geophysical methods in archaeology began around 1950 in the Dorchester-on-Thames site (England) where the electrical Wenner device was used to detect a perimeter wall (Aikten 1958 and 1961). Due to the success of this pioneering work, the use of geophysical surveys applied to archaeology began to spread to other sites. In just over a decade, electrical and magnetic surveys were being used at Iron Age sites, recording important anomalies due to iron objects and the remains of forges (Gaffney and Gater 2003). From the 1980s onwards, the evolution of geophysical equipment made a wider application possible at Prehistoric (Gibson 1986), Bronze Age (Clark 1986) and Historical Classic sites (Fassbinder 1994). These days, geophysical prospecting is a common practice at all types of sites.

Right from these early beginnings, the feature that has made geophysical modelling useful is the fact that the models are obtained using non-destructive methods and cause no permanent alteration of any buried materials. Usually, data acquisition is carried out from the surface. These two attributes are also responsible for the first constraint of geophysical imaging: the models will always be approximate.

Another feature to consider when a geophysical survey is planned is that each method has a different capacity to analyse the subsurface. For example, the electrical resistivity tomography method (ERT; Loke 2015) is able to provide detailed 2D models of geological strata (Berge and Drahor 2011; Teixidó et al. 2013a, b; Martínez-Moreno et al. 2014); ground penetrating radar (GPR; Daniels 2004) can generate accurate models of the geometry of buried structures (Novo et al. 2012); the 2D-gradient magnetic method is a useful tool for exploring extensive areas (Becker 1997) due to the velocity of acquisition data, and so on. Each of these capabilities is determined by the particular methodological *corpus* of each technique, and the resulting geophysical models are images showing how a given physical property of the materials (e.g., resistivity, dielectric constant, imannation,…) is distributed in the subsurface. Therefore, the second constraint to consider when choosing the most suitable method for a particular study involves knowing how the physical properties vary between the various buried materials and how to operate the geophysical acquisition devices.

Considering the above two points, and in a very general way, we could say that geophysical models are like "eyes" that look into the subsurface from different viewpoints; and from a different perspective than human perception. Hence the principal objective of geophysical prospection is to provide the appropriate "glasses" so that the geophysical images obtained can be read archaeologically. That is to say, the usefulness of geophysical models lies in establishing the proper relationship between the geophysical and archaeological systems.

To understand this relationship (or synergy), it is necessary to establish certain methodological points on which the geophysical models are based:

- The measurements for each geophysical model are conditioned by the contrast between the studied body and the surrounding materials. In each particular study, the physical property that presents the greatest contrast determines the most appropriate method to be used: the higher the contrast, the more accurately the body is detected.
- Each geophysical method has a specific resolution that is imposed in different ways: by the order of magnitude of the measured data, the accuracy during the field acquisition stage, and the numerical approaches employed in the data processing step.
- Geophysical anomalies have no chrono-cultural ascription. This means that on a single site, geophysical anomalies generated by different sources, such as modern structures, water level, lithology changes, and so on, may coexist.

Figure 8.1 provides an example illustrating the particular viewpoint and experimental resolution of a GPR model. The aim of the study (Teixidó et al. 2013a, b) was to evaluate the extent to which 3D-GPR models could describe a cork oak root system. The upper left image shows the cork oak studied and the rectangular sector explored with 400 and 900 MHz antennas (GSSI, Inc.). The right upper image is the tree root system after its excavation. The two middle images correspond to the 3D-GPR models obtained using each of the antennas (Figs. 8.1a and b) and Fig. 8.1c is a 3D-laser scanner image

Fig. 8.1 Geophysical models versus reality. The 3D-GPR models were obtained using 400 MHz (**a**) and 900 MHz (**b**) antennas. (**c**) This image was obtained with a 3D-laser scanner tool. The two upper images are the cork oak with the rectangle explored (left) and the same tree after its root system was excavated. Bottom images are the geometry acquisition of two GPR sets

obtained after digging out the roots. The bottom images show the GPR profiles performed in each case; note that they are undertaken in two directions in order to provide the best cover. In the first case (a) the profiles are spaced 0.25 m apart, and in the second case (b) the spacing was 0.15 m. Of course, the best approximation is the 3D-laser scanner model with a sampling distance of close to 0.02 m, but this technique requires direct measurements and only works when the root system is visible, damaging the tree.

Now consider the two GPR models; these were obtained in the same way: by firing pulsed electromagnetic waves through the ground and receiving the reflections. The main cause of the subsurface reflections was the dielectric contrast between the roots ($\alpha = 7.4$) and the soil consisting of organic materials and silty sands ($\alpha = 14$). Therefore, differences between them result only from the resolution concept (Irano et al. 2009). Two contributing factors must be taken into account: the central frequency value that determines the

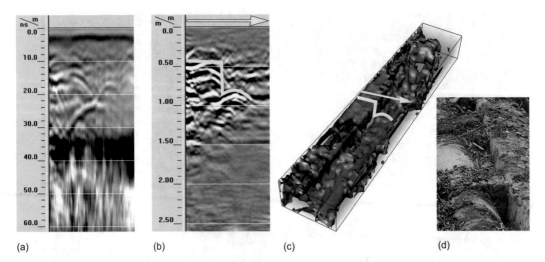

(a)   (b)   (c)   (d)

Fig. 8.2 Data processing is also crucial for obtaining good geophysical models. (**a**) This image is a raw radargram obtained with a 400 MHz GPR antenna. (**b**) The same profile after a careful processing flow has been applied. (**c**) This image is a 3D-GPR model resulting from the spatial interpolation of all GPR profiles. (**d**) This photo shows a view of a similar pipeline located near the Roman site

minimum size of objects which can be detected as well as their depth; and the space between the profiles. Both factors affect the horizontal and vertical resolution (Rayleigh and Fresnel criteria; Daniels 2004), following the rule that higher frequencies mean higher resolution but less penetration depth. Having established the limits of each model, we can assess which is best suited to our objective, because it may be that the coarsest model (a), with faster acquisition, amply fulfils our purpose.

Two additional points should be mentioned in this overview. The first refers to the data processing stage, because although we can apply basic data processing techniques, to obtain good geophysical models it is necessary to apply expert processing; even if this takes more time. Figure 8.2 shows the same GPR profile before (a) and after (b) a careful processing flow has been applied. This profile is part of a GPR survey carried out to locate a water pipeline running through the Roman site of Ciavieja, Spain (Teixidó et al. 2014). The aim of the study was to visualise the damage to this pipe and how it could affect the Roman remains. Special measures were taken during the field data acquisition stage, and profiles were obtained transversally across the structure in

order to capture the best sections, rather than longitudinally, which would have been easier. Furthermore, the carefully processed flow provides a good 3D-GPR pipeline image (c) obtained by a spatial interpolation of all the GPR profiles processed. The image is a 3D-isosurface obtained using the spatial distribution of a given trace amplitude value. It is not difficult to imagine the resulting 3D-model model if we had worked parallel to the pipeline direction and a basic filter were applied to the raw data set.

Finally, the second additional point involves the methodological strategy: whenever possible, it is preferable to use more than one geophysical method in order to obtain better subsurface interpretation. Doing this, we have different sets of measurements from the same materials that react in different ways; correlating these different responses provides improved information on the buried structures.

This situation was true in the following case study (Peña et al. 2008). The figure shows a magnetic gradient survey carried out where a Roman villa was thought to be (Fig. 8.3a). Let us analyse only the two significant sets of anomalies that are marked. Both are circular bipolar anomalies with high remanent magnetism

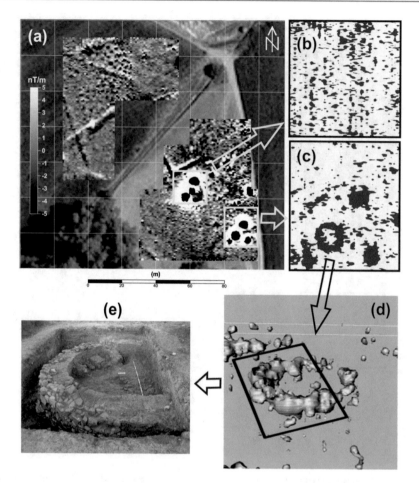

Fig. 8.3 (a) Magnetic gradient anomaly map overlying the corresponding orthophoto. The magnetic data was acquired from a potassium-vapour magnetometer (GEM System, Inc.) sweep of the area in parallel profiles spaced 1 m apart. (b) and (c) are representative GPR depth slices at 0.6 m depth (0.15 m thick) for two rectangular areas. The circles correspond to Roman kilns. (d) 3D-GPR model for the marked circle. (e) The archaeological dig reveals the existence of a Roman kiln. Roman villa of Cortijo de Quintos, Spain

suggesting the possibility of Roman kilns. A 3D-GPR prospection was carried out in two rectangles using a 400 MHz antenna with parallel profiles spaced 0.25 m apart. The depth-slice in Fig. 8.3b shows no coherent reflections, suggesting there is no dielectric contrast between the supposed body and the surrounding materials. However, the depth-slice of Fig. 8.3c shows some circular reflections indicating that there is a dielectric contrast. An archaeological excavation was planned using the 3D-GPR model generated for the selected circle (Fig. 8.5d) and the result is displayed in Fig. 8.5e. A possible explanation of

why two sectors with similar magnetic responses have different dielectric responses may be that in the GPR sector (b) the kilns were destroyed and only the heated surrounding material has remained; this therefore has remanent magnetism but does not present changes in dielectric constant.

8.2   Geoarcheological Approach

In the same way that historical knowledge of a particular site is decisive for its archaeological

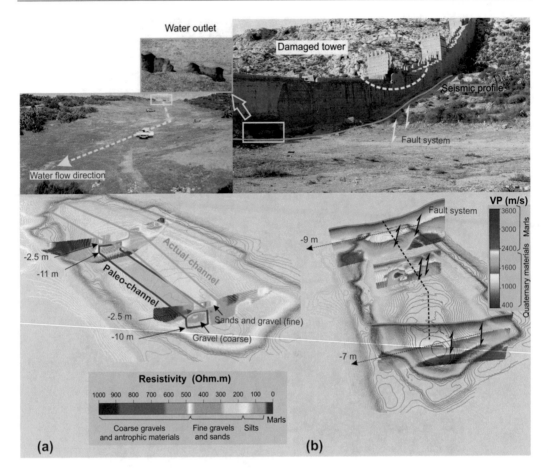

Fig. 8.4 Geoarchaeological study carried out on the Arab wall section of Almería Alcazaba (Spain). (**a**) ERT profile interpretation provides information on the basin drainage and shows the barrier effect of the wall over the palaeo-channel. (**b**) The seismic refraction profiles (first-time reading, geoarchaeological information provides tomography method) detected an unconsolidated Quaternary layer over the marl basement. At the bottom of the Quaternary, two jumps (arrows) are detected that may indicate a possible fault system

reading, geoarchaeological information provides a framework for understanding the geological setting and some of the reasons that motivated the choice of geographical location and construction at a particular site. Geophysical models can be used to characterise the morphologies and lithologies of geological strata, and detailed geological or geomorphological maps can be generated to obtain information on the geological history of the site. For example, we can determine ancient watercourses (palaeo-channels) or perform archaeo-seismic work. In addition, the study of the subsurface allows us to understand general aspects of the site such as the stability of

the terrain where the structures are settled, preferred water runoff, erosion effects, and so on. Thereby, they provide valuable information for planning the valorisation and conservation of the site. Moreover, if geoarchaeological work is combined with other environmental disciplines such as bioarchaeology, it can also be applied to the study of small territories.

Figure 8.4 is an example of a geoarchaeological study (Peña and Teixidó 2008) carried out on the Arab wall of Almería Alcazaba (Spain) constructed by Abd al-Rahman III in 995 (A.D.). This Arab wall is an important fortification, 3 m wide and 5 m high, with a perimeter

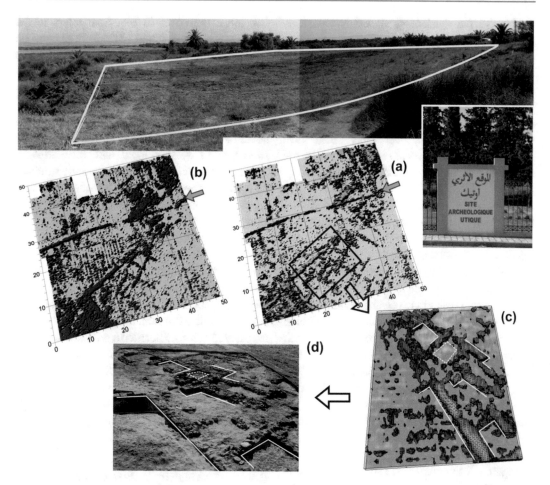

Fig. 8.5 GPR survey at the Phoenician-Roman Utica site (Tunisia). The upper image is the studied area. It is a 50 × 50 m square. The two models (**a**) and (**b**) are cover surfaces relating to the superficial archaeological level (between 0.15 and 0.50 m depth) and the basal level (between 0.5 and 1 m depth). The arrangement of the buried structure reveals the archaeological potential of this area and provides guidance on its possible functions. Image (**c**) is a portion of the first cover surface (**a**) that was excavated; (**d**) shows the first stage of field excavation

length of 1430 m and defensive towers distributed along it. One wall section is located over a small basin through which a stream flows. Two electrical ERT profiles and three seismic refraction profiles (tomography method, Lecome et al. 2000) were acquired in this area, where one of the towers is damaged.

The interpretation of the electrical profiles (Fig. 8.4a) reveals the geological effect produced when the Arab wall was constructed over the basin. From that time, it acted like a dam and widened the natural drainage course of the torrent, reducing the energy of the water circulation. We can see this effect in the ERT profiles as a discontinuity at 2.5 m depth, over the first palaeo-channel. The bottom of the first palaeo-channel is narrower and has higher resistivities, indicating a time where the stream flow was stronger (with more energy) and the strata were excavated, with coarse materials being deposited. However, above 2.5 m depth we can see the drainage of the current channel. It is wide with lower resistivities according to the decrease in water flow energy and increased fine deposits due to the barrier effect of the Arab wall. On the other hand, the seismic refraction profiles (Fig. 8.4b)

distinguish the Quaternary materials characterised by lower velocities, which vary according
to a vertical gradient and the marl basement with
higher velocities. The Quaternary formation in
contact with the marls shows two morphological
jumps (arrows) that may indicate a possible fault
system under the damaged tower. These unconsolidated materials, over time, could have
induced differential ground settlement which has
then been transmitted to the damaged tower.

8.3   Looking at Buried Structures

Geophysical prospection for discovering the spatial distribution of buried structures is the most
widespread use of geophysical methods in archaeology. These types of surveys have a number of
advantages over other non-destructive techniques,
e.g., geophysical models are more accurate in their
placement of structures than the study of the surface distribution of archaeological artefacts. Furthermore, they tend to have better resolution than
remote sensing prospection, and, when it is not
possible to dig, geophysical interventions become
the most appropriate techniques for establishing a
site's potential.

Figure 8.5 summarises the study carried out in
the northern part of the Utica site (Tunisia). This
site is remarkable as both Phoenician and Roman
remains coexist. The objective of the exploration
was to spatially locate the buried structures
(Teixidó et al. 2010). In this case a 400 MHz
antenna (GSSI, Inc.) was used and the data was
acquired through parallel GPR profiles spaced
0.5 m apart. The resulting models (a) and (b) are
the covered surfaces calculated from both depth-
slice sets (Peña and Teixidó 2012). These
surfaces represent the depth at which the first
maximum amplitude reflection is detected at
each point of the 3D-GPR set. The result is a
kind of "topographic surface" for the buried
structures that bears some similarity to stratigraphic archaeological excavation (or excavation
by natural levels). The covered surface of
Fig. 8.4a corresponds to accumulative reflectors
for the superficial level (between 0.15 and 0.50 m
depth) and shows the first level of buried

structures; the cover in Fig. 8.5b refers to the
basal level of the structures, at a depth of between
0.5 and 2 m. The spatial distribution of these
GPR-reflections reveals the archaeological potential of this area and provides information on the
possible functions of the buildings. In this context, there is a lineal reflector detected in two
cover surfaces which crosses the study zone
from east to west (marked with blue arrows).
This reflector corresponds to a hydraulic structure
that comes from a thermal artesian well and
supplies Roman baths located 2 km from this
study area. We can therefore infer that the
neighbouring structures located in the northeast
corner will be related to this and that they had
hydraulic functions (cisterns, bathing areas, etc.).
To the south of this Roman pipe, in the central
part and southwestern corner, we can see
structures with a different constructive orientation
that may indicate a Phoenician origin. The bottom
image (Fig. 8.5c) is a portion of the superficial
cover surface of Fig. 8.5a, displayed as a 3D
iso-surface image, and Fig. 8.5d is the corresponding photo showing the first stage of excavation; note the good coincidence.

8.4   Analysing a Particular
         Structure

The size and settlement of a particular structure
are the two principal aspects that define the geophysical methods and tools to be used. This
choice determines how detailed the models of
the structures will be. The case study presented
(Fernández et al. 2015) is a peculiar structure
located outside the perimeter wall of the
Celtiberian Segeda I site, Spain (from 153 to
133 A.D.). Several questions have arisen regarding the functionality of this unusual building. If it
had been built for ritual purposes, a comparison
with other sacred buildings would suggest the
presence of foundation deposits or an inhumation
urn. But if it were a defensive building, it would
probably have a different constructive pattern and
special foundations compared with the other
constructions on the site (Fig. 8.6).

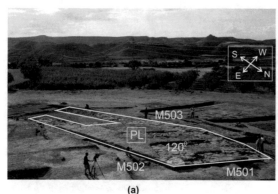

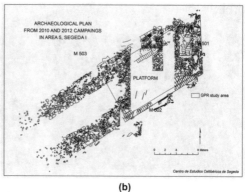

(a)   (b)

Fig. 8.6 (**a**) Semi-aerial view of the monumental structure. It comprises different constructive elements, principally: the central platform (PL) part of which is currently visible, representing 200 m^2 of the building, three walls limit this structure (M501, M502 and M503). (**b**) Archaeological plan

In order to collect more information on the internal structure of the building, non-destructive geophysical methods were applied to survey the buried inner parts of the building (architectonic target), and investigate the terrain under and around it (geological target). A 2D-magnetic gradient survey was undertaken over the structure to locate possible foundation elements. 3D-laser scanner topography and 3D- GPR survey were combined to ascertain the internal arrangement of ashlars and blocks. Finally, 2D-ERT profiles were performed to investigate the geological materials underneath the structure.

Figure 8.7a shows the resulting 2D magnetic anomaly map placed over the 3D-laser scanner surface. We selected a non-classic colour scheme to emphasise the most extreme anomalies. Three dipolar anomalies were detected in the western sector; these were checked and shown to correspond to modern iron (labelled Fe). Two negative anomalies were also detected beside wall M503 (A_1^-, A_2^-). These were excavated and proved to be residues of burnt wood of Celtiberian age. The anomalies within the building are weak and the boundaries between the positive and negative values appear to coincide with the limits of the walls (dotted lines). Finally, we detected one positive anomaly (A^+) coinciding with the cornerstone of building. This is very likely due to the presence of a ritual object.

Figure 8.7b shows the three ERT-profiles in their true positions, beneath the scanned surface in order to see the fit of the structure. In these electrical models the anomalies with high resistivity correspond to the walls and foundations, meaning the dimensions of the structure can be inferred. A broad band of resistivity surrounding the structure was detected (marked with dotted lines). Geological evidence on the surface indicates that this band is an anthropogenic deposit, principally comprising gravels and blocks. It is 5–6 m wide in transverse section and approximately 1.5–2 m thick. From an architectural point of view, this has a double function: (1) it regulates and contains the clayey terrain, and (2) it provides drainage; as described in the archaeological record, although it is larger than predicted. Analysis of profiles ERT2 and ERT3 reveals that the contact between the platform (PL) and walls (M501, M502 and M503) comprises thick materials; probably arranged slabs, once again coinciding with the archaeological description.

The GPR interpretation was followed by the recognition of the constructive elements described in the archaeological record. Firstly, the individual radargrams obtained were analysed in order to comprehend the internal elements of the structure, and then 3D GPR images were reviewed to help understand the spatial distribution of the monument.

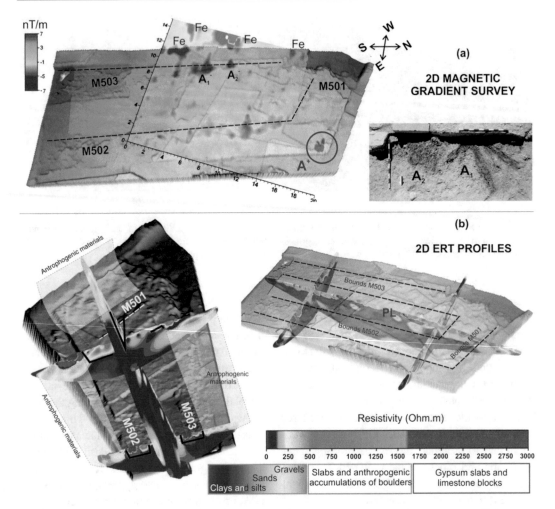

Fig. 8.7 Resulting geophysical models. (**a**) The 2D magnetic gradient map has been superimposed over the 3D-laser scanner surface. The anomaly A^+ coincides with the cornerstone, the negative anomalies (A_1^-, A_2^-) are due to burnt wood residues. (**b**) The three 2D-ERT profiles have been placed in their subsurface position, below the 3D-laser scanner map. These show the anthropogenic arrangement of the underground construction elements

In agreement with the archaeological record, Fig. 8.8 shows that mud bricks form a superficial layer (level A), which previously covered the entire structure but is now only preserved in certain sectors due to the effect of the agricultural work that the structure has suffered. Under this horizon, a gypsum slab layer (B) is found with varying thicknesses; in the central part of the building the gypsum slabs are arranged in rows (generally two levels) and are an average of 60 cm thick, although toward the sides the thickness of the slabs increases up to 80–150 cm, defining the perimeter walls. It can be seen that the transition between the central reflectors and the wall reflectors is continuous as there are individual slabs that form parts of both sections, providing good cohesion between the two structural elements. The GPR profiles also show the base of the foundations in the walls, characterised by smaller slabs that may be interleaved limestone blocks. Therefore, the walls progress downwards from carved, regularly arranged slabs to a coarser base where waste material was used. This implies that the contact between the bottom of the gypsum with the top of level C is not clearly defined under the walls; the limit is set considering that

Fig. 8.8 GPR profile interpretation of the inner structure of the Platform. The yellow line shows the part of the GPR profile analysed

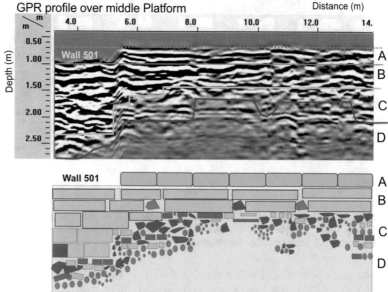

the decreasing size of the reflectors must match the size of the reflectors located under the slab rows in the central part. Layer C, filled with gypsum, limestone boulders and gravel, is a foundation layer levelling the palaeoground. Lastly, a thin layer (level D) was detected; this layer is characterised by a homogeneous band with short reflections corresponding to levelled gravels arranged as a drainage bed placed on top of the impermeable silts and clays.

The 3D-GPR images were analysed to help understand the spatial distribution of the monument. The images in Fig. 8.9a show the cover surfaces. The uppermost covering surface (between 0–0.4 m depth) is the level containing the remains of mud bricks and some gypsum

slabs. The successive covering surfaces clearly show that the M501 and M503 walls are the deepest and wall M501 is more solid than the rest of the structure. These walls become deeper towards the western boundary of the structure, confirming that originally the land sloped downwards in this direction, where the monument was sited and its foundations dug. Fig. 8.9b shows the final synthesis of the 3D-GPR models.

The conclusion is that special measures were taken in the construction of the monumental building as, even though the structure was adapted to the palaeorelief and not vice versa, in the same way as the other buildings excavated at the Segeda I site, the ground was slightly regularised prior to the placement of the

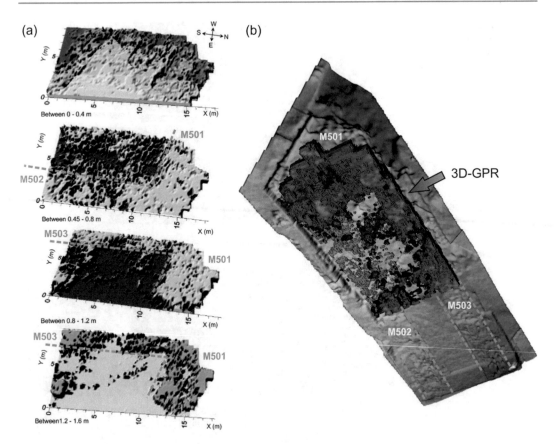

Fig. 8.9 (**a**) Cover surfaces at different depth intervals. The cover surfaces represent the cumulative reflectors between defined stretches. (**b**) Combined geophysical view from underneath the 3D-GPR model; for display purposes the western 2D-ERT profile is shown

constructive materials. Furthermore, a 5–6 m wide and 1.5–2 m deep ditch was built around the structure and we also detected a possible foundation element at the cornerstone of building. This gives relevance to the monument, but only by excavating the deposit and discovering its nature, would it be possible interpret its function.

8.5   Final Comments

Using just the few examples above, we hope to have shown the value of geophysical methods as non-destructive techniques for exploring the subsurface.

Interpretation of archaeological sites involves collaboration between researchers from different scientific disciplines where each looks at the evidence from their own epistemological framework.

Consequently, modern archaeology not only includes classic perspectives such as historiography, processes of excavation and restoration and constructive methods, but is also a multidisciplinary science that accommodates technological approaches. Geophysical exploration involves technologies from outside archaeology for which the fundamental objectives can be synthetized into three main areas: (1) understanding the geo-archaeological context of a site, (2) obtaining subsurface models containing the spatial distribution of buried structures; and (3) making a detailed study of a particular structure. These three categories can also be referred to as geological, planimetric and architectonic approaches to an archaeological site.

Obviously, like any indirect measuring technique, geophysical models never reach the level of detail of an archaeological excavation, but neither is this their mission. Their usefulness should be seen as a complementary tool that is used to collect information prior to excavation. Geophysics provides valuable data on the morphology, local lithology, palaeo-topographic relief, and preferred direction of runoff, as well as the spatial distribution of structures and their constructive features. All this information is invaluable in planning archaeological actions at a site.

Acknowledgements All the geophysical models presented in this chapter were generated by Applied Geophysics in the Andalusian Institute of Geophysics at the University of Granada (IAG-UGR). Both the geophysical instruments for acquiring the data and the processing software were purchased thanks to a European Regional Development Fund (ERDF).

References

Aitken J (1958) Magnetic prospecting I. The Water Newton Survey. Archaeometry 1:24–29

Aitken MJ (1961) Physics and archaeology. Interscience, New York

Becker H (1997) Basilika Kaiser Konstatins in Ostia Antica mit modernsedtem Magnetometer entdeckt? Denkmalpflege Informationen, hrsg. V. Bayer. Landersamt für Denkmalpflegem Ausgabe B. Nr. 106/22. August 1997-2

Berge MA, Drahor MG (2011) Electrical resistivity tomography investigations of multilayered archaeological settlements: part I – modelling. Archaeol Prospect 18:159–171

Clark A (1986) Archaeologycal geophysics in Britain. Geophysics 51(7):1404–1413

Daniels DJ (ed) (2004) Ground-penetrating radar, 2nd edn. The Institution of Electrical Engineers, London, p 726. isbn:0-86341-360-9

Fasbindert JWE (1994) Die magnetishen Eigenschaften und die Genese ferromagnetique Minerale in Bödem im Hindblick auf die magnetsche Prospektion archäelogischer Boderndenkmäler, Buxh am Erlbach

Fernández G, Teixidó T, Peña JA, Burillo F, Claros J (2015) Using shallow geophysical methods to characterise the monumental building at the Segeda I site (Spain). J Archaeol Sci Rep 2:427–436. https://doi.org/10.1016/j.jasrep.2015.04.0062352-409X/2015

Gaffney C, Gater J (2003) Revealing the buried past: geophysics for archaeologists stroud. (Tempus)

Gibson TH (1986) Magnetic prospection on prehistoric sites in western Canada. Geophysics 51(3):553–560

Irano Y, Dannoura M, Aono K, Igatashi T, Ishii M, Yamase K, Makita N, Kanazawa Y (2009) Limiting factors in the detection of tree roots using ground-penetrating radar. Plant Soil 319:15–24

Lecome I, Gjoystdal H, Dalhe A, Pedersen OC (2000) Improving modeling and inversion in refraction seismics with a first order Eifonal solver. Geophys Prospect 48:437–454

Loke MH (2015) 2-D and 3-D ERT surveys and data interpretation, Piedmont regional Order of geologists Pasi geophysics. Turin, Italy, 10th Sept. 2015. Geotomo Software Pvt Ltd. www.geotomosoft.com

Martínez-Moreno FJ, Galindo-Zaldívar J, Pedrera A, Teixido T, Ruano P, Peña JA, González-Castillo L, Ruiz-Constán A, López-Chicano M, Martín-Rosales W (2014) Integrated geophysical methods for studying the karst system of Gruta de las Maravillas (Aracena, Southwest Spain). J Appl Geophys 107:149–162. https://doi.org/10.1016/j.jappgeo.2014.05.021

Novo A, Lorenzo H, Rial FI, Solla M (2012) From pseudo-3D to full-resolution GPR imaging of a complex Roman Site. Near Surf Geophys 10:11–15. https://doi.org/10.3997/1873-064.2011016

Peña JA, Teixidó T (2008) Intervención arqueológica punctual, prospección superficial y prospección geofísica con radra del subsuelo modelaidad 3D en la Hoya Nueva, junto a la Alcazaba de almería. Internal document of Andalusian Institute of Geophysics (Granada University). Ref. AGA/2008-56

Peña JA, Teixidó T (2012) Cover surfaces as a new technique for 3D GPR image enhancement: Archaeological aspplications. Repositorio Institucional de la Universidad de Granada. http://hdl.handle.net/10481/22949

Peña JA, Teixidó T, Carmona E, Sierra M (2008) Prospección magnética y radar 3D como métodos para obtener información a priori en la planificación de una excavación arqueológica. Caso de estudio: Yacimiento del Cortijo de Quintos (Córdoba, España). http://hdl.handle.net/10481/23460

Teixidó T, Peña JA, Lopez Castro JL (2010) Prospecció Geofísica en el Yacimiento Arqueológico de Utica, Túnez. Campaña arqueológica de 2010. Internal document of Andalusian Institute of Geophysics (Granada University). Ref. AGA-ID/2010-78

Teixidó T, Artigot EG, Peña JA, Molina F, Nájera T, Carrión F (2013a) Geoarchaeological context of the Motilla de la Vega Site (Spain) based on electrical resistivity tomography. Archaeol Prospect 20:11–22. https://doi.org/10.1002/arp.1440

Teixidó T, Peña JA, Ruso JA, Alcalá S (2013b) Assessment of experimental device and the GPR resolution suitable to characterize the cork oak root system. Case of study la Alcaidesa Natural Park, Cádiz (SPAIN). Internal document of Andalusian Institute of Geophysics (Granada University). Ref. AGA-ID/2013-19

Teixidó T, Peña JA, Moya L, Alemán B, Claros J (2014) Archaeologic superficial prospection and GPR exploration in the Roman Site of Ciavieja (Elejido, Spain). Internal document of Andalusian Institute of Geophysics (Granada University). Ref. AGA/2014-104

Ground Penetrating Radar Resolution in Archaeological Geophysics

9

David C. Nobes and Juzhi Deng

Abstract

Ground penetrating radar (GPR) is now a common tool for archaeological imaging. However, difficulties arise in choosing the right antenna. Do we choose high-frequency antennas to yield lots of detail? Or do we choose lower frequency antennas to see larger scale features that provide site context? Often we are tempted to use a high-frequency signal, in order to see all the detail. However, this can be counter-productive. Lots of detail may actually obscure the features that are the primary targets. Conversely, lower frequency antennas can locate the features of interest, but may not provide as much detail as desired. In general, choosing a lower frequency antenna yields better results. The optimum choice of antenna depends on the site conditions; using a range of antennas for initial tests helps establish the "best" signal frequency to use. The imaging may be best done in two stages: an initial stage using low frequency antennas; followed by high-frequency imaging to yield greater detail over the areas where it is useful.

In addition, conditions can change around a site, both laterally and with depth. For example, if there is a void—such as crypt beneath an old church site or a cave of archaeological interest—then the GPR velocity will be radically different across that void. If the void is air-filled, then the velocity will be much faster than in the surrounding soil or rock; any void reflections will arrive much sooner than those from within the surrounding material. If the void is water-filled, then the velocity will be much slower, causing a significant time delay when compared with the reflections from within the surrounding material. In addition, voids can generate multiple reflections because of the strong velocity contrast at the void top and bottom, and near-vertical voids, such as cracks, can generate stacks of diffractions caused by scattering from the rough walls of the crack.

Keywords

Ground penetrating radar · Resolution · Frequency · Voids

9.1 Introduction

One of the difficulties in using ground penetrating radar (GPR) for imaging of archaeological sites, is the choice of the antenna to use, e.g. Chamberlain et al. (2000). The temptation is to use a higher-frequency antenna, to obtain greater detail, e.g. Pérez Gracia et al. (2000) or Leucci (2002). However, sometimes the greater detail comes from geological "noise", that is

D. C. Nobes (✉) · J. Deng
School of Geophysics and Measurement-Control Technology, East China University of Technology, Nanchang, Jiangxi, China
e-mail: david.nobes@canterbury.ac.nz

© Springer International Publishing AG, part of Springer Nature 2019
G. El-Qady, M. Metwaly (eds.), *Archaeogeophysics*, Natural Science in Archaeology,
https://doi.org/10.1007/978-3-319-78861-6_9

small near-surface or surficial features, that mask the underlying features that are our real targets, e.g. Nobes (1999). Often, choosing a lower frequency antenna is the better option, but the question then still remains: How do we choose the right frequency antenna? It is usually a compromise between depth of penetration and the detail desired, e.g. Shaaban et al. (2008).

There is a fundamental trade-off between the depth of penetration of the radar signal and the resolution (see, e.g., Jol 1995). The issue of depth of penetration depends strongly on the properties of the materials in the ground, and are highly site specific. Resolution is tied up with depth of penetration, but some fundamental issues can be addressed independently of depth of penetration. Thus, the purpose of this chapter is to discuss issues of resolution, particularly as they relate to imaging of archaeological sites.

When we use the term "resolution", we mean it in the technical sense—the scale or dimensions of objects that can be resolved using an antenna of a given frequency. Some older papers, e.g. Leckebusch and Peikert (2001) or Conyers (2004), who used the term "accuracy", use the term to also mean "detection", i.e. the ability to observe the desired feature in the data. That is not the sense with which "resolution" is used here.

There are various considerations associated with resolution and the choice of the appropriate antenna, in particular the dependence of the radial (and vertical) resolution and the lateral resolution on the antenna frequency. We use examples of imaging, particularly archaeological imaging, to illustrate the different aspects of the resolution and the choice of antenna, including how a lower frequency antenna was the better choice for imaging a burial site.

Finally, we deal with a special case associated with resolution—voids. Voids can be characterised as caves, cracks, or crypts. Because a void is filled with a material, air or water, with properties radically different from the surrounding rock and soil, its interpretation can be problematic. In particular, strong reflections from the top can overlap with strong reflections from the base to produce a "bright spot" that is characteristic of a void. In addition, because of

the different travel times in the voids, the reflections from the base and below will not align with reflections from the surrounding materials. Some simple numerical models and examples illustrate the "bright spot" phenomenon.

9.2 Resolution

9.2.1 Radial/Vertical Resolution

There are two aspects to the resolution: the *radial resolution* and the *lateral resolution* (Fig. 9.1). The radial resolution is what can be resolved perpendicular to the wavefront, and thus depends on the width of the pulse. As the signal moves out and down into the ground, attenuation tends to reduce the higher frequency content, thus broadening the pulse. However, in the very near surface, the attenuation is still small, and the transmitted pulse width is still adequate for describing the resolution. If W is the pulse width, then we can determine the radial (or vertical) resolution, Δr, as (Annan 2005):

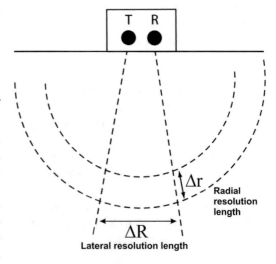

Fig. 9.1 The two aspects to GPR resolution are: the *radial resolution*, Δr, which determines the vertical resolution, and is related to the pulsewidth, as discussed in the text; and the *lateral resolution*, ΔR, which is related to the spread of the wavefront with depth (distance from the source) and its interaction with any boundary or target. (Modified from Annan 2005)

$$\Delta r \geq WV/4 \qquad (9.1)$$

where W is the pulse width, in ns, and V is the velocity in the medium, in m/µs, m/ns, or cm/ns, taking care to note the units used. The pulsewidth and velocity depend, in turn, on the frequency.

The resolution of a radar antenna depends on the frequency of the antenna and the geometry of the radiated wave. Ideally, the radar pulse would be a spike, but in reality, we try to have as clean and short a pulse as possible. The Ricker wavelet (Sheriff 2002) is an example of an idealised pulse that is realisable in practice. The Ricker wavelet is what is called a *zero phase* wavelet, that is, the wavelet is centred and symmetric in time. The wavelet is generated in time, t (in nanoseconds, ns), for a mean frequency, f_M (in megahertz, MHz), and is defined by:

$$f(t) = \left(1 - 2\pi^2 f_M^2 t^2\right) e^{-\pi^2 f_M^2 t^2}. \qquad (9.2)$$

The shape of the 100 MHz pulse is shown in Fig. 9.2. It has the characteristic large central positive, T_R, with two small negative sidelobes, and a pulse width from minimum to minimum of T_D, where:

$$T_D = \sqrt{6}/\pi f_M \text{ and } T_R = T_D/\sqrt{3}$$
$$= \sqrt{2}/\pi f_M. \qquad (9.3)$$

The pulsewidth, W, is commonly equal to T_R. How the radial resolution depends on the velocity and $\log_{10}$ of the frequency is illustrated in Fig. 9.3b. Note the rapid decline with frequency.

9.2.2 Lateral/Fresnel Resolution

The lateral resolution (Fig. 9.1) depends instead on the interaction of the wavefront with a boundary or the edge of a target. This is called Fresnel scattering (Sheriff 2002). The resolution from the first Fresnel scattering zone is ΔR, which depends on the both the wavelength of the signal and the depth as:

$$\Delta R^2 = (D + \lambda/4)^2 - D^2$$
$$= D\lambda/2 + \lambda^2/16, \qquad (9.4)$$

where D is the depth and λ is the wavelength, equal to V/f, which follows from the fundamental

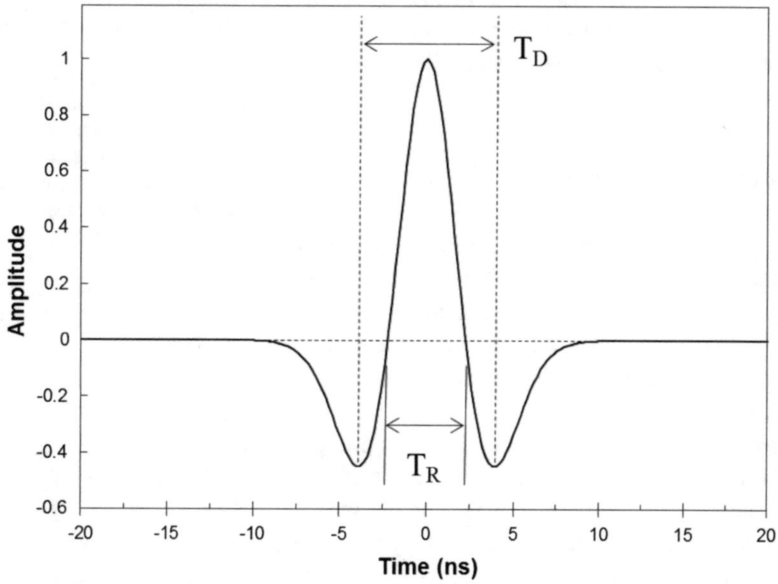

Fig. 9.2 The Ricker wavelet is a pulse that can realistically be generated, and is characterized by a large positive central peak that has a pulsewidth T_R, and two small negative sidelobes, with a separation from minimum to minimum of T_D. See text for more discussion

relationship between frequency, wavelength and velocity: $V = \lambda f$. If we take the limit as depth becomes significantly greater than the wavelength, then we obtain a standard relationship for the resolution at depth (Annan 2005):

$$\Delta R_{D \to \infty} \cong \sqrt{(D\lambda/2)} = \sqrt{(DV/2f)}. \quad (9.5a)$$

On the other hand, if we let D become small, then the resolution approaches the limit:

$$\Delta R_{D \to 0} \cong \lambda/4 = V/4f. \quad (9.5b)$$

This is the usual standard used to decide the optimum trace spacing in a GPR imaging survey: the trace spacing, Δx, is less than or equal to the Fresnel resolution at the surface, ΔR. That is $\Delta x \leq \Delta R$. This is also often called the *Nyquist sample interval* (see, e.g., Novo et al. 2008). The dependence of the surficial limit for the lateral resolution, $\Delta R_{D \to 0}$, on velocity and the $\log_{10}$ of the frequency is shown in Fig. 9.3a. Note the similarity in shape with the radial resolution in Fig. 9.3b, but the lateral resolution is of the order of twice the radial resolution.

The change in the resolution as a function of frequency is more clearly seen by plotting the logarithms of the resolutions as a function of velocity and $\log_{10}$ of the frequency (Fig. 9.4). Note the linear relationship for $\log_{10} (\Delta R_{D \to 0})$ and $\log_{10} (\Delta r)$ with $\log_{10} (f_M)$.

9.2.3 Diffractions

There is another manifestation of resolution, and that is the presence of diffractions in GPR profiles. A diffraction is due to the reflection of energy from a discrete feature, such as offset or truncated stratigraphy due to a fault, a crack, or an excavation, or a buried object of some sort. The reflected energy travels out and back at the speed of radar in the medium around and above the feature in question (Fig. 9.5), but because we don't know where the reflected energy originates, we plot it as part of a normal trace, expressed as amplitude versus travel time. As we approach the feature, the travel time there and back is less, and

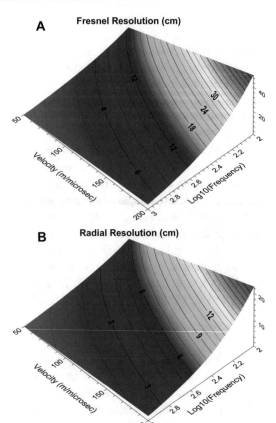

Fig. 9.3 (a) The Fresnel (lateral) resolution as a function of the logarithm of the frequency (front axis) and the velocity (left axis, in m/μs). The frequency covers the usual range for archaeological imaging, from 100 MHz ($\log_{10} f = 2$, at the right) to 1000 MHz ($\log_{10} f = 3$, at lower left). Note the sharp decline as a function of frequency, and the gentler dependence on velocity. (b) The radial/vertical resolution as a function of the logarithm of the frequency and the velocity. The scales are the same as for the Fresnel resolution in (a). Red indicates resolutions of half metre to metre or greater. Green is for resolutions of order of 10s of centimetres. Blue is for resolutions of the order of centimetres. Note that the Fresnel resolution is about twice that as for the radial resolution. See text for more discussion

we reach a minimum travel time immediately over top of the feature (Fig. 9.5).

A diffraction only occurs if the feature is about the same size as the lateral (Fresnel) resolution, ΔR, as defined in Eq. (9.4). Thus, if we use different frequency antennas, we can get some sense of the size of the features involved. An example is shown in Fig. 9.6. A profile across

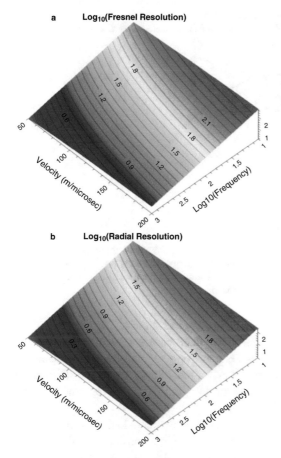

a   Log₁₀(Fresnel Resolution)

b   Log₁₀(Radial Resolution)

Fig. 9.4 The logarithm of the Fresnel resolution (**a**) and the radial resolution (**b**) as a function of the logarithm of the frequency (front axis) and the velocity (left axis, in m/μs). The frequency range covered is a more complete range, from 10 MHz ($\log_{10} f = 1$, at the right) to 1000 MHz ($\log_{10} f = 3$, at lower left). Note the linear decline in $\log_{10}$ of the resolutions versus the $\log_{10}$ of the frequency. The scales are the same for both images. Red indicates resolutions of metres or greater. Green is for resolutions of the order of 10s of centimetres. Blue is for resolutions of the order of centimetres

the Fox Glacier, in New Zealand, crossed a feature that generated stacked multiple diffractions at depth that were clear in the 50 MHz profile, but not as numerous or clear in the 25 MHz profile. Similarly, a dipping feature (also highlighted in Fig. 9.6) was clear in the 50 MHz profile, but not in the 25 MHz profile. Using a velocity determined for the site and the equation for ΔR, the 50 MHz diffractions were caused by features of the order of 0.8 m, whereas the features that

generated the diffractions in the 25 MHz profile were of the order of 1.6 m. Keep in mind that at lower frequencies, we might be seeing diffractions from the feature as a whole, whereas at higher frequencies we would be seeing smaller discrete aspects of the feature. For example, at a lower frequency, a burial as a whole might yield a diffraction, whereas at higher frequencies, diffractions might originate from truncated beds or some smaller part of the burial.

The diffraction shape or curvature is also useful. The curvature of the diffraction hyperbola depends on the velocity as $1/V^2$. Thus the curvature yields an estimate of the average velocity in the area near and above the feature causing diffractions. We can then determine the physical properties of the ground. We can also distinguish between features that are below ground, and have velocities that are characteristic of the materials present, and features that are above ground, but for which the radar energy is travelling through the air. The radar velocity in air is 300 m/μs (0.30 m/ns), and all other material velocities are much less. Thus we have a clear discriminator for distinguishing surface geological or cultural noise from any targets of interest. In addition, the subsurface radar velocity allows us to carry out additional processing, specifically *migration* (see, e.g., Annan 2005) which collapses diffractions down to the point from which they originate and which places dipping features into their correct geometric positions, and time to depth conversion, where $d = tV/2$. Migration in particular is required before doing time slices of 3D GPR data sets, otherwise the diffractions won't have been collapsed and the slices will be diffuse and not properly focussed.

9.3   Lower Versus Higher Frequency

Two examples clearly show how a higher frequency signal may have so much clutter or unwanted signal as to obscure or partly obscure the features we wish to see. The first example is from Nobes (1999), which described the geophysical imaging of a Maori burial site. A set of

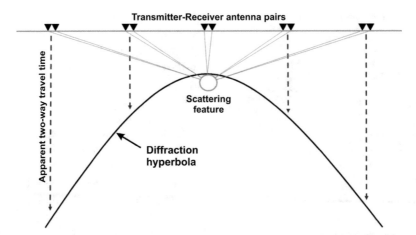

Fig. 9.5 A discrete object that is about the same size as the lateral resolution can scatter energy in a distinctive way, giving rise to a *diffraction hyperbola*, or a *diffraction* for short. The reflected energy is simply recorded as a function of travel time, without regard for the direction from which the energy originates. As we cross over top of the discrete feature, we obtain a characteristic diffraction, as illustrated

test profiles were carried out using both 200 and 400 MHz antenna (Fig. 9.7). The high frequency profile (Fig. 9.7a) has many diffractions scattered from the numerous tree roots. The lower frequency profile on the other hand (Fig. 9.7b) sees the graves, but the tree roots are too small to be observed. The resolution of the 200 MHz signal is much greater than the dimensions of the roots.

Nuzzo et al. (2002) and Yalçiner et al. (2009) made similar observations. Nuzzo et al. noted that 500 MHz was close to the "clutter frequency" (Annan and Cosway 1994, as cited in Nuzzo et al. 2002) where soil heterogeneities can scatter the GPR signal, whereas the 200 MHz signal was able to resolve the soil-bedrock interface and features within the soil profile. Yalçiner et al. (2009) similarly noted that the 250 MHz GPR signal clearly showed the features of interest, whereas the 500 MHz profile had many diffractions from olive tree roots (Fig. 9.8). What the diffractions did allow Yalçiner et al. to do was to determine the most appropriate velocity with which to convert time to depth, and to carry out additional processing.

Pipan et al. (1999) used three separate frequencies: 100, 200, and 400 MHz. After conducting site test surveys, they concluded that the 200 MHz results yielded the best compromise between depth of penetration and resolution of

their archaeological targets. The site conditions determined their choice of survey parameters, thus emphasising again the importance of variations in conditions from site to site in choosing the optimum survey modes and methods.

In a more complex situation, Novo et al. (2008) wanted to locate a sarcophagus buried beneath a convent. They used three frequencies: 800, 500, and 250 MHz. The data were acquired using trace and line spacings less than the Nyquist sampling distance (Eq. 9.5b), to yield ultra-dense images in three dimensions (3D). The high-frequency signals yielded too much near-surface clutter and could not resolve any features. The 500 MHz antennas yielded clear images of the cross-beams that supported the floor above the burial chamber. Finally, the 250 MHz signals yielded images of the suspected target. This reinforces the point that we need to recognise what is our target, and what imaging settings are required to observe it in our data.

A similar test was carried out by Barton and Montagu (2004), for a completely different purpose: the imaging of tree roots. They constructed a "sand box" to test the GPR imaging of tree roots of various sizes at a range of depths, and used 500, 800, and 1000 MHz antennas to test the imaging of the tree roots of different sizes

Fig. 9.6 A
segment of a GPR profile
from the Fox Glacier,
New Zealand, shows the
different diffractions
generated for 25 MHz
(**a**) versus 50 MHz
(**b**) signals. The scales of
the features causing the sets
of diffractions will be
different. The dipping
feature that is clear in the
50 MHz profile (**b**) is barely
visible in the 25 MHz
profile (**a**), and the
diffractions are much more
numerous and stacked in
the 50 MHz profile (**b**)
compared to the 25 MHz
profile (**a**)

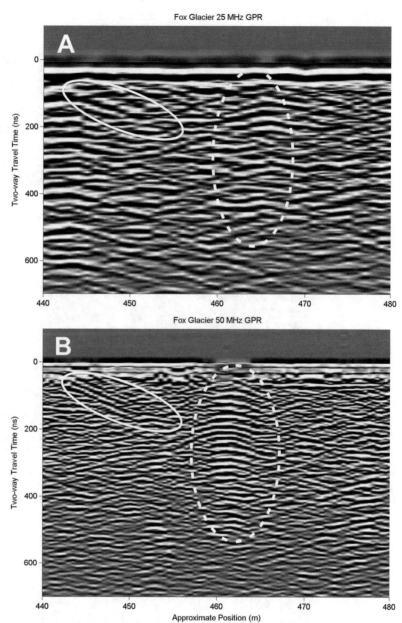

(Fig. 9.9) and different depths (Fig. 9.10). The
800 MHz images yielded the clearest profiles, but
the resultant waveforms were still "confused"
(Barton and Montagu 2004). The 500 MHz
results yielded parameters that best correlated
with the root diameters. The clutter and numerous
diffractions in the 1000 MHz profiles yielded
little information that was useful. Again, the

lower frequency signals yielded the most useful
information. The higher frequency results were
too often too cluttered for clear interpretations
and parameter determinations. In addition, the
diffractions in Figs. 9.9 and 9.10 overlap, making
interpretation difficult. When the profiles were
migrated, 8 of the 9 roots could be clearly
identified in the 500 MHz transect across the

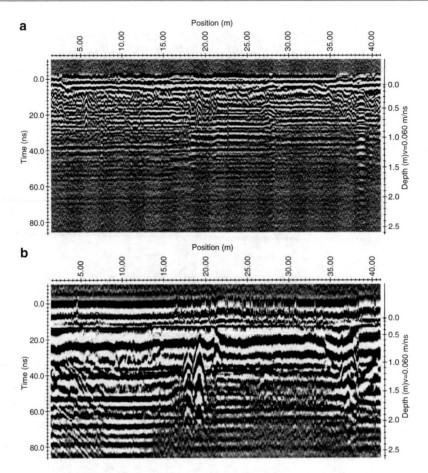

Fig. 9.7 Comparison of filtered 450 MHz (**a**) and 200 MHz (**b**) GPR profiles across a set of indigenous Maori graves. The 450 MHz profile (**a**) has numerous shallow diffractions arising from the roots of nearby large flax bushes, partly obscuring the deeper events from the graves. The graves themselves are better imaged in the 200 MHz profile (**b**). (From Nobes 1999)

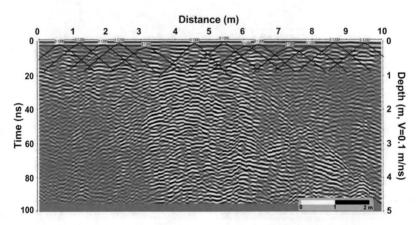

Fig. 9.8 The 500 MHz profile from a test site at Nysa, Turkey, is dominated by diffractions from the roots of olive trees. The deeper stone wall targets are less clear. The tree root diffractions did allow velocities to be determined for the site (superimposed curves), which yielded a consistent velocity of 0.10 m/ns (100 m/μs or 10 cm/ns). (Adapted from Yalçiner et al. 2009)

Fig. 9.9 Radar transects across the centres of nine roots of various diameters (1–10 cm) buried 50 cm deep in a sand pit. (Modified from Barton and Montagu 2004)

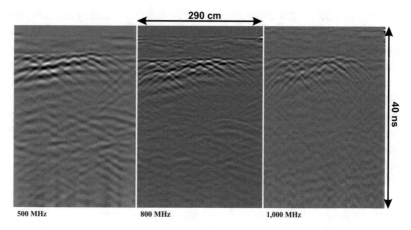

Fig. 9.10 Radar transects across the centre of nine roots buried at depths ranging from 15 to 155 cm. (Modified from Barton and Montagu 2004)

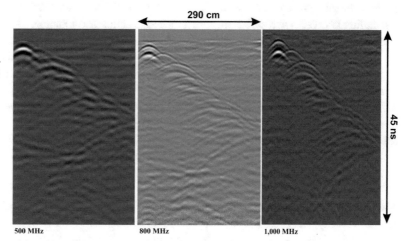

roots buried at different depths, and all nine could be clearly identified in the 500 MHz transect across roots of different sizes.

Conditions vary from site to site, and often we cannot pre-judge what results we will obtain, nor which antennas are best used. For example, Porsani et al. (2010) used both 200 and 400 MHz antennas, but the site conditions allowed them to obtain all the information they needed using the 400 MHz results. Testing different frequency antennas allows us to choose the optimal signal for a given site. This should simply be part of our normal operating procedures.

We also should be following a hierarchy of imaging work, from large scale to finer scale, and from mapping mode to cross-sectional imaging,

as recommended by McCoy and Ladefoged (2009), amongst many others. A multi-parameter approach is always best. The results from the different techniques complement and reinforce each other, as illustrated in numerous studies (e.g., Nobes 1999; Nobes and Lintott 2000; Field et al. 2001; Chianese et al. 2004; de Domenico et al. 2006; Keay et al. 2009; Papadopoulos et al. 2009; Selim et al. 2014; Nobes and Wallace 2018).

9.4   Resolution of Buried Objects

In order to decide which antenna frequency may be the best choice in a given situation, we need to assess: (1) what are the dimensions of the buried feature, and (2) what are the ground conditions. For example, Novo et al. (2008) estimated the dimensions of their target, a sarcophagus that was about 0.6 m wide, about 1 m high, and about 2 m long. They calculated the required Nyquist trace spacings and proceeded accordingly. Similarly, Conyers (2004) showed how important the ground conditions can be, specifically wet vs dry, in determining what system configuration may be best in a given situation.

If we take a grave as an example, which is approximately 1 m wide and 2 m long when excavated, then by the sampling theorem, we need at least 2 samples per metre to minimally observe the grave when we cross it. Thus we need a sampling interval of at least 0.5 m, which indeed was the sampling interval used by Nobes (1999) for acquisition of magnetic and electromagnetic data. A smaller interval would have been preferred, to yield more detail and have some redundancy in the readings, but any increase in the number of readings is then offset by the additional time required to carry out those readings.

The sampling interval used for the GPR profiles at the Oaro burial site (Nobes 1999) was 10 cm for a 200 MHz system, and 5 cm for a 450 MHz system. Often we use an initial estimate of the velocity, e.g., 0.10 m/ns (100 m/μs or 10 cm/ns). This yields a Nyquist sampling (Eq. 9.5b) of 12.5 cm for 200 MHz. Thus the sampling interval of 10 cm appeared to be adequate and was chosen for ease of implementation. However, the velocity was found to be 0.06 m/ns (60 m/μs or 6 cm/ns), which yields Nyquist sampling intervals of $\Delta x \leq 7.5$ cm for 200 MHz, and 3.3 cm for 450 MHz. Thus the data were slightly undersampled. The results were, nonetheless, adequate for imaging the burial sites. The survey design and layout was prepared before the velocities could be determined, which illustrates that survey designs need to be done conservatively, and over-sampling of GPR profiles is generally useful, as illustrated by Novo et al. (2008).

We face similar issues when attempting to design surveys to obtain images of buried walls or trenches. When imaging a possible kumara (Polynesian sweet potato) storage pit at an archaeological site on Okoura farm (Gordon et al. 2004), a GPR trace spacing of 10 cm was used, and the profiles were placed 20 cm apart. The velocity was 0.12 m/ns (120 m/ms or 120 cm/ns), which yields a Nyquist sampling interval of 15 cm. Thus the profiles were oversampled along each profile, and slightly undersampled across the profiles. The net results, however, are able to resolve what appear to be central post holes to support the roof (Fig. 9.11), and possibly the entry sump to trap cold air and maintain the conditions necessary for adequate storage of the kumara.

Knowledge of ground conditions, possible dimensions, and other aspects of the targets and the setting are needed to plan an optimised campaign for imaging. As noted earlier, a staged hierarchical campaign of imaging is optimal (McCoy and Ladefoged 2009), and as described by Böniger and Tronicke (2010) in their work on tomb detection, the survey designs and interpretations need to be designed for the specific target in mind.

9.5 Voids

A special case of resolution occurs when we consider voids. Examples of voids in archaeology include tunnels, crypts beneath churches, caves of archaeological interest (especially those of archaeological interest), and small voids within ancient constructions such as walls and bridges (e.g., El-Fouly 2000; Pérez Gracia et al. 2000; Beres et al. 2001; Leucci 2002; Leucci et al. 2003; Arias et al. 2007; Novo et al. 2008; Shaaban et al. 2008; Böniger and Tronicke 2010). The detection of voids of archaeological interest could be summarised as the search for caves, cracks, and crypts. There have been a number of papers about detecting such voids, but most don't address the mechanisms behind the GPR response from voids. An example is illustrated in Fig. 9.12. A GPR profile had been acquired above a known cave in South Australia, and the researcher asked why he couldn't see the cave (A. White, pers. comm., 2002). It was noted that the radar velocity is much faster than in the

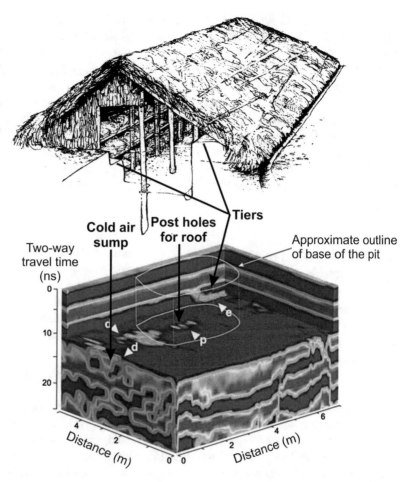

Fig. 9.11 The 3D GPR image of a kumara storage pit (**b**) shows features consistent with post holes to support a roof over the storage pit (**a**), and a possible entry sump used to trap cold air to maintain good thermal conditions for storage. (Adapted from Gordon et al. 2004)

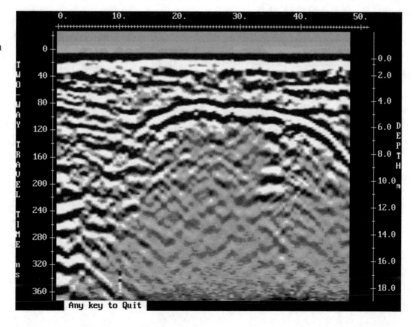

Fig. 9.12 GPR profile acquired on the surface above a cave, located from about 15 m along the line onwards. Note the bright reflection at about 80–120 ns. (Profile courtesy of A. White, deceased)

surrounding rock, and hence the two wavelets almost coincide. The result was an enhanced positive-negative pulse.

Similarly, results from a GPR investigation of a concrete construction that appeared to be failing showed a strong "bright spot" response (King et al. 2003). The bright reflections were due to voids beneath the concrete base. Repairs were done by injecting grout into the voids. Follow-up GPR imaging allowed quality control checks to be done on the repairs because grout has similar physical properties to that of concrete.

9.5.1 Modelling of GPR Void Response

The response of GPR to the presence of a void, whatever the mechanism by which it formed, is frequently attributed to the strong velocity contrast. In fact, it is due as much to the superposition of the wavelets from the top and bottom of the void as it is due to the strong velocity contrast. By examining the interactions of the top and bottom reflected wavelets, we can see that the point at which amplitude reaches its peak, the point at which the wavelets can be separately identified, and the point at which the wavelets are no longer superimposed, all depend on the wavelet frequency, which thus in turn governs the wavelet resolution, as discussed earlier.

The GPR response to a void was examined using a simple model constructed using an Excel™ spreadsheet. A Ricker wavelet Eq. (9.2) was used as the basic pulse. The model was constructed using a simple background material with properties similar to limestone, with a dielectric coefficient, ε_r, equal to 6.25. The radar velocity is then 120 m/μs (0.12 m/ns). Within the medium is a void of variable thickness; the void

is filled with air or with water. The radar velocity in air is 300 m/μs (0.30 m/ns or 30 cm/ns), and in water is ~33 m/μs (0.033 m/ns or 3.3 cm/ns). Then using Eqs. (9.1) and (9.3), we can define the pulsewidths and the resolutions as derived from T_R, as shown in Table 9.1.

This approach is far simpler than that of Carcione (1996), who used the full Maxwell's equations to model the propagation of the electromagnetic waves and their response to a coffin-like stone structure, and developed a process for reconstructing the shape of the void from the observed GPR response. Here, instead, we use a very simple three-layer rock-void-rock model in order to illustrate the influence of the superposition of the waves reflected from the top and bottom of the void to reinforce the strong reflections due to the void.

Using standard wave propagation theory, the reflection coefficient from the top of the void will be R_{12}:

$$R_{12} = (V_2 - V_1)/(V_2 + V_1) \qquad (9.6)$$

where V_1 is the velocity of the host medium, in this case 0.12 m/ns, and V_2 is the velocity in the void, either air- or water-filled. The amount of energy reflected is equal to $R_{12}{}^2$, and the transmission coefficient is thus T_{12}, given by:

$$T_{12}{}^2 = 1 - R_{12}{}^2 \text{ so } T_{12} = \sqrt{(1 - R_{12}{}^2)}. \quad (9.7)$$

The reflection coefficient at the base of the void will similarly be R_{23}, defined by:

$$R_{23} = (V_3 - V_2)/(V_3 + V_2) \qquad (9.8)$$

where V_2 is the velocity in the void, as before, and V_3 is the velocity in the host medium, so $V_3 = V_1$. The reflected wavelet amplitude is then equal to the transmission coefficient for the wavelet transmitted through the top of the void, times the

Table 9.1 Ricker wavelet pulsewidths and resolutions

Frequency (MHz)	T_D (ns)	T_R (ns)	Δr_{rock} (m)	Δr_{air} (m)	Δr_{water} (m)
50	15.6	9.0	0.270	0.675	0.074
100	7.8	4.5	0.135	0.338	0.037
200	3.9	2.3	0.068	0.169	0.019
400	1.9	1.1	0.034	0.084	0.009

reflection coefficient from the bottom of the void. The net composite superimposed wavelet is thus:

$$W(t) = R_{12} W_0(t) + T_{12} R_{23} W_0(t + \Delta t) \quad (9.9)$$

where $W(t)$ is the net composite pulse recorded as a function of time, and W_0 is the Ricker wavelet. $W_0(t)$ is the wavelet reflected from the top of the void, which is used as the zero time reference, and $W_0(t + \Delta T)$ is the wavelet reflected from the bottom of the void. $W_0(t + \Delta t)$ is simply the Ricker wavelet delayed by Δt, i.e. the delay determined by the travel time in the void. The time delay is then comparable to a void thickness, d, where $d = V_2 \Delta t$. The reflection and transmission coefficients for air- and water-filled voids are shown in Table 9.2.

Using the simple spreadsheet design, Eq. (9.9), and the coefficients listed in Table 9.2, wavelets with increasing time delays were added to the reflection from the top of the void. The increasing time delays are equivalent to increasing void thicknesses. The superimposed composite wavelets were then plotted as a function of travel time and of the time delay. The top of the void was used as the time reference. The time sampling interval was 0.4 ns for 50, 100, and 200 MHz signals, and 0.2 ns for 400 MHz.

For these idealised layered models, no noise was introduced, and no geological or morphological complexities were added. The void response can thus be examined in "ideal" conditions, i.e. these models represent the best that we could do for ascertaining the different aspects of the GPR void response. The simple models were constructed to highlight the essential elements of the void responses, especially the superposition of the reflected wavelets that serve to enhance the strong reflections already present.

The 100 MHz composite wavelet results are representative; the results are structurally similar for all of the model frequencies. The results differ only in the sharpness and duration of the wavelet used. The composite wavelet amplitudes are shown for an air-filled void (Fig. 9.13) and a water-filled void (Fig. 9.14). Note that the amplitude is about 50% higher at the "bright spot" than for the peak amplitude of a single wavelet (solid line highlights). Notice also the transitions (dashed boxes) where there at first appears to be only one large wavelet, with a large ± pair of peaks, until the wavelets separate (arrows and dotted lines).

When we compare the composite wavelets in air- vs water-filled voids, we see some similarities and some important distinctions. First of all, both have peak amplitudes (positive and negative) that far exceed the peak amplitudes for a single reflected wavelet. Secondly, there is a sequence of time delays (indicated by the dashed rectangles in Figs. 9.13 and 9.14) over which the superimposed wavelet response appears to be one large wavelet. The point at which the two wavelets can be seen to be separate entities are the same (between 5 and 10 ns).

The distinctions arise from the obvious differences between the radar velocities in air vs water. The water radar velocity is almost 1/10th of the air velocity. Thus whereas the wavelets cannot be separated until the air-filled void is more than 1 m thick, the wavelets in the water-filled void can be separated when the void is of the order of 15 cm thick or greater.

When we look at suites of individual wavelets, we obtain additional insights (Fig. 9.15). For the air-filled void (Fig. 9.15a), the minimum amplitude occurs for the minimum time delay, 0.4 ns. The maximum amplitude occurs when the time

Table 9.2 Reflection and transmission coefficients

Boundary	Reflection coefficient	Transmission coefficient
Air: Top of void	$R_{12} = +0.4286$	$T_{12} = 0.9035$
Air: Bottom of void	$R_{23} = -0.4286$	
Air: Base composite	$T_{12} R_{23} = -0.3872$	
Water: Top of void	$R_{12} = -0.5686$	$T_{12} = 0.8226$
Water: Base of void	$R_{23} = +0.5686$	
Water: Base composite	$T_{12} R_{23} = +0.4678$	

Fig. 9.13 Superposition of 100 MHz wavelets reflected from the top and bottom of an air-filled void. There is a strong amplitude "bright spot" (arrow at top) due to superimposed positive and negative peak amplitudes. A sequence of time delays where the top wavelets cannot be separated are highlighted (dashed rectangle). A point where the two wavelets can be separately distinguished is indicated by an arrow in the upper centre of the figure. This separation point corresponds to a time delay (dotted line, centre) close to 10 ns, and corresponds to an air void thickness greater than 1 m

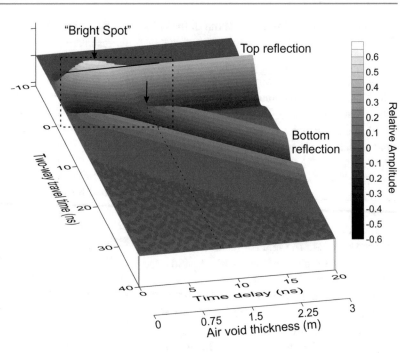

Fig. 9.14 As for the air-filled void (Fig. 9.10), but instead for a water-filled void. The separation between the wavelets again occurs for time delays close to 10 ns, but this then corresponds to a water void thickness of about 15 cm (0.15 m). The wavelet superposition is shown from a different perspective here, compared to Fig. 9.10, to show both the negative and positive peaks, i.e., the top and bottom reflections, respectively

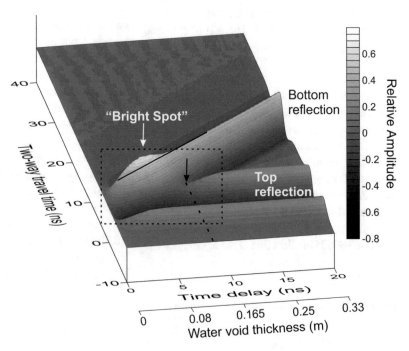

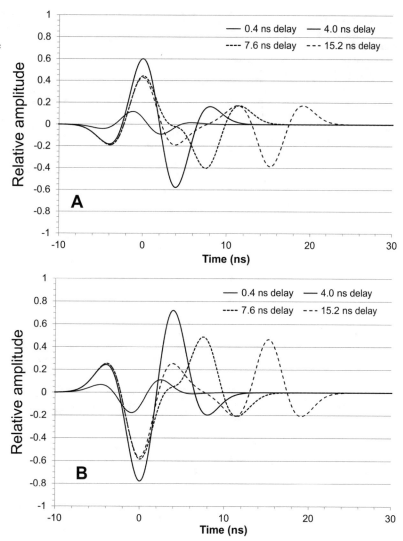

Fig. 9.15 100-MHz wavelets at delays of 0.4, 4.0, 7.6, and 15.2 ns for the (**a**) air-filled void and (**b**) water-filled void

delay is 4.0 ns, which is close to the value of T_R (Table 9.1). The wavelets appear to be distinct when the peaks are separated by a clear "plateau", which develops when the time delay is 7.6 ns, which is roughly equal to the value for T_D (Table 9.1) when we consider the sampling interval is 0.4 ns. Finally, the wavelets are fully separated, with no overlap, when the time delay is 15.2 ns, about the same as twice T_D.

For the water-filled void (Fig. 9.15b), the minimum amplitude again occurs for the minimum time delay, 0.4 ns. The maximum amplitude again occurs for a time delay equal to 4.0 ns,

which as noted previously is close to the value for T_R (Table 9.1). Note the phase reversal of the water-void wavelet relative to the air-void wavelet, due to the low velocity in water. The wavelets again appear to be distinct when the peaks are separated by a clear "plateau", again for a 7.6 ns delay. Finally, the wavelets are again fully separated, with no overlap, for a delay of 15.2 ns.

When we examine in greater detail the transition from what appears to be one wavelet with large positive and negative pulses (Fig. 9.16), we can see that when the time delay is 6.8 ns, the curve goes smoothly from the maximum positive

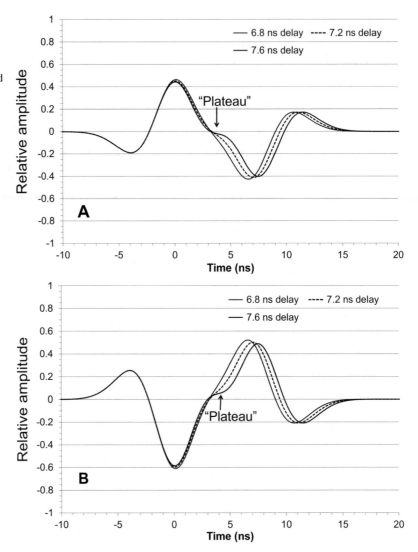

Fig. 9.16 A closer examination of the transition from a single composite wavelet for voids filled with air (**a**) and water (**b**), shows a large positive-negative pair of peaks, changing to peaks separated by a flatter "plateau" with increasing time delays

value to the maximum negative value. There is a subtle divergence from a simple linear change when the delay is 7.2 ns, but in the presence of noise, both system and geological noise, such a subtle feature would likely be easily missed. It isn't until the delay is 7.6 ns that the wavelets appear to be separated by a clear small "plateau", that is, a segment with a flattened response, as labelled in Fig. 9.15.

We can compile the parameters as a function of frequency (Table 9.3), and we observe some simple relationships. The peak amplitude of the composite superimposed wavelets, the composite peak amplitude (CPA) occurs at a time almost equal to the pulsewidth, T_R. We could alternatively say that the CPA is equal to half the T_D, which is where we see the maximum superposition of the respective $-/+/-$ and $+/-/+$ lobes and peaks of the first and second reflected wavelets. The "plateau" then occurs for the time delay, approximately equal to T_D, when peaks no long align well enough to enhance the peak amplitudes. Finally, the full separation of the wavelets occurs when the wavelets no longer interact at all, at twice T_D.

Table 9.3 Computed parameters vs frequency

Frequency (MHz)	T_R (ns)	T_D (ns)	CPA[a] (ns)	Plateau[b] (ns)	Full[c] (ns)
50	9.0	15.6	8.0	15.2	31
100	4.5	7.8	4.0	7.6	15.6
200	2.3	3.9	2.0	4.0	8.0
400	1.1	1.9	1.0	1.8	4.0

[a]"CPA" is the composite peak amplitude of the superimposed wavelets
[b]"Plateau" is the time delay at which the composite superimposed response clearly flattens
[c]"Full" is time delay at which the superimposed wavelets no longer overlap at all

Table 9.4 Rock, air and water wavelet separation vs frequency

Frequency (MHz)	Δr_{rock} (cm)	Δr_{air} (cm)	Δr_{water} (cm)
50	91.2	228	25.1
100	45.6	114	12.5
200	24.0	60.0	6.6
400	10.8	27.0	3.0

These time delays represent different radial (i.e. vertical) resolutions when looking at rock vs air vs water. The "plateau" separation can be translated into a void thickness (Table 9.4) by using Δr and the simple relation noted earlier, $d = tV/2$. However, instead of using T_R for the pulsewidth, W, in Eq. (9.1), we use T_D, which is the time delay required to separate the wavelets. The ability to differentiate the two wavelets represents only centimetres to 10s of centimetres in rock or ice, but 2.5 times that in air. That means, for example, that the minimum thickness of an air void for which we can begin to separate the wavelets (the "plateau" in Table 9.3) is 4 ns or 60 cm for a 200 MHz wavelet.

One issue not addressed so far in this section, but which will also be prevalent in voids, are multiple reflections or reverberations (e.g., Kofman et al. 2006). The strong velocity contrasts at the top and bottom of a void will act to reflect any energy back into the void, generating multiple reflections, usually simply called *multiples*. Thus another feature of a void will frequently be reverberations or multiples. In addition, multiple diffractions will often be generated in near vertical cracks, such as crevasses in glaciers, cracks in rock, and cracks in building stone and foundations. Such cracks will be common in karst terrains, where caves of archaeological interest may be commonplace. Because the sides of the crack are not smooth,

any rough protuberance will generate a diffraction. So multiple, sometimes "stacked" diffractions will appear as a result. These can be additional evidence of voids that are near-vertical.

9.5.2   Void Examples

This brings us back to the specific topic of resolution. Diffractions will occur at different frequencies for different scales of roughness, just as diffractions occur at different frequencies for different sizes of roots, as shown in Figs. 9.6, 9.8, and 9.9. We see a similar set of stacked reverberations in the results from a sinkhole in Poland (Fig. 9.17, from Marcak et al. 2008). The Chinese have numerous examples of karstic terrains, and have often observed strong reflections from voids and multiples in voids in karstic environments (e.g. Deng 2001; Deng et al. 2001; Leucci and De Giorgi 2005). An example from Wuha Xuancheng in Anhui illustrates nicely both the strong reflection and the occurrence of multiples with karstic voids (Fig. 9.18, from Deng 2001). Beres et al. (2001) observed a mix of strong reflections and diffractions from La Grande Rolaz cave in Switzerland (e.g., Fig. 9.19). Similar multiples and reverberations were observed in other studies of archaeological sites (e.g., Pérez Gracia et al. 2000; Shaaban et al. 2008). Leucci

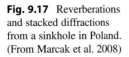

Fig. 9.17 Reverberations and stacked diffractions from a sinkhole in Poland. (From Marcak et al. 2008)

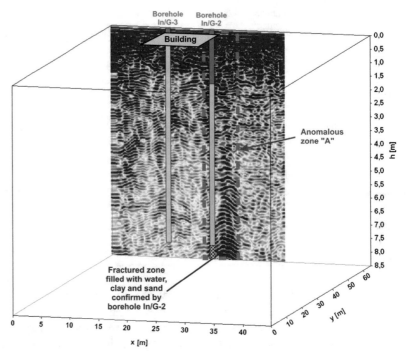

(2002) has a nice example of strong reflections from a structure beneath an old church (Fig. 9.20).

Voids can also be used to identify structures that may be in need of repair and remedial action (e.g., Arias et al. 2007; Leucci et al. 2003; Ranalli et al. 2004; Diamanti et al. 2008). As mortar or building stone decays, gaps can form, either in the building coating, such as the outer decorative lamination, or in the building structure itself, thus weakening the coating or the building. Leuci et al. (2003) devised a model to represent such voids (Fig. 9.21), and sand box tests yielded both the void response and a multiple. Such voids are regularly detected in tunnel maintenance scans, and are diagnostic of the sort of response one might expect from a void (e.g., Fig. 9.22, from Deng et al. 2015), especially a small elongate void as depicted in Fig. 9.21. The resolution needed to "see" these cracks will then determine what frequency signal to use.

Arias et al. (2007) used GPR not only for imaging cracks, but also for imaging the structure of a medieval bridge and compared it with digital photogrammetry images to construct a three-dimensional (3D) model of the bridge. The aim

was both to assess the structural integrity of the bridge but also to match the GPR images to the photogrammetric reconstructions, creating a 3D model of the bridge in its entirety. At every stage, the question of scale and resolution was addressed through the use of the complementary techniques. They used three separate GPR frequencies—250, 500, and 800 MHz antennas—to evaluate the homogeneity of the interior of the bridge and to detect any holes or cracks. The GPR results were then cross-referenced to the photogrammetry and to the mapping of surface cracks.

9.6   Conclusions

It is tempting to choose a high frequency ground penetrating radar (GPR) antenna for archaeological imaging. However, lower frequency signals are usually better at delineating the targets of interest. Higher frequency signals are prone to "clutter" from smaller features, and can obscure the features in which we are interested. Such clutter can appear as diffractions, but our targets

Fig. 9.18 The strong reflection from the top of a void and the multiples generated in the void are illustrated by two profiles, Line 9 (**a**) which is 100 m long and Line 30 (**b**) which is about 32 m long, across a segment of the Wuha Xuancheng karst terrain in the Anhui province, China. The time window is 400 ns for both profiles. (Adapted from Deng, 2001)

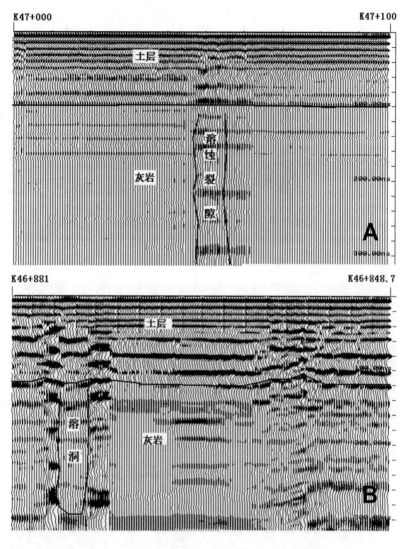

Fig. 9.19 La Grande Rolaz cave in Switzerland generates both strong void reflections (far right) and numerous diffractions (to the right of the strong void reflections). (From Beres et al. 2001)

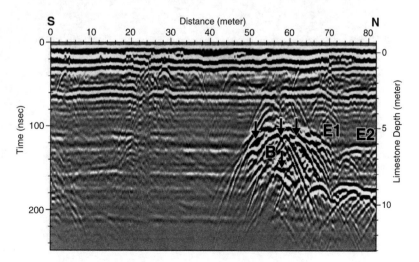

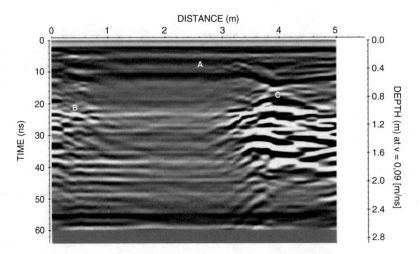

Fig. 9.20 Multiples generated in a structure beneath an old church (at the right side of the profile), likely due to a crypt. (From Leucci 2002)

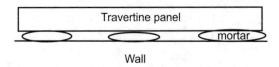

Fig. 9.21 An idealised model of the sort of voids that might be encountered in old structures, especially ones built using mortar. When the mortar decays, gaps remain, weakening at least the outer coating of the building, and potentially weakening the structure itself. (Modified from Leucci et al. 2003)

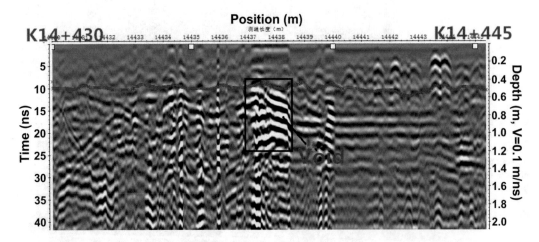

Fig. 9.22 Strong reflections and multiple reflections caused by a void behind a tunnel wall. The primary void is outlined by the solid rectangle. (Adapted from Deng et al. 2015)

may yield diffractions as well, but at the lower frequencies. When in doubt, or in complex situations, multiple frequencies are helpful in delineating targets of different dimensions.

Voids due to caves, cracks, or crypts are a special case when dealing with resolution, because the strong velocity contrasts yield strong reflections. The strong reflections are made even stronger by the superposition of reflections from the base of the voids arriving soon after the reflections from the top of the voids. The separation of the top and bottom void reflections may not occur for a significant distance, and may be substantially larger than the resolution of the antenna in use. In addition, multiples (reverberations) can further obscure the interpretation of the GPR profile.

Vertical cracks, which can occur in karst terrains and in buildings that are under mechanical stress, are a special case of voids that may give rise to diffractions due to the roughness of the crack edges. Stacked diffractions are then the equivalent of multiples or reverberations in the general case of a void.

Finally, the best results are obtained when GPR imaging is carried out as part of a multi-parameter imaging program.

References

Annan AP (2005) The principles of ground penetrating radar. In: Butler DK (ed) Near-surface geophysics, vol 13. Society of Exploration Geophysicists, Investigations in Geophysics, Tulsa, OK, pp 357–438

Annan AP, Cosway SW (1994) GPR frequency selection. In: GPR '94: Proceeding of the Fifth International Conference on Ground Penetrating Radar, Kitchener, Canada, vol. II, pp 747–760

Arias P, Armesto J, Di-Capua D, González-Drigo R, Lorenzo H, Pérez-Gracia V (2007) Digital photogrammetry, GPR and computational analysis of structural damages in a mediaeval bridge. J Eng Fail Anal 14:1444–1457. https://doi.org/10.1016/j.engfailanal.2007.02.001

Barton CVM, Montagu KD (2004) Detection of tree roots and determination of root diameters by ground penetrating radar under optimal conditions. Tree Physiol 24:1323–1331

Beres M, Luetscher M, Olivier R (2001) Integration of ground penetrating radar and microgravimetric methods to map shallow caves. J Appl Geophys 46:249–262

Böniger U, Tronicke J (2010) Improving the interpretability of 3D GPR data using target–specific attributes: application to tomb detection. J Archaeol Sci 37:360–367. https://doi.org/10.1016/j.jas.2009.09.049

Carcione JM (1996) Ground radar simulation for archaeological applications. Geophys Prospect 44:871–888

Chamberlain AT, Sellers W, Proctor C, Coard R (2000) Cave detection in limestone using ground penetrating radar. J Archaeol Sci 27:957–964. https://doi.org/10.1006/jasc.1999.0525

Chianese D, D'Emilio M, Di Salvia S, Lapenna V, Ragosta M, Rizzo E (2004) Magnetic mapping, ground penetrating radar surveys and magnetic susceptibility measurements for the study of the archaeological site of Serra di Vaglio (southern Italy). J Archaeol Sci 31:633–643. https://doi.org/10.1016/j.jas.2003.10.011

Conyers LB (2004) Moisture and soil differences as related to the spatial accuracy of GPR amplitude maps at two archaeological test sites. In: Slob E, Yarovoy A, Rhebergen J (eds) GPR 2004: Proceedings of the 10th International Conference on Ground Penetrating Radar, Delft, Netherlands, vol. II, pp 435–438

de Domenico D, Giannino F, Leucci G, Bottari C (2006) Integrated geophysical surveys at the archaeological site of Tindari (Sicily, Italy). J Archaeol Sci 33:961–970. https://doi.org/10.1016/j.jas.2005.11.004

Deng J (2001) Research on the data processing system and measurement parameters setting of ground penetrating radar, Report from East China geological Institute [in Chinese], 23p

Deng J, Mo H, Liu Q (2001) The application of ground-penetrating radar to karst detection. Geophys Geochem Explor 25(6):474–477

Deng X, Du T, Yuan Q, Zhong X (2015) Tunnel lining thickness and voids detection by GPR. Electron J Geotech Eng 20:2019–2030

Diamanti N, Giannopoulos A, Forde MC (2008) Numerical modelling and experimental verification of GPR to investigate ring separation in brick masonry arch bridges. Non-Destr Test Eng Int 41(5):354–363. https://doi.org/10.1016/j.ndteint.2008.01.006

El-Fouly A (2000) Voids investigation at Gabbari Tombs, Alexandria, Egypt using ground penetrating radar technique. In: Proceedings of ICEHM2000, Cairo University, Egypt, pp 84–90

Field G, Leonard G, Nobes DC (2001) Where is Percy Rutherford's grave? In: M. Jones M, Sheppard P (eds) Australasian Connections and New Directions: Proceedings of the 7th Australasian Archæometry Conference, Research in Anthropology and Linguistics, University of Auckland, 5, pp 123–140

Gordon HW, Bassett KN, Nobes DC, Jacomb C (2004) Gardening at the edge: documenting the limits of tropical Polynesian kumara horticulture in southern New Zealand. Geoarchaeology 19(1):185–218

Jol HM (1995) Ground penetrating radar antennae frequencies and transmitter powers compared for penetration depth, resolution and reflection continuity. Geophys Prospect 43:693–709

Keay S, Earl G, Hay S, Kay S, Ogden J, Strutt KD (2009) The role of integrated geophysical survey methods in the assessment of archaeological landscapes: the case of portus. Archaeol Prospect 16:154–166. https://doi.org/10.1002/arp.358

King ML, Wu D, Nobes DC (2003) Non-invasive ground penetrating radar investigation of a failing concrete floor slab. In: Proceedings of NDT-CE 2003, the international conference on non-destructive testing in civil engineering, Deutsche Gesellschaft für Zerstörungsfreie Prüfung, Berlin, Proceedings BB85-CD. http://www.ndt.net/article/ndtce03/papers/p043/p043.htm

Kofman L, Ronen A, Frydman S (2006) Detection of model voids by identifying reverberation phenomena in GPR records. J Appl Geophys 59(4):284–299

Leckebusch J, Peikert R (2001) Investigating the true resolution and three-dimensional capabilities of ground-penetrating radar data in archaeological surveys: measurements in a sand box. Archaeol Prospect 8:29–40

Leucci G (2002) Ground-penetrating radar survey to map the location of buried structures under two churches. Archaeol Prospect 9:217–228. https://doi.org/10.1002/arp.198

Leucci G, De Giorgi L (2005) Integrated geophysical surveys to assess the structural conditions of a karstic cave of archaeological importance. Nat Hazards Earth Syst Sci 5:17–22

Leucci G, Negri S, Carrozzo MT (2003) Ground penetrating radar (GPR): an application for evaluating the state of maintenance of the building coating. Ann Geophys 46(3):481–489

Marcak H, Golebiowski T, Tomecka-Suchon S (2008) Geotechnical analysis and 4D GPR measurements for the assessment of the risk of sinkholes occurring in a Polish mining area. Near Surf Geophys 6:233–243

McCoy MD, Ladefoged TN (2009) New developments in the use of spatial technology in archaeology. J Archaeol Res 17:263–295. https://doi.org/10.1007/s10814-009-9030-1

Nobes DC (1999) Geophysical surveys of burial sites: a case study of the Oaro urupa. Geophysics 64(2):357–367. https://doi.org/10.1190/1.1444540

Nobes DC, Lintott B (2000) Rutherford's "Old Tin Shed": Mapping the foundations of a Victorian-age lecture hall. In: Noon DA, Stickley GF, Longstaff D (eds) GPR 2000: Proceedings of the 8th international conference on ground penetrating radar, Gold Coast, Australia, Society of Photo-Optical Instrumentation Engineers (SPIE), 4084, pp 887–892

Nobes DC, Wallace LR (2018) Geophysical imaging of an Early nineteenth century colonial defensive blockhouse: applications of EM directionality and multi-parameter imaging. In: El-Qady G, Metwaly M (eds) Archaeogeophysics. Springer, Berlin (in press)

Novo A, Lorenzo H, Rial FI, Pereira M, Solla M (2008) Ultra-dense grid strategies for 3D GPR in Archaeology. In: Proceedings of GPR 2008: 12th international conference on ground penetrating radar, Birmingham, UK

Nuzzo L, Leucci G, Negri S, Carrozzo MT, Quarta T (2002) Application of 3D visualization techniques in the analysis of GPR data for archaeology. Ann Geophys 45:321–337

Papadopoulos N, Sarris A, Yi M-J, Kim J-H (2009) Urban archaeological investigations using surface 3D ground penetrating radar and electrical resistivity tomography methods. Explor Geophys 40:56–68

Pérez Gracia V, Antonio Canas J, Pujades LG, Clapes J, Caselles O, Garcia F, Osorio R (2000) GPR survey to confirm the location of ancient structures under the Valencian Cathedral (Spain). J Appl Geophys 43:167–174

Pipan M, Baradello L, Forte E, Prizzon A, Finetti I (1999) 2-D and 3-D processing and interpretation of multifold ground penetrating radar data: a case history from an archaeological site. J Appl Geophys 41:271–292

Porsani JL, de Matos Jangelme G, Kipnis R (2010) GPR survey at Lapa do Santo archaeological site, Lagoa Santa karstic region, Minas Gerais state, Brazil. J Archaeol Sci 37:1141–1148. https://doi.org/10.1016/j.jas.2009.12.028

Ranalli D, Scozzafava M, Tallini M (2004) Ground penetrating radar investigations for the restoration of historic buildings: the case study of the Collemaggio Basilica (L'Aquila, Italy). J Cult Herit 5:91–99

Selim EI, Basheer AA, Elqady G, Hafez MA (2014) Shallow seismic refraction, two-dimensional electrical resistivity imaging, and ground penetrating radar for imaging the ancient monuments at the western shore of Old Luxor City, Egypt. Archaeol Discov 2:31–43. https://doi.org/10.4236/ad.2014.22005

Shaaban FA, Abbas AM, Atya MA, Hafez MA (2008) Ground-penetrating radar exploration for ancient monuments at the Valley of Mummies -Kilo 6, Bahariya Oasis, Egypt. J Appl Geophys 68(2):194–202. https://doi.org/10.1016/j.appgeo.2008.11.009

Sheriff RE (2002) Encyclopedic dictionary of exploration geophysics, 4th edn. Society of Exploration Geophysicists, Tulsa, OK, p 429

Yalçiner CÇ, Bano M, Kadioglu M, Karabacak V, Meghraoui M, Altunel E (2009) New temple discovery at the archaeological site of Nysa (western Turkey) using GPR method. J Archaeol Sci 36:1680–1689

The Standardized Pricking Probe Surveying and Its Use in Archaeology

10

S. Szalai, I. Lemperger, Á. M. Pattantyús, and L. Szarka

Abstract

In this paper we present the so-called standardized pricking probe surveying technique and demonstrate its usefulness in an archaeological study. The buried target is a Paleochristian sepulchral chapel, which had already been excavated 82 years ago, then re-buried and forgotten.

By applying this technique, it was possible to locate the buried remnants of the chapel in a large field, in spite of the dense undergrowth, where classical geophysical methods could have hardly been applied. When the area was mopped-up, a detailed and systematic pricking probe surveying was carried out. The pricking-probe results have been compared to geoelectric, magnetic and georadar mapping results. The standardized pricking probe images, at least in this field experiment, proved to be competitive to the geophysical maps.

The optimum pricking probe parameters such as horizontal interval, pricking depth, observable quantity and the way of presentation were optimized through field experiments.

For a detailed investigation a rectangular grid with an interval of 50 cm (i.e. a grid interval, corresponding to the wall thickness) is recommended, while for reconnaissance measurements a two times larger horizontal interval (1 m in this case) proved to be sufficient. In this case study the optimum pricking depth was 20–30 cm; in general it depends on the burial depth of the investigated object. For the presentation of the results a suitable running average of a two-valued observable quantity is defined.

The merits of the standardized pricking probe technique are as follows: its field procedure and data processing are simple, it is cheap and relatively quick; it does not need any electronic instrument, therefore there are no investment costs and there is no risk of technical failures; the technique can be applied even among the most unfavourable field conditions like e.g. bad weather, extreme topography, dense undergrowth, etc.; At the same time, the standardized pricking probe method should be applied only in areas, where the possible damaging of the buried structures is excluded.

S. Szalai (✉) · I. Lemperger · L. Szarka
MTA CSFK GGI, Sopron, Hungary

University of West-Hungary, Sopron, Hungary
e-mail: szalai@ggki.hu; szarka@ggki.hu

Á. M. Pattantyús
Eötvös Loránd Geophysical Institute of Hungary,
Budapest, Hungary
e-mail: miklos@elgi.hu

Keywords

Pricking probe · Geophysics · Archaeological survey · Geoelectrics · Geomagnetics · GPR · Mapping

10.1 Introduction

The Palaeochristian cemetery chapel in Sárisáp (Hungary, Fig. 10.1.) was already excavated in the beginning of the twentieth century, and then it was re-buried, and later forgotten. Recently, some ideas have emerged about its re-excavation and exhibition. As a first step, the chapel is to be re-found again. It has been known from an old document (Gerevich 1979, Fig. 10.2) that the area of the chapel including the apsis is 11.2 × 5.5 m, and it is situated somewhere in an area of 25 × 30 m.

In 2007, because of the dense vegetation onsite, it was impossible to locate the chapel by applying common geophysical methods. This failure led us to the idea of the systematic pricking probe (PP) technique. By using it, we easily found the chapel. Later on, when the area was mopped-up from undergrowth, we carried out detailed pricking probe experiments and, for comparison and verification, we also carried out geophysical measurements by applying three standard methods. Due to the fact that the pricking probe map proved to be competitive to the standard geophysical maps, additional test measurements were undertaken, in order to reveal the full capacities of the standardized PP technique.

In this paper at first we give an archaeological overview, then we present the pricking probe technique. Then we provide a detailed comparison between the PP data and the results of standard geophysical measurements. Based on detailed field experiments, we elaborated the optimum pricking probe parameters for archaeological prospecting: (1) the penetration depth, (2) the horizontal grid distance, (3) the observable quantity, and (4) the optimum way of presentation of observed values. Finally, the strengths and weaknesses of the PP technique are summarized, together with the archaeological results obtained from the PP measurements.

10.2 Archaeological Background

About some 82 years ago in Sárisáp (Fig. 10.1.), along the so-called Quadriburg accommodation road, a local historian Albin Balogh excavated the walls of a sepulchral chapel, made of travertine in Palaeochristian ages (Gerevich 1979). Balogh found remnants of frescos, pots and glass rings, two complete pots and an iron knife. The results were first released in a report from the year 1927. A description of the building and of the frescos was published in 1934. The sepulchral chapel is

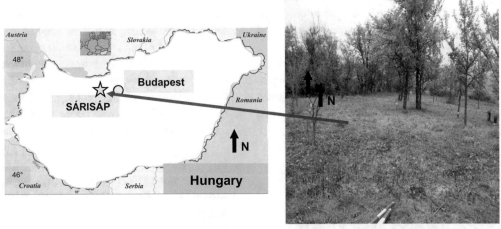

research area: 25*30 m

Fig. 10.1 Location of the Sárisáp archaeological site

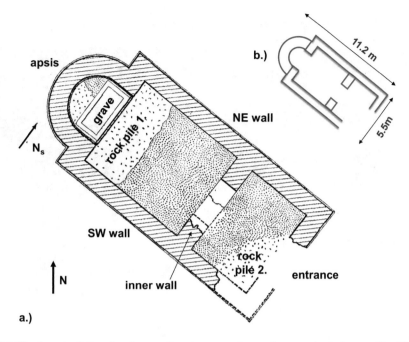

Fig. 10.2 (**a**) The layout of the chapel, according to Gerevich (1979), with identification of the main features of the chapel, (**b**) a schematic picture of the chapel, to be superimposed on geophysical maps; N_s—North direction as given by Gerevich (1979), N—the true North direction

very typical of the Roman epoch (Fig. 10.2). According to the original description, the chapel should have a width of 5.5 m, and a length of 11.2 m including an apsis on the opposite side of the entrance. The building is divided by an inner crossing wall into two parts. By the time of the excavation the entrance had already been destroyed. In the middle part of the apsis there was an empty grave, hardly distinguishable from the floor level. The frescoes were characteristic of the fourth century. No other graves were found around the chapel. On the surface limestone debris, embedding mortar and small fresco remnants were found. (82 years later we also found such remnants.) Nowadays, the wall remnants are supposed to be at a depth of 0–1 m covered by clayey-sandy sediments. According to the original document, the chapel is W-E oriented (mistakenly, see in Sect. 10.3.1), as shown by N_s in Fig. 10.2a. The topography is relatively flat (with a maximum dipping of 4° in SE direction), and there are no geomorphologic indications for the buried chapel.

10.3   The Standardized Pricking Probe Technique

It is an old experience that the electrodes used in geoelectrical prospection sometimes hit some debris in the soil. Although some pricking techniques have been known for a long time (e.g. in tomb looting and in searching for survivors of avalanches), as far as we know, the first attempt to get objective and quantitative information by using this apparently intuitive searching approach was performed by Szalai et al. (2009).

Systematic PP measurements can be carried out either along a profile, or over an area. A metal rod with a sharp peak, which can be an ordinary electrode used in geoelectric prospecting (as shown in Fig. 10.3.), is pushed into the soil to a given constant (e.g. 20–30 cm) depth. It either penetrates into the soil gently or sticks in. In the first case we assign to the pricking probe location a value of $k = 0$, while in the second case a value of $k = 1$ is assigned.

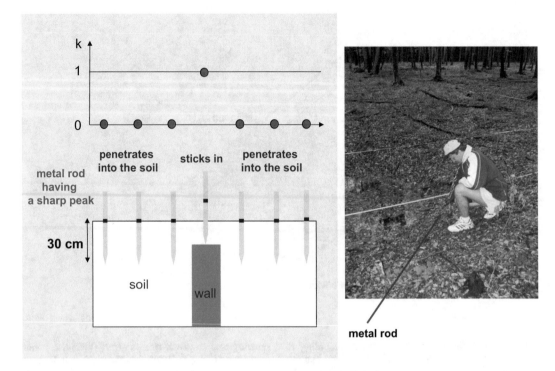

Fig. 10.3 Scheme of the pricking probe and a photo showing its field application

This technique is sensitive, of course, to localized, random features. A more robust parameter can be obtained evidently by using a suitable average of the k value. As we found, the arithmetic mean of the obtained k values in five/nine neighbouring points (k_5, in the case of profile-, k_9 in the case of areal measurements) largely eliminates random features. In order to reduce the effect of roots of living trees, it is advisable to divert the profiles around such trees for a distance of at least 0.5–1 m, if it is possible at all. In the case of very detailed areal measurements (e.g. in detailed archaeological investigation) this is not an option. Instead of doing this, the coordinates of living trees should be carefully recorded, and the results over the critical area (close to the tree) should be critically taken into account.

10.3.1 Pricking Probe Measurements in Sárisáp

Knowing that the chapel had already been excavated and then re-buried more than eight decades ago, one may assume that its walls are situated very close to the surface. We applied a penetration depth of 30 cm, which is a reasonable compromise. A larger depth of penetration would have needed higher pushing force.

Through some test measurements both the optimum interval between the neighbouring pricking probes and the optimum depth were determined. In the case of a reconnaissance (overview) mapping, a grid distance of 1 m was applied. In Fig. 10.4a, showing the whole area, the black squares indicate the rock hits. Based on this test it was possible to reduce the area for further investigation (see Fig. 10.4b).

During the first, more detailed measurements a grid distance of 50 cm and a depth of 30 cm were applied, the results of which are shown in Fig. 10.4b. A 9 point average, that is the mean value of the point and its 8 nearest neighbours, in Fig. 10.5 is even more informative. This figure already contains the tree root correction: $k = 0$ values were put inside of a 0.5 m radius around the trees.

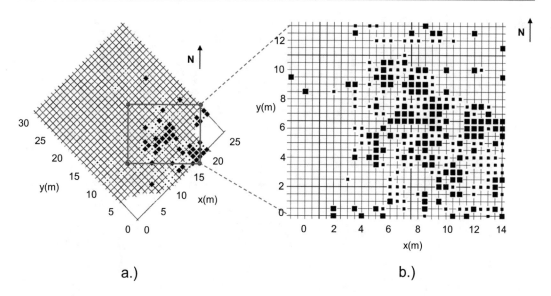

a.)   b.)

Fig. 10.4 The pricking probe results. (**a**) the whole area with a horizontal interval of 1 m, (**b**) the reduced area with a horizontal interval of 50 cm. Filled square: rock-hit sites (where $k = 1$)

As it is shown in Fig. 10.5, the pricking probe technique provides information rather about the whole area occupied by the chapel than just about its walls, because the effect of standing walls cannot be distinguished from that due to rock- and debris piles. The presence of debris was

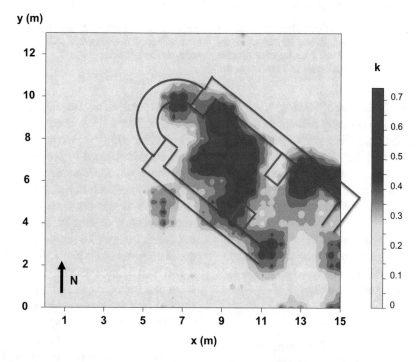

Fig. 10.5 Filtered isoline maps of the root-corrected PP results applying a penetration depth of 30 cm and horizontal grid distance of 50 cm, together with the layout of the chapel. As shown in Fig. 10.4, $+y$ in Figs. 10.5–10.9 points into North direction

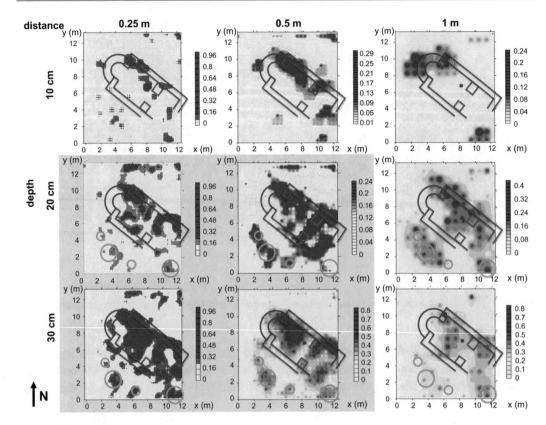

Fig. 10.6 Filtered isoline maps of the PP results applying various penetration depths (10, 20 and 30 cm) and horizontal grid distances (0.25, 0.5 and 1 m), together with the layout of the chapel and major tree- and bush colonies (large circles) and individual trees (small circle). The map is not corrected for roots. The scaling depends of depth- and horizontal distance

most likely within the area of the chapel, therefore the chapel could be correctly delineated, with the exception of the apsis and a part of the entrance.

Gerevich (1979) found the entrance already partially destroyed, but he did not note the state of the apsis. We are convinced of that the pricking probe technique was able to detect the yet standing walls.

Anyway, according to the PP survey—in conflict with the original document—the orientation of the chapel is NW-SE (and not N-S) directed, as it is shown (as N) in Fig. 10.2a.

10.3.2 Recommended Parameters for Detailed PP Survey

Further, systematic field studies were carried out in this area in order to be able to recommend

optimum parameters, lateral grid distance and penetration depth for detailed archaeological explorations. (In the reconnaissance survey, 1 × 1 m grid seemed to provide unambiguous indication about the presence and the approximate location of such objects.) We applied 0.25, 0.5 and 1 m as grid distances and, in each case, three penetration depths: 10, 20 and 30 cm were considered. Figure 10.6 shows the k_9 maps in the case of each variants without tree root correction.

The least favourable trial, namely the case of 10 cm penetration depth and 1 m grid resolution provided little information about the subsurface. Increasing either the pricking depth or sampling density, the subsurface features are gradually fading in. The NE wall becomes more or less visible in all other maps, but really useful results are only obtained if the grid distance is not larger than 50 cm and the pricking depth is not less than

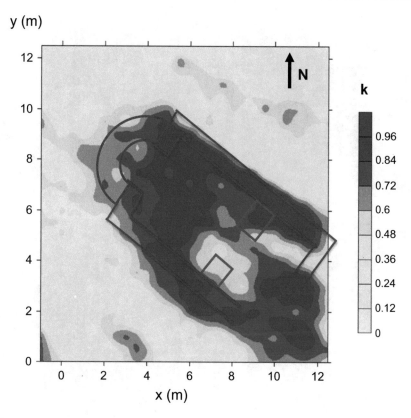

Fig. 10.7 Filtered isoline map of the PP results applying a penetration depth of 60 cm and horizontal grid distance of 50 cm, together with the layout of the chapel

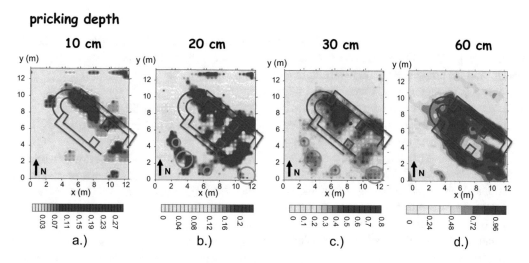

Fig. 10.8 Sounding with the PP technique with pricking (penetration) depths of 10, 20, 30 and 60 cm. The scaling is depth-dependent

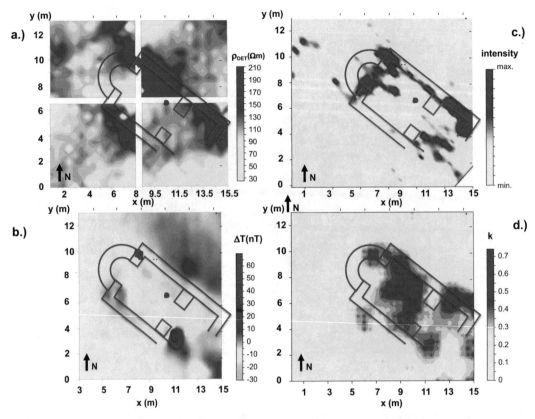

Fig 10.9 Results of three standard geophysical methods and of the pricking probe: (**a**) geoelectric method, (**b**) magnetic method, (**c**) georadar, (**d**) pricking-probe (same as in Fig. 10.5)

20 cm (see the four pictures in the left-bottom corner of Fig. 10.6). Further increase of the grid density and the pricking depth may result provide more detailed imaging, but it would need much more physical effort and more measuring time.

The optimum value of the pricking depth depends on the given conditions. In normal soils it is easy to reach a pricking depth of 30 cm, and usually it gives sufficient information. If the investigated object were laying at significantly larger depths, the uppermost part of the soil would be worthwhile physically removing, as it is usually done in all detailed archaeological investigation.

In the four lower left maps in Fig. 10.6 (25 cm and 50 cm grid distances, 20 cm and 30 cm penetration depths, with highlighted background) there is an anomaly-free zone inside the chapel area, elongating in SW-NE direction. This zone is

probably a former excavation trench, where the rocks were removed from.

10.3.3 Sounding Experiments with the PP Technique

We carried out a PP measurement with a pricking depth of 60 cm and grid distance of 50 cm (see Fig. 10.7, where root correction was applied). In Fig. 10.7 the contours of the chapel are clearly visible, including the apsis, which cannot be seen in any other map so distinctly. However, in the case of pricking depth of 60 cm the inner wall cannot be already seen as clearly as in the case of smaller pricking depths, since its effect becomes negligible compared to that due to the debris pile. Penetrating down to a depth of 60 cm needed already enormous physical effort. For such a deep or even deeper penetration, the application of a machinery is deemed more suitable.

Figures 10.6 and 10.7 tell us that the SW wall has remained; only its height is less than that of the NE wall. It is also clearly indicated, where the rock piles 1 and 2 (shown in Fig. 10.2) are located.

The standardized pricking probe technique in the same area with a few increasing pricking depths provides an approximate depth-dependent information (so-called sounding information), as shown in Fig. 10.8a–d. The visual information of the image set as a whole may enable one even to estimate the present state of the existing wall remnants.

10.3.4 About the Objectivity of the PP Measurements

The basic parameters are evaluated man-made, without any instrument implying a risk of subjectivity. In the most often, simplest case, when the mechanical resistance of the soil is low enough and there are no roots and ice pieces, it can be unambiguously distinguished, whether the probe hits rocks or not. Debris, ice pieces and tree roots should be handled as restrictive (but mostly not prohibitive) noises.

Subjectivity might be present if the pricking probe survey in the same area is done by more than one person, especially, when the mechanical resistance of the soil is relatively high. In our field of investigation the survey was done altogether by six persons, and there is no indication of subjectivity.

Fig. 10.5 was obtained with the same field parameters as one of the maps in Fig. 10.6 (pricking depth of 30 cm, horizontal interval of 50 cm), but only Fig. 10.5 is root-corrected. In spite of the fact that data displayed in Fig. 10.5 was acquired by a different team than those presented on maps in Fig. 10.6, there are no significant differences between the images.

It should be noted that such slight differences may occur with standard geophysical methods as well: the geophysical anomaly maps are not free of random and systematic noise. The objectivity of any technique should be judged by comparing

the resulting anomaly and investigated objects. Although the walls has not been unearthed, from the fact that the standardized PP technique provided more or less similar anomalies as the geophysical methods (see their results in details in Sect. 10.4.2), the objectivity of the standardized PP technique cannot be questioned.

10.4 Perspectives of the Pricking Probe Survey

10.4.1 Comparison with Standard Penetration Test (SPT[1])

The standardized pricking probe survey can be considered as a very special case of the so-called standard penetration test (SPT), a useful method in engineering geophysics (Lunne et al. 1997). The standard penetration test (SPT) is a dynamic penetration test designed to provide information on the geotechnical engineering properties of soil by recording among others the peak- and cone pressure values with depth. At the same time, the PP technique gives one single value about the peak pressure. Namely, it tells if the peak pressure is lower or higher than a certain threshold value. In the Sárisáp area this single parameter proved to be sufficient to achieve our goals. The PP technique is much more mobile, simpler and cheaper than the SPT, which may have limitations in accessing of field sites due to the truck and machinery. It is needless to use complicated techniques, if a simpler and cheaper one gives satisfactory results.

10.4.2 Comparison of the PP Technique with the Applied Geophysical Methods

For comparison, detailed geophysical (geoelectric, magnetic and georadar) measurements were also carried out. The technical details of the applied geophysical methods are summarized in the Appendix. We note that

---

[1] SPT = Standard Penetration Test

electromagnetic method such us e.g. the planshet induction electromagnetic survey (Atya et al. 2010) could have also been used in archaeological investigations which may even give higher resolution than GPR.

In the detailed geoelectric investigation two main, high-resistivity anomalies were found (Fig. 10.9a). The elongated one corresponds to the long NE wall, and a section of the apsis, while the shorter one corresponds to a part of the SE wall. The corner of NE and SE (entrance) walls is well-expressed, but in this map there are unfortunately some pseudo-anomalies, most likely due to geological inhomogeneities. Figure 10.9b shows the magnetic anomaly map. The contours of the chapel appear very ambiguously; the highest, small-size anomaly is likely due to some object in the soil having ferromagnetic properties. The georadar proved to be perfect in localisation of the chapel, and it provided a reliable picture on the NE wall and a section of the apsis (Fig. 10.9c, which presents one of the depth-slices), while the entrance wall, the SW wall and the SW part of the apsis are clearly shown. The georadar map was able to detect the grave as well, while the inner wall is not visible at all. In Fig. 10.9d the PP map is shown (the same as in Fig. 10.5).

Due to fact that the susceptibility contrast between the soil and the wall-building material (travertine limestone) is very weak, the least information was obtained from the magnetic method (Fig. 10.9b). The geoelectric maps (in Fig. 10.9a) indicated more or less clearly the main walls, but the results are strongly distorted by geological noise, i.e. local inhomogeneities. The georadar (in Fig. 10.9c) provided excellent, noise-free pictures, but it lacks some details. At the same time the PP map (Fig. 10.9d) contains a lot of details (see below).

The three geophysical maps and also the PP results confirm the existence of the NE wall and a part of the SW wall, thus providing a correct orientation of the chapel. The georadar and PP maps are less distorted, while the electric and magnetic maps are more distorted by noise.

Some further details are exclusively visible in the PP maps (Fig. 10.9d) such as: (1) the SW wall

is in a much worse condition than the NE wall: it is discontinuous, and not as high as the NE wall, (2) the large part of the apsis with exception of its northern section is destroyed, (3) there is a substantial amount of debris between the apsis and the inner wall, (4) the chapel could have been crossed by excavation, where the rocks were removed from. By applying a larger pricking depth (see Fig. 10.7), even more details would have been revealed.

It is worth mentioning that while the geoelectric and georadar maps indicate primarily the main subsurface bodies (the wall remnants), the PP indicates the near-surface rock distribution, independently of the fact whether it belongs to an intact wall or to a rock pile. Such rock piles below a certain size are hard to detect by geoelectric or georadar mappings. Therefore the pricking probe provides complementary information to the results obtained by the conventional geophysical methods.

In addition, a direct comparison between the PP technique and the georadar, the two most informative tools on the study area, was performed. The field work with the GPR method is evidently more elegant and faster than that with the PP technique. At the same time the PP method has also several advantages over GPR: it does not require a mopped-out area; it can be used among extreme field conditions including bad weather and topography; wet or clayey soil does not make difficulties; field work can be done without any instrument; data acquisition and interpretation is very simple. Finally the PP maps are completely free of the measuring direction. Note that the georadar anomalies in Fig. 10.9c are somewhat elongated in the same (measuring) direction.

10.4.3 Weaknesses of the PP Technique

After presenting the strengths of the PP technique, its weaknesses and possibilities for the elimination of these weaknesses are summarized.

1. A major demerit of the standardized PP method is a possible damaging of sensitive structures below the surface.
2. Sub-horizontal roots of living trees may produce pseudo-anomalies. Elimination of such problems requires either a slight diversion of the measuring profiles (from the trees), or omitting the suspected values.
3. Subjectivity of data acquisition, which may eventually occur when large physical efforts are needed. The subjectivity can be handled as a signal/noise problem.
4. Penetration into compact, hard soils or penetration to larger, e.g. 60 cm depth may need too large physical efforts. This problem could be solved only by using machinery designed to perform PP measurements. However the discussion of this is beyond the scope of the present paper.
5. Posterior debris displacements may distort the interpretation. Very significant posterior debris displacement may, of course, prohibit any reasonable interpretation.

10.5   Summary

The PP technique and its application to archaeological prospecting have been discussed in the present study. Successful reconnaissance measurements were carried out in the case of dense undergrowth, which would have been hardly possible with standard geophysical methods. In the detailed investigation the pricking probe proved to be also competitive. Only the georadar was able to provide high quality images about the chapel walls, but some details are not clearly seen, or can hardly be recognised in contrast to the PP maps. The imaging power of the PP technique, especially with a 60 cm pricking depth, proved to be superior to that of the applied geophysical methods.

It is also possible to carry out PP measurements with a few different pricking depths in the same area. It is called PP sounding, and, as we have found, it is especially useful.

Among common soil conditions the method can be universally applied, except in presence of significant posterior debris distribution. For a deeper insight it is advisable to remove the uppermost soil layer.

For archaeological reconnaissance measurements, i.e. to find buried elongated objects with a width of about 50 cm a rectangular grid with an interval of 1 m is recommended, while for a detailed survey a rectangular grid with an interval of 50 cm is advisable.

The main advantages of the pricking probe method are as follows: field procedure and data processing are simple, cheap and quick; it does not need electronic instrument, just a metal rode, therefore there are no investment costs, and there is no chance of technical failure; the method can be applied even among the most unfavourable field conditions as bad weather, extreme topography, dense undergrowth, etc.; it is possible to elaborate its sounding variant; it provides complementary information to the geoelectric and georadar maps; it may show some details remained hidden using standard geophysical methods

In this paper altogether four independent methods/techniques are inter-compared over the buried object in a given field. The archaeological results, which can be truly verified only after the chapel will be unearthed, are as follows: (1) localization and orientation of the chapel is not N-S as it stands in the old documents, but it is NW-SE oriented; (2) there is a still existing massive NE wall; (3) the SE wall is not as high as the NE wall; (4) the large part of the apsis except of its northern section has been destroyed; (5) the northern part of the inner wall still exists; (6) there can be found a large amount of debris between the apsis-closing wall and the inner wall and also in the SE corner of the chapel; (7) the chapel was most probably crossed by a shallow excavation trench, where the rocks were removed from; (8) there are no other archaeological objects in the investigated area. Most of these results come from the PP measurements, and sometimes they can be best recognised in PP results.

On the basis of our results the standardized PP technique in archaeological prospection is a useful complementary tool, of standard geophysical methods. It is important to add that it should be applied only in areas, where the possible damaging of the buried structures is excluded.

Acknowledgements Hungarian Research Fund (K049604 and 61013), Bolyai Scholarship of the Hungarian Academy of Sciences (Sándor Szalai). Field assistance: András Koppán, Attila Novák, Krisztina Rokob, János Túri, Árpád Kis, Mihály Varga; also Gábor Gombás and Kitti Szokoli (students of University of West-Hungary). Georadar data processing was made by Boriszláv Neducza (ELGI); geoelectric data processing was made by Attila Novák. Local technical support was provided by Péter Vincze, Károly Kollár and János Csicsmann. Comments by Antal Ádám, Gábor Újvári and the Reviewers were very helpful.

Appendix

Description of the Applied Standard Geophysical Methods

Geoelectric Mapping

We used the tensor-invariant method by Varga et al. (2008), which had already been successfully applied in another archaeological investigation. In this three-dimensional approach the detection of walls is not influenced by the current direction, an inherent problem of directional (two-dimensional) measurements. Somewhat similar three-dimensional techniques were described by Papadopoulos et al. (2006). Di Fiore et al. (2002) realised another tensor-invariant technique based on probability tomography measurements. Drahor et al. (2007) carried out a full three-dimensional electrical resistivity tomography.

In the field, the current electrode distance was fixed to 15 m. Two perpendicular AB directions were used, and $16 \times 15 = 240$ potential electrodes with an equidistant space of $\Delta x = \Delta y = 50$ cm were put in the central nearly square, 7.5 m $\times 7$ m area between the current electrodes. Due to a four-channel measuring system, it was possible to determine both components of a horizontal electric vector simultaneously. The time needed to measure all potential differences between the neighbouring potential electrodes, thus to obtain

$15 \times 14 = 210$ resistivity tensors, was about 40 min. The data processing method is described in details by Varga et al. (2008). In this paper the authors presented the areal distribution of one of the invariants of the apparent resistivity tensor, namely the determinant (Fig. 10.9a.). The actual area of investigation in Sárisáp could be completely covered by four layouts.

Magnetics

The magnetic mapping is a classical technique of archaeological explorations (Clark 1990). For a recent example see e.g. Piro et al. (2007). Magnetic and geoelectric techniques were jointly used by Neubauer and Eder-Hinterleitner (1997).

The measurements were carried out by GSM-19 v7.0 Overhauser magnetometer. The time variation of the geomagnetic field was corrected by using a base station.

Georadar

The geo-probing radar (GPR) has been recently successfully applied in an archaeological study by Bonomo et al. (2009). Capizzi et al. (2007) compared two-dimensional electrical resistivity tomography (2D ERT) and GPR results, while in the archaeological study by Cardarelli et al. (2008) magnetic, geoelectric and georadar results were inter-compared.

The first georadar measurements were carried out by means of a Noggin 250 MHz instrument. Based on these measurements a high-resolution georadar map was prepared by using a 450 MHz PulseEKKO-1000 instrument with 5 cm sampling along the measuring profiles, and 1 m distance between the neighbouring profiles.

Time Requirements of the Methods

For the detailed measurements, the pricking probe technique needed less than 1 day for two people (0.5 m lateral distance, 30 cm depth). The geoelectric measurements, with a staff of four persons needed 1 day, the georadar measurements were completed by two geophysicists within half a day, and the magnetic measurements needed 1 day for two.

References

Atya MA, Khachay OA, Soliman MM, Khachay OY, Khalil AB, Mahmoud G, Shaaban FF, Hemali IA (2010) CSEM imaging of the near surface dynamics and its impact for foundation stability at quarter 27, 15th of May city, Helwan, Egypt. Earth Sci Res J 14(1):76–87

Bonomo N, Cedrina L, Osella A, Ratto N (2009) GPR prospecting in a prehispanic village, NW Argentina. J Appl Geophys 67:80–87

Capizzi P, Cosentino PL, Fiandaca G, Martorana R, Messina P, Vassallo S (2007) Geophysical investigations of the Himera archeological site, northern Sicily. Near Surf Geophys 5:417–426

Cardarelli E, Fischanger F, Piro S (2008) Integrated geophysical survey to detect buried structures for archeological prospecting. A case history at Sabine Necropolis (Rome, Italy). Near Surf Geophys 6:15–20

Clark AJ (1990) Seeing beneath the soil. B.T. Batsford, London

Di Fiore B, Mauriello P, Monna D, Patella D (2002) Examples of application of tensorial resistivity tomography to arcitectonic and archeological targets. Ann Geophys 45:417–429

Drahor MG, Göktürkler G, Berge MA, Kurtulmus TÖ, Tuna N (2007) 3D resistivity imaging from an archeological site in South-Western Anatolia, Turkey: a case study. Near Surf Geophys 5:195–202

Gerevich L (ed) (1979) Archaeological topography of Hungary 5. Akadémiai Kiadó, Budapest. (in Hungarian)

Lunne T, Robertson PK, Powell JJM (1997) Cone penetration testing in geotechnical practice. Blackie Academic/Chapman & Hall, E&FN Spon, London, 312 pp

Neubauer W, Eder-Hinterleitner A (1997) Resistivity and magnetics of the Roman town Carnuntum, Austria: an example of combined interpretation of prospection data. Archaeol Prospect 4:179–189

Papadopoulos NG, Tsourlos P, Tsokas GN, Sarris A (2006) Two-dimensional and three-dimensional resistivity imaging in archeological site investigation. Archaeol Prospect 13:163–181

Piro S, Sambuelli L, Godio A, Taormina R (2007) Beyond image analysis in processing archeomagnetic geophysical data: case studies of chamber tombs with dromos. Near Surf Geophys 5:405–416

Szalai S, Lemperger I, Pattantyús ÁM, Szarka L (2009) Pricking probe as a complementary technique in archeological prospecting, Proceedings of Near-Surface Conference 2009, A03, Dublin

Varga M, Novák A, Szarka L (2008) Application of tensorial electrical resistivity mapping to archeological prospection. Near Surf Geophys 6:39–47

Geophysical Imaging of an Early Nineteenth Century Colonial Defensive Blockhouse: Applications of EM Directionality and Multi-parameter Imaging

11

David C. Nobes and Lynda R. Wallace

Abstract

In 1845, the French navy built three blockhouses as part of their defence of French settlers in Akaroa, located on Banks Peninsula, near Christchurch, New Zealand. In the late 1860s, the blockhouses were removed and the timber used for other purposes. Two of the blockhouses were situated at either end of Akaroa township; their locations are well known and documented. The position of the third, in the village of Takamatua, near Akaroa, was not well known, but was thought to have been situated in what became a public reserve, first known as the Blockhouse Domain and more recently as the Takamatua Domain.

To support local archaeological studies, non-invasive, non-destructive geophysical imaging was carried out across the Takamatua Domain. We used a complementary combination of horizontal loop electromagnetic (HLEM), total field magnetic, ground penetrating radar (GPR), and tomographic elec-trical imaging (EI) techniques. We expected that little if any of the blockhouse itself would remain. However, the nature of the construction was such that we expected to find the defensive trench that enclosed the blockhouse.

Like many geophysical techniques, HLEM methods exhibit directionality in their responses, i.e. the response depends on the orientation of the instrumentation relative to the target orientation. This directionality can be used to enhance the detection of linear features. When we combined our suite of com-plementary geophysical methods, we identified coincident linear anomalous responses. The clear and unequivocal results of the geophysical surveys indicate that the blockhouse and its surrounding trench or moat have been found.

11.1 Introduction

In the 1840s, the French navy built three blockhouses as part of their defence of French settlers in the Akaroa area (Fig. 11.1 from Maling 1981; Jacobson 1940; Walton 2003; Tremewan 1990, 2010), which is located on Banks Penin-sula, near Christchurch, New Zealand (Fig. 11.2, inset). In the 1860s, the blockhouses were removed and the timber used for other purposes. Two of the blockhouses were situated in the

D. C. Nobes (✉)
School of Geophysics and Measurement-Control Technology, East China University of Technology, Nanchang, Jiangxi, China
e-mail: david.nobes@canterbury.ac.nz

L. R. Wallace
Akaroa Museum, Akaroa, New Zealand

© Springer International Publishing AG, part of Springer Nature 2019
G. El-Qady, M. Metwaly (eds.), *Archaeogeophysics*, Natural Science in Archaeology,
https://doi.org/10.1007/978-3-319-78861-6_11

Fig. 11.1 Rough historical sketches of the French defensive blockhouses constructed in the mid-nineteenth century at Akaroa, New Zealand (From Maling 1981)

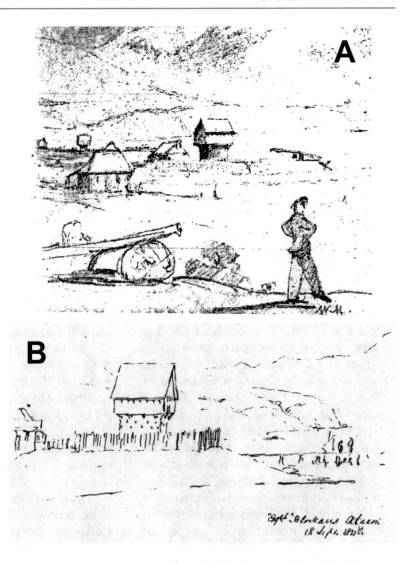

Akaroa township, and the locations are well known and documented (Jacobson 1940; Maling 1981; Walton 2003). The position of the third, in the village of Takamatua, is not well known, but is thought to have been sited in what became a public reserve, first known as the Blockhouse Domain and more recently as the Takamatua Domain (Fig. 11.2), near Akaroa.

We expected that little if any of the blockhouse itself would remain. However, the nature of the construction was such that we expected to find the defensive trench that enclosed the blockhouse. The first storey of the blockhouse would have been of the order of 6 m², and the second storey overhung the first by 1–2 m. The trench was 1.5–2.5 m across, and the soil removed was

banked up against the blockhouse (Walton 2003). Thus the blockhouse footprint would have been 8–10 m, and the distance from trench centre to trench centre would have been of the order of 10–13 m. The entire site would have been about 15 m square.

To aid local archaeological studies, non-invasive, non-destructive geophysical imaging was carried out across the Takamatua Domain (Fig. 11.3). A combination of horizontal loop electromagnetic (HLEM), total field magnetic, ground penetrating radar (GPR), and tomographic electrical imaging (EI) techniques were used. The HLEM results had to be processed to allow for the directionality of the response, which yielded a clear target. The combined multi-parameter

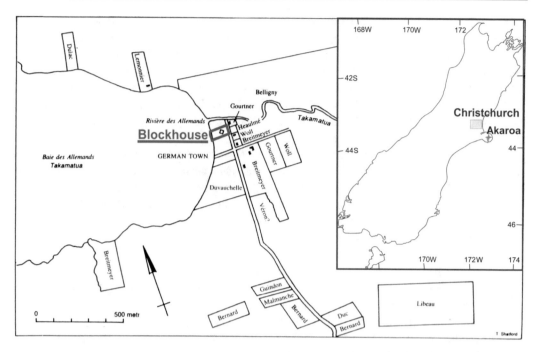

Fig. 11.2 Map showing the Takamatua blockhouse location as inferred from contemporary accounts. The blockhouse orientation as shown in the figure was arbitrary (Tremewan, personal communication).The area shown in Fig. 11.2 (labelled "Blockhouse") is outlined. Adapted from Tremewan (1990). (Inset) Akaroa township is on the east coast of the South Island of New Zealand

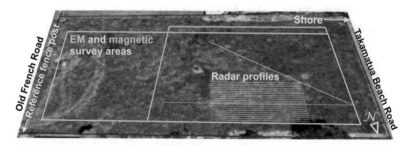

Fig. 11.3 Perspective view of the Takamatua survey, as viewed from the north. The shore is at the right. The circular track from Old French Road (at the left) is obvious. Residents drove their vehicles in at one corner, dropped rubbish (e.g. tree cuttings), and exited from the other corner. The areas covered by the electromagnetic (pale orange) and magnetic (pale blue) surveys are marked in the corresponding colours. The 3D ground penetrating radar profiles are shown (marked "Radar profiles"). The two long ground penetrating radar lines and one of the short ground penetrating radar lines were also used for electrical tomographic imaging

geophysical data sets yielded coincident linear anomalous responses that are consistent with the targeted blockhouse features. The individual anomalous responses are clear, however the combination of all of the anomalous geophysical responses yields unequivocal results that indicate that we have located the blockhouse and its surrounding trench or moat.

11.2 Survey Design and Results

Ultimately, the geophysical response arises primarily from soil disturbance, as in the construction of trenches or of a significant timber building. The stratigraphy and soil structure are destroyed, as different soils are mixed or compacted, and layer continuity is truncated. The soil porosity and consequently its water content are altered, and while the average water content can eventually recover over time, the natural variability of the water content in the soil column is irrevocably altered.

By considering the physical properties that can be affected by the blockhouse and its surrounding trench, we can choose the most appropriate geophysical methods. In general, finer-grained soils, i.e., clays, have greater concentrations of magnetic minerals and elevated electrical conductivities; if clayey soils are then mixed in with other materials, the bulk magnetic and electrical properties of the existing layers will be affected. The water content and water salinity are also two of the primary influences on the electrical conductivity and dielectric properties of the soil.

Thus, magnetic, electrical, electromagnetic (EM) and ground-penetrating radar (GPR) methods are the most likely to yield anomalous responses, and these were the four methods used in the blockhouse survey (Fig. 11.3):

1. Horizontal loop electromagnetic (HLEM) profiles were first gathered across the site. HLEM is a fast and efficient way to cover a large area and quickly identify those areas that may be anomalous. The results show the anomalous responses in map view.
2. Then a total field magnetic survey was completed on the portion of the site where the EM response appeared to be anomalous. As with HLEM, the results are shown in map view.

3. Ground penetrating radar (GPR) cross-section profiles and three-dimensional (3D) imaging were acquired across the area that had an anomalous EM response.
4. Finally electrical imaging (EI) was completed along selected GPR lines, for comparison and correlation with the other results.

Together, the collective interpretation of the individual results yields a more complete picture of the target features, and provides greater confidence in the interpretation of the results and the location of the Takamatua blockhouse.

11.2.1 Horizontal Loop Electromagnetic Survey

Electromagnetic (EM) methods are non-contact techniques, unlike resistivity methods which require direct physical contact with the ground. Thus, EM methods are well suited to rapid non-invasive surveys of archaeological sites. The specific instrument selected was a Geonics EM31™, which is a horizontal loop EM (HLEM) system. The basic principles of HLEM are illustrated in Fig. 11.4, and are described in more detail by McNeill (1980, 1990) and Reynolds (1997). A transmitter loop antenna (coil) sends a time-varying magnetic field (the "primary field") into the ground, usually at a specific-frequency. The time-varying field induces electric currents to flow; these currents generate a secondary field which in turn generates a response

Fig. 11.4 Schematic diagram illustrating the basic concepts of electromagnetic induction. A transmitting coil produces a time-varying EM field which induces electric currents to flow in the ground. These subsurface currents in turn generate a secondary EM field which is measured by the receiving coil. (From Kearey and Brooks 1991)

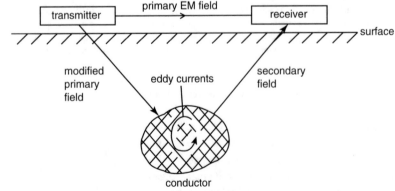

in a receiving coil. This secondary response arises from a volume of the subsurface immediately beneath and surrounding the HLEM instrument, and the strength of the response depends on the electrical conductivity of the ground. The depth of penetration is a function of the transmitter-receiver coil spacing and the electrical properties, but for most surveys, and indeed for the Takamatua site, the electrical conductivity is small enough so that the determining factor is the coil spacing. For the EM31, the coil spacing of 3.66 m yields a depth of penetration of approximately 5.5 m, with a peak sensitivity at about 1 m depth.

The secondary response can be separated into two components or modes. The quadrature response yields a measure of the bulk electrical properties of the ground, usually expressed as an apparent electrical conductivity in millisiemens per metre (mS/m). The in-phase or real response is, in effect, a metal detector mode, and is usually expressed in units of parts per thousand (ppt) of the primary field. As the electrical conductivity increases, so too, initially, do the quadrature and in-phase responses. At a conductivity that varies for each instrument, the quadrature response peaks and then decreases, whereas the in-phase response continues to increase to the so-called "inductive limit" (McNeill 1980, 1990). Thus, it is important for any survey to measure both quadrature and in-phase responses. In the presence of metal, the in-phase response will always be high, but the quadrature response can yield high or low apparent conductivities, depending on the volume and proximity of metal present. In the case of the Takamatua site, there was little metal present, except in fences or power lines along the site boundaries, and thus only the quadrature response yielded useful information.

The EM31 readings were acquired using 1 m line spacings with 1 m station spacings, and were taken using two orientations, one parallel to the survey line and the other perpendicular. This yielded a 1 m by 1 m grid. The readings were compiled using Excel™ and plotted using Surfer™. The oriented readings are usually averaged to yield an overall image of the variations of the bulk electrical conductivity (Fig. 11.5), which reveals a large-scale trend across the site. The conductivity is highest at the shore, where the pore waters have the highest salinity and thus the highest conductivities. The conductivity decreases away from the shore. There are subtle patterns that may or may not indicate potential targets of interest, and larger anomalous responses due to power lines and metal fences along the northern boundary. The real response is superimposed as contours on the colour image of the quadrature response. As for the quadrature response, there are subtle features, but nothing that stands out as obviously anomalous, aside from the power lines and fences.

11.2.2 HLEM Directionality

The horizontal-loop electromagnetic (HLEM) response is sensitive to the orientation of the instrument relative to buried objects. Linear features have

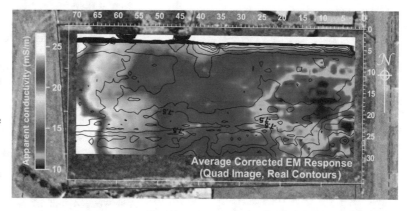

Fig. 11.5 The average EM31 quadrature readings taken with the instrument oriented parallel and perpendicular to the survey lines. The map is rotated to have north at the top, by convention, so the shore is now to the left. The average conductivity is dominated by increasing conductivity towards the shore, because of the higher salinity of sea water

maximum responses when they are oriented perpendicular to the long axis of the HLEM instrument (McNeill 1990; Frischknecht et al. 1991), especially for features with relatively sharp boundaries, such as pipes and archæological targets. The EM31 response is maximal for linear features with relatively sharp boundaries and that are perpendicular to the orientation of the instrument (Nobes 1999, 2007). HLEM responses are thus often acquired using two orientations, parallel and perpendicular to the survey line direction, as described in the previous section, and the responses are then combined, usually by summing or averaging the bidirectional responses (e.g., Fig. 11.5).

If we define $\sigma_\parallel$ as the apparent conductivity when the EM31 instrument is oriented parallel to the survey lines, and $\sigma_\perp$ is the apparent conductivity when the EM31 instrument is oriented perpendicular to the survey lines, then we normally take the average of the two:

$$\langle\sigma\rangle = \sigma_\parallel + \sigma_\perp. \qquad (11.1)$$

If the ratio or the difference between the two responses are used instead, the contrast between the directionality of any linear anomalies is emphasised. If we take the difference for example, then we can define it in any one of three ways:

$$\delta\sigma_1 = \sigma_\parallel - \sigma_\perp \quad \text{or} \qquad (11.2a)$$

$$\delta\sigma_2 = \sigma_\perp - \sigma_\parallel \quad \text{or} \qquad (11.2b)$$

$$|\delta\sigma| = |\delta\sigma_1| = |\delta\sigma_2|. \qquad (11.2c)$$

Alternatively, we can define the ratio of the quadrature response as:

$$R_1 = \sigma_\parallel/\sigma_\perp \quad \text{or} \qquad (11.3a)$$

$$R_2 = \sigma_\perp/\sigma_\parallel \quad \text{or.} \qquad (11.3b)$$

It doesn't matter which of these definitions we choose, because we note that $\delta\sigma_2 = -\delta\sigma_1$, and $R_2 = 1/R_1$. The only difference will be the sign of the difference, or whether the anomalous ratio is significantly greater or less than 1.

The ratio of the quadrature response between the parallel and perpendicular readings has been found to be particularly sensitive to linear features (Nobes 2007). An example is shown in Fig. 11.6. A soccer field at the University of Canterbury Recreation Centre was crossed in one corner by what may be an old buried stream channel and by a concrete conduit at about midfield. At each survey point, EM31 readings were taken parallel and perpendicular to the survey line, yielding responses optimally for linear objects oriented respectively perpendicular and parallel to the line. The parallel and perpendicular responses (Fig. 11.6a, b) show the possible buried channel and indications of criss-crossing features. The average (Fig. 11.6c) emphasises the larger scale features; the metal goal posts are also obvious in the top left of Fig. 11.6. In contrast, the quadrature ratio (Fig. 11.6d) highlights the criss-crossing buried irrigation pipes, and the large scale features have essentially disappeared because they have gradational boundaries, like along an old buried stream channel for example. The responses in the two directions are thus the same (Nobes 1999), and any difference or ratio essentially sees those large scale responses as part of the background.

When we apply this process to the EM31 results from Takamatua, the quadrature ratio reveals a rectilinear feature (Fig. 11.7) that became the focus of our work.

11.2.3 Total Field Magnetic Survey

Total magnetic field surveys use the Earth's magnetic field as the source. Local variations in the content of magnetic minerals, such as magnetite, can cause anomalous responses to the background magnetic field (e.g., Milsom and Eriksen 2011). A local concentration of magnetic minerals will have a field aligned to and superimposed on that of the Earth's field (Fig. 11.8). Thus a survey of the local variations in the total magnetic field can indicate the locations of significant anomalous concentrations of magnetic minerals. At the Takamatua site, the magnetic field inclination is steep, of the order of 68 degrees, and the positive anomalous response will be more centred over the anomalous target. However, steel fences and power lines can also cause anomalous magnetic responses, and to minimise the effects of such external influences, the magnetic survey was restricted to the western half of the Takamatua Reserve site, that is the portion of the site closer to the shore.

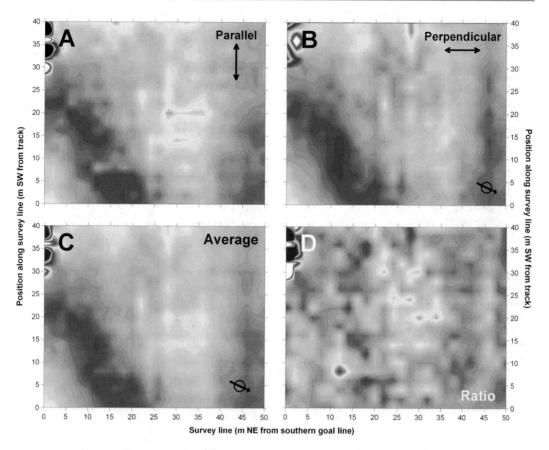

Fig. 11.6 Parallel (**a**, upper left), perpendicular (**b**, upper right), average (**c**, lower left), and ratio (**d**, lower right) EM31 quadrature responses for the Recreation Centre soccer pitch. The oriented and average responses (**a–c**) are dominated by the goal posts (upper left) and the subsurface paleochannel (lower left). The grid of irrigation pipes is visible but largely masked in the raw results. The ratio (**d**) subdues the large scale features and enhances the response of the buried grid of pipes

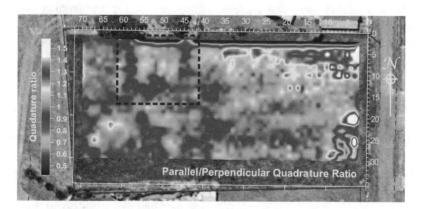

Fig. 11.7 The EM31 quadrature ratio highlights what appears to be a square feature (bold dashed box) that was investigated further using magnetic, ground penetrating radar, and electrical imaging. The conductive sea response is gone because the oriented readings are essentially the same

Fig. 11.8 Schematic diagram illustrating the basic concepts of magnetic surveying. The Earth's magnetic field interacts with local concentrations of magnetic minerals (**a**, top), generating a magnetic dipole field that is aligned with the Earth's field. This localised anomalous dipolar field yields, at the surface, changes in the total magnetic field magnitude that can be both aligned with and thus enhance the Earth's field, and elsewhere be opposite to the Earth's field. At mid-latitudes, measurements of the local total magnetic field (**b**, bottom) yield a response in the presence of an anomalous subsurface feature that has a dipolar (+/−) character. At higher magnetic latitudes, the positive anomaly will be more centred over the anomalous target

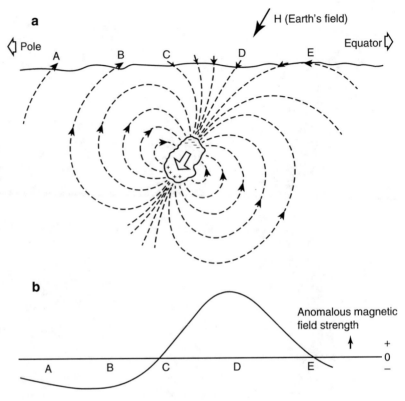

Readings were taken every metre in the east-west direction and 0.5 m in the north-south direction, using a Geometrics G856 proton-precession magnetometer/gradiometer. Each line was surveyed twice, once along the line and again in reverse to the start of the line. The repeated readings allowed us to check for short-period diurnal variations in addition to the base station readings taken at regular intervals. As for the EM readings, the results were compiled in Excel™, the diurnal variations removed, and the results were plotted using Surfer™. The total magnetic field results are coarse but show the relatively anomalous responses in the same locations as for the HLEM results (Fig. 11.9). The EM anomaly aligns with slightly positive magnetic responses, as indicated by the thin black line in Fig. 11.9a.

11.2.4 Ground Penetrating Radar

The basic principles of ground-penetrating radar (GPR) are well covered elsewhere (e.g. Davis and

Annan 1989; Field et al. 2001; Jol and Bristow 2003), and will only be summarised here. The essential elements of GPR are schematically represented in Fig. 11.10a. A laptop is used for system control and data storage. A radar pulse generated at the transmitting antenna travels through the air (the direct air wave), the ground (the direct ground wave), and the subsurface, where reflections ("echoes") are generated at changes in the physical properties, e.g. as at the water table (Fig. 11.10b). The direct and reflected signals are detected at the receiving antenna, and are recorded in the control computer. The speed at which the signals travel are primarily a function of the water content. (Davis and Annan 1989; Field et al. 2001). The subsurface boundaries arise due to changes in the velocity of propagation of radar, and to a lesser extent due to changes in the electrical properties.

Two long GPR profiles were gathered across the anomalous features to confirm their presence within a stratigraphic context, and then a set of closely spaced profiles were acquired to yield a

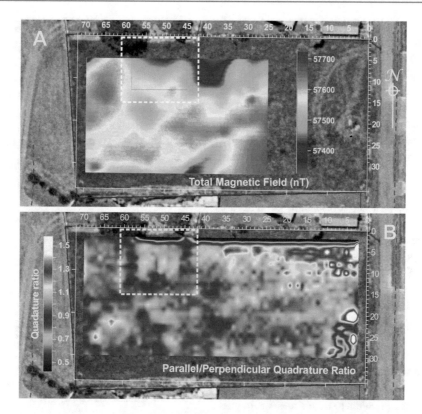

Fig. 11.9 The total magnetic field results (**a**), corrected for diurnal drift, superimposed on the Takamatua airphotos and compared with the EM31 quadrature ratio (**b**). The magnetic field has many features but in particular there is an anomalous response in approximately the same location as the EM ratio anomaly (dashed rectangle). The EM anomaly location is shown as a thin black line on the magnetic results (**a**)

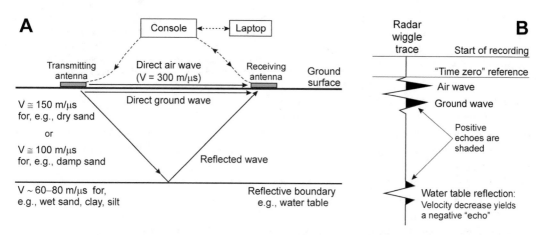

Fig. 11.10 (**a**) Basic principles and components of a ground penetrating radar (GPR) system. (**b**) The signal wavelet is not a perfect impulse "spike" but rather has a structure to it. The most common wavelet has a small negative, large positive, and small negative, and closely spaced signals can superimpose, as for the direct air and ground waves. The water table is a common strong sub-surface reflector, where the radar velocity decreases sharply, yielding a reversed "echo". Individual traces are plotted side-by-side to yield a radar cross-section or profile

three-dimensional (3D) image of the subsurface in an area containing the anomalous EM and magnetic responses. The line spacing was 0.5 m, and the trace spacing was 0.1 m using a Sensor & Software pulseEKKO 100A GPR system with 200 MHz antennas. The subsurface velocity was found to be approximately 100 m/μs (0.10 m/ns), which is typical of partly saturated sands and silts, the soils present at the site. The profiles were migrated and compiled into a 3D data cube (Fig. 11.11). Migration is a processing step that collapses scattering due to truncated bedding, for example at the edges of trenches, and places any dipping beds into their proper geometric positions, that is, deeper and steeper than the beds appear in the raw profiles. The beds appear shallower due to the geometry of reflections from dipping beds (e.g., Reynolds 1997; Yilmaz 2001).

The 3D data cube is a compilation of profiles so that subsurface reflections can be readily traced across the site. Such a data cube can be "sliced" in a variety of ways to reveal subsurface anomalous structures. The most common mode of "slicing" is to take cuts equivalent to time slices, i.e. to peel back the data cube layer by layer from the surface down. Each time slice is equivalent to a depth, once the travel time is corrected for the radar velocity of propagation.

The results for two different time slices (Fig. 11.11) reveal linear features that are coincident with the anomalies identified in the HLEM quadrature ratio (Fig. 11.7). The time slices are shown in "chair" mode to put them into the context of the data cube as a whole. The velocity used to convert time to depth was 0.10 m/ns, which is typical for partly saturated sands and silts. Note the linearly trending sub-parallel anomalous features near the left- and right-hand edges of the 3D cube (highlighted by the arrows in Fig. 11.11), which we interpret to be the defensive blockhouse trenches. The linear trends are coincident with features apparent in the cross-sections at the fronts of the data cubes (circled in Fig. 11.11). There also appears to be an almost square response in the middle of the GPR slices. The anomalous features appear to be at slightly different depths because material was dumped on the surface in the latter part of the twentieth century. The change in thickness of the fill material is

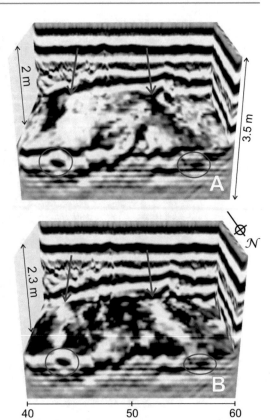

Fig. 11.11 The 3D GPR data cube, as viewed from the north, sliced at echo two-way travel times of (**a**, top) 40 ns, equivalent to a depth of 2 m, and (**b**, bottom) 46 ns, equivalent to a depth of 2.3 m. The data cube has not been corrected for the topography of the site, which dips gently to the west towards the shore (to the right). At the left side (east) of the cubes, the reflectors immediately above the linearly trending anomalies are actually almost horizontal but appear to trend downward because of the fill added to the site over the years

apparent along the back planes adjacent to the time slices, which are left visible to put the slices into their 3D context.

11.2.5 Electrical Resistivity Tomography

Finally, a tomographic electrical imaging (EI) survey was performed. EI is based on the simple principles of electrical resistivity surveying (e.g. Reynolds 1997; Milsom and Eriksen 2011). One pair of electrodes is used to inject current through the ground, thus turning the ground into

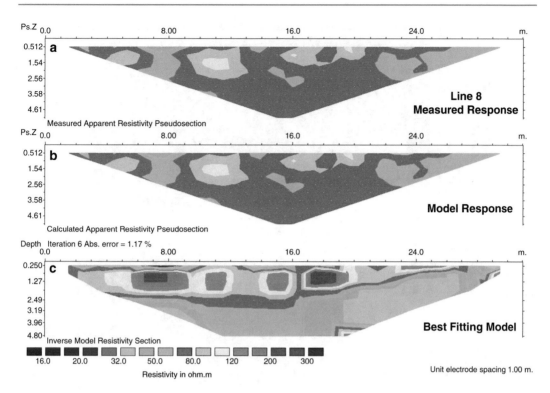

Fig. 11.12 Representative electrical imaging (EI) cross-sections for Line 8 across the site, showing (**a**) the raw measured response, (**b**) the response of the best-fitting model for comparison, and (**c**) the model that best fits the measured response. The best fit model minimises the misfit between the measure response and the model response, in this case approximately 1.2%, which indicates an excellent degree of fit. Note one set of anomalous features centred approximately at locations 7 and 18 m, at depths of approximately 1–2 or more m, and a second set centred approximately at 11 and 15 m. The results for the other lines were similar

a low-power electrical circuit. A second pair of electrodes is used to measure the voltage in that circuit, that is the force required to move the current in the ground, and by Ohm's law, the ratio of the voltage to the current yields the resistance in the circuit. That resistance, however, depends on the size of the circuit. Thus a scaling factor is applied to normalise the measurements and yield a value of the electrical properties, called the electrical resistivity, that is independent of the size of the circuit. The electrical resistivity can thus be directly compared from one material to another. Different electrode geometries can be used; the simplest is the Wenner array, where all the electrodes are evenly spaced, the two outer electrodes are the current electrodes, and the two inner electrodes are the voltage electrodes.

The modern variant of electrical resistivity surveying is EI. A line of electrodes is set out, and connected to a console switching unit, which

in turn is controlled by a computer, usually a laptop. In this case, a Campus Tigre multi-electrode resistivity system was used for EI imaging of selected GPR lines, for comparison and correlation with the other results. Each line consisted of 32 electrodes, using 1 m electrode separation, in a Wenner array configuration. The shallowest data are acquired using four adjacent electrodes, switched across the electrode string in sequence. The next set of data is acquired using every second electrode, as in 1-3-5-7, then 2-4-6-8, etc. The two long GPR lines were used for EI; an additional EI line was acquired to confirm and test repeatability of the EI results from line to line.

The raw EI results are averages of the electrical properties over the volumes through which the electric current flows. The results for Line 8 (i.e. 8 m from the northern boundary fence) are typical and are shown as an example (Fig. 11.12a). The deeper the current flows, the

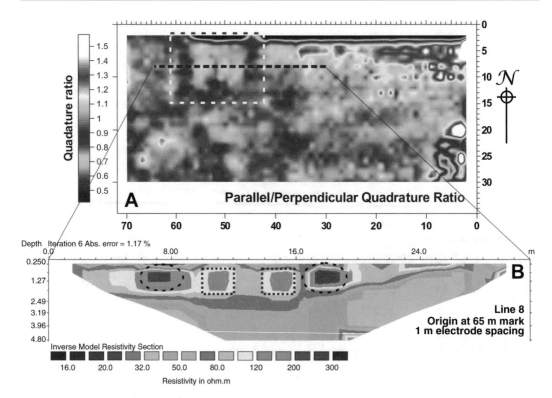

Fig. 11.13 When the electrical imaging model (**b**) is compared to the quadrature ratio (**a**), we can see that the large electrical anomalies (red, at 7 and 18 m), correspond well with the linear features seen in the EM results. These features likely correspond to the outer trench. The inner set of electrical anomalies (brown, at 11 and 15 m) likely correspond to the remnants of the blockhouse foundations

larger the volume. The data must therefore be modelled to obtain the subsurface electrical properties, using a process called inversion. The data were inverted using Res2DINV, which is based on an algorithm developed by Loke and Barker (1996). A robust inversion with no directional bias was used. The effects of the edges, where the electrode arrays end and the data are truncated, have been minimised. The model response (Fig. 11.12b) is almost identical to the measured response (Fig. 11.12a), with only a small misfit (about 1.2%). The model derived through the inversion (Fig. 11.12c) shows clear anomalous features which, again, are coincident with the linear trends identified in the results for both the EM and GPR. A direct comparison between the best-fitting electrical imaging model and the EM results (Fig. 11.13) illustrate the good correlation between the two sets of results.

11.3 Discussion

While individual results can be persuasive, it is when multi-parameter results are compared that we obtain the maximum impact. Thus when we directly compare the EM quadrature ratio (Fig. 11.14a) which also has contours of the total magnetic field superimposed, with a GPR 3D time slice (Fig. 11.14b) and the model for one of the EI lines (Fig. 11.14c), that we can see that all pieces of evidence align and strengthen the conclusion that we have successfully determined the location of the French naval defensive blockhouse.

The EM response ratio shows a feature that is approximately 14 m square. At the same location, the GPR 3D image shows subparallel features that are about 12–14 m across, and the electrical image cross-section shows two sets of subsurface

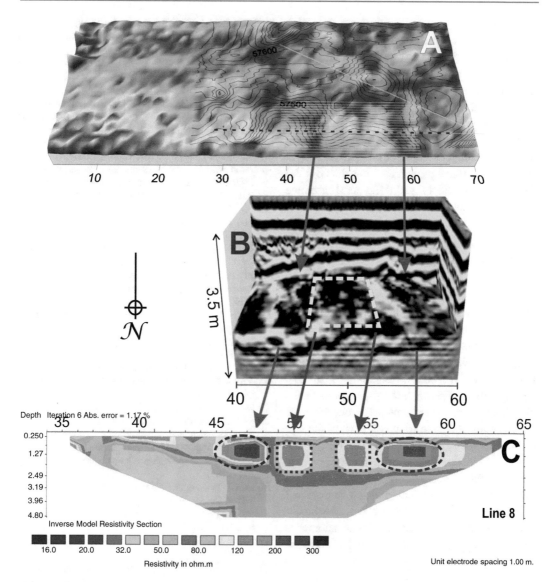

Fig. 11.14 The comparison of a perspective view, looking from the north to the south, of the quadrature ratio (**a**, top), the GPR data cube (**b**, middle) and EI line 8 (**c**, bottom), illustrates the consistency of results from one method to another and reinforces the interpretation based on each technique in turn. Contours of the magnetic field response are superimposed on the quadrature ratio results (**a**). The locations of the GPR lines used in constructing (**b**) (middle) are shown relative to the EM results in (**a**) (top). The EI lines shown in (**c**) (bottom) was coincident with one of the long GPR lines, as indicated by the black dashed line in (**a**) (top). The GPR and EI results suggest the presence of trench-like features such as would have surrounded and enclosed the blockhouse; the features are marked (dashed ovals) in (**c**) (bottom). In addition, the squarish feature seen in the GPR time slice [yellow dashed box in (**b**)] corresponds with the second, weaker set of electrical anomalies (dotted boxes in (**c**)). See text for more discussion

features, one about 9–12 m across and the other about 5 m across.

When all the sets of results are compared, as in Fig. 11.14, we can see the very good correspondence between the results for the different techniques. The different techniques respond to different properties that represent different parts of the subsurface features, particularly the edges vs. the central portions of the defensive trenches. The EM image provides a map view of the subsurface feature, and is an average over a volume of the subsurface; similarly the EI image is a cross-section view of the anomalous response due to the subsurface features, and the response is an average over some volume of the subsurface. The 3D GPR image ties these different responses together, yielding an image of the contrasts in the subsurface physical properties. Thus the GPR 3D data cube will likely yield the most accurate dimensions of the site, in this case about 12–14 m.

11.4   Conclusion

The comparison of the complementary geophysical results are strongly indicative of features that are relatively linear and almost square. The strongest anomalous responses are from a larger feature, most likely corresponding to the outer trench, with a weaker anomaly that like corresponds to what were the foundations of the blockhouse. The dimensions are about what we would anticipate given what is known of the size of the blockhouse and the moat-like trench that surrounded it. The distance from the centre of the trench on the west side to the centre on the east side is about 12–14 m, within the range expected based on contemporary accounts of the blockhouse construction. The inner set of anomalies likely correspond to remains of the blockhouse foundation, and are 4–6 m across.

Acknowledgements DCN thanks his field assistants, dubbed "Team Radar" by one of its members: Richard Cooksey and Jon Lapwood. The entire survey team wish to thank the Akaroa Museum and the Takamatua community for their support and hospitality, as well as Gill and Trevor Bedford, who provided beds and meals, John Roe, who also assisted in the field, and Jean and Neville Rogatski, who supplied us with many cups of tea and trays of biscuits.

References

Davis L, Annan AP (1989) Ground penetrating radar for high-resolution mapping of soil and rock stratigraphy. Geophys Prospect 37:531–551

Field G, Leonard G, Nobes DC (2001) Where is Percy Rutherford's grave? In Jones M, Sheppard P (eds) Australasian connections and new directions: Proceedings of the 7th Australasian Archæometry Conference. Research in anthropology and linguistics, vol 5. University of Auckland, pp 123–140

Frischknecht FC, Labson VF, Spies BR, Anderson WL (1991). Profiling methods using small sources. In Nabighian MN (ed) Electromagnetic methods in applied geophysics, vol 2. Applications. Society of Exploration Geophysicists, Tulsa, pp 105–270

Jacobson WEM (1940) Akaroa and Banks Peninsula, 1840–1940: Story of French Colonising venture and early whaling activities. Akaroa Mail, Akaroa

Jol HM, Bristow CS (2003). GPR in sediments: advice on data collection, basic processing and interpretation, a good practice guide. In Bristow CS, Jol HM (eds) Ground penetrating radar in sediments. Geological Society, London, Special Publications 211, 9–27.

Kearey P, Brooks M (1991) An introduction to geophysical exploration, 2nd edn. Blackwell Scientific Publications, Malden

Loke MH, Barker RD (1996) Rapid least squares inversion of apparent resistivity pseudosections by a quasi-Newton method. Geophys Prospect 44:131–152

Maling PB (1981) Early sketches and charts of banks Peninsula, 1770–1850. Reed, Wellington

McNeill JD (1980) Electromagnetic terrain conductivity measurement at low induction numbers. Technical Note TN-6, Geonics Ltd, Mississauga

McNeill JD (1990) Use of electromagnetic methods for groundwater studies. In Ward SN (ed) Geotechnical and environmental geophysics, I: review and tutorial. Society of Exploration Geophysicists, Tulsa, pp 191–218

Milsom J, Eriksen A (2011) Field geophysics, 4th edn. Wiley, Chichester

Nobes DC (1999) How important is the orientation of a horizontal loop EM system? Examples from a leachate plume and a fault zone. J Environ Eng Geophys 4:81–85

Nobes DC (2007) Detecting linear features using the directionality of the HLEM response. In: Proceedings of Near Surface 2007: 13th European Meeting of Environmental and Engineering Geophysics, Paper A23, Istanbul

Reynolds JM (1997) Applied and environmental geophysics. Wiley, Chichester

Tremewan P (1990) French Akaroa. Canterbury University Press, Christchurch

Tremewan P (2010) French Akaroa: an attempt to colonise Southern New Zealand, 2nd edn. Canterbury University Press, Christchurch

Walton A (2003) New Zealand Redoubts, Stockades and Blockhouses, 1840–1848. Department of Conservation Science Internal Series No. 122, Wellington

Yilmaz O (2001) Seismic data analysis: processing, inversion, and interpretation of seismic data. Society of Exploration Geophysicists, Tulsa

Geophysical Assessment and Mitigation of Degraded Archaeological Sites in Luxor Egypt

12

Ahmed Ismail

Abstract

This study demonstrates the use of geophysics in the field of archaeology not for archaeological exploration but for site assessment and mitigation of degraded archaeological sites because of rough environmental or cultural hazards. Accelerated deterioration of the stone foundations of many temples and monuments at Luxor, Egypt has been documented and is causing global concern for their long-term safety and serviceability. These stone foundations appear to be degrading due to the rise in level and increase in salinity of groundwater. Groundwater transported into the stone foundations by capillary rise through the underlying soil is thought to cause a loss of cohesion and rigidity of these foundations. Moreover, the capillary waters deliver salts into the stone foundations. Pressure developed during crystallization and hydration of these salts exfoliates the outer layers of the stone foundations, allowing them to be easily eroded by wind and other physical processes. The rise in level and increase in salinity of groundwater is thought to be the main problem behind the antiques degradation scenario. We conducted integrated geophysical survey in the form of resistivity and seismic refraction and collected surface water samples for chemical analysis in order to determine the reasons of rise in level and increase in salinity of groundwater. The results showed groundwater is flowing from the east toward the temples area and the groundwater salinity is increasing in the direction of groundwater flow. Our proposed solution is to interrupt or reverse the groundwater flow to stop rise in level and increase in salinity of groundwater.

Keywords

Geophysics · Archaeology · Stone foundation · Luxor · Deterioration · Mitigation

12.1 Introduction

The monuments of the east bank of Luxor study site (Fig. 12.1) are mostly made of sandstone and built on the silty clay layer of the Nile Valley flood plain. Water and soluble salts move upward from the saturated zone into the porous stone foundations through the silty clay. This layer is characterized by high capillary attraction of about 16 m (Sevi 2002). Within the porous stone foundations of some monuments at Luxor, water rises to elevation of about 1.5 m above the ground. As the water evaporates, salts are precipitated on the surface or within the soil and stone foundations.

A. Ismail (✉)
Boone Pickens School of Geology, Oklahoma State University, Stillwater, OK, USA
e-mail: Ahmed.ismail@okstate.edu

© Springer International Publishing AG, part of Springer Nature 2019
G. El-Qady, M. Metwaly (eds.), *Archaeogeophysics*, Natural Science in Archaeology,
https://doi.org/10.1007/978-3-319-78861-6_12

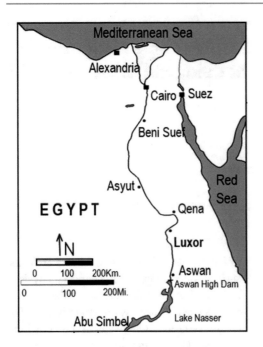

Fig. 12.1 Location map showing the modern city of Luxor

The precipitation of salts in the fine-grained soils accumulates sufficiently to form an impervious hard pan (Fig. 12.2). This hard pan represents serious problems for future archeological excavation. The hard pan also influences drainage and seepage when the area is flooded by irrigation water (Sowers and Sowers 1970). Extensive accumulated salts (efflorescence) are visible on the surface of the soil and the foundations of many temples at Luxor (Fig. 12.2). Salt crystallization on the surface and inside the pore spaces of the stone foundations is believed to be the causative factors of the stone deterioration.

Deterioration of porous stones (Fig. 12.3) is mainly attributed to the presence of salts, which are brought into the sandstone while dissolved in water (Charola and Herodotus 2000). The soluble salts are transferred from the soil into the sandstone through capillary rise of groundwater (Behlen et al. 1997). This may help to relate the results obtained from this study to the problem of deterioration of the monuments' stone foundations.

Although the antiquities degradation seems to be an engineering problem, engineering fixes

such as cementations, water pumping, etc. did not seem to be working. We strongly recommend exploring the root cause of the problem, which is the reason of the rise in groundwater level and increase in salinity in the Luxor area. Non-invasive geophysics seem to be the ideal technique to use in such sensitive archaeological site to study this problem. We propose integrating two geophysical technicities, resistivity and seismic refraction to achieve the goals of this study. We also propose acquiring surface and borehole water samples from the site and analyze the chemical contacts of these samples to aid in tracing the ground water flow direction.

12.2 Site Description

Luxor or ancient Thebes with the High Priest of Amun at its head, formed a counterbalance to the realm of the 21st and 22nd Dynasty Kings, who ruled from Tanis in the delta (Baines and Malek 1980). The city proved difficult to rule and it rebelled against the 22nd Dynasty rulers, and again under the Ptolemies and Romans. The Ptolemies built widely in the east and west bank. During the 25th Dynasty Kings, Thebes was sacked twice, in 671 and 663 BC. In the later Egyptian history the power of Thebes steadily declined and its influence ended since when the political capital moved northward. Even though Thebes remained the administrative center of the south, its imperial glory had departed (Kamil 1989).

During the early Christian times, Luxor was an important center as several shrines were converted into churches. The general trend was one of decline, though it was never completely deserted. After the Muslim conquest, the district was called al-Tibah, and the name al-Uqsur was applied to the town. The importance of Luxor became obvious after Napoleon's survey expedition in 1799. In 1896, Luxor was established as the chief town of the district, which increased its importance, and in 1985 it was given the status of a province (Kamil 1989).

The modern city of Luxor lies within one of the largest archaeological sites in the world, that

Fig. 12.2 Visible salt accumulation on the soil surface inside the temples

of ancient Thebes. Luxor has been a United Nations World Heritage Site since 1979. Few locations on the earth have yielded such a large number of archaeological treasures. These include temples (which were the most important and the wealthiest in the land), tombs (which were the most luxurious Egypt ever saw), colossal statues, coffins and funerary figures, pieces of jewelry, and the pottery of everyday life (Strudwick and Strudwick 1999).

Luxor has Karnak Temples complex which is arguably the largest historical religious complex in Egypt, and perhaps, in the world. The name Karnak, from a modern nearby village (El-Karnak), is used to describe a vast collection of ruined temples, chapels and other ancient buildings, measuring 1.5 by at least 0.8 km. The whole area was known as Iput-Isut, "The Most Select of Places", and represented the main place of worship of the Theban Triad, the great god Amun-Re, his wife Mut and their son Khonsu (Strudwick and Strudwick 1999). Since Thebes became prominent (at the beginning of the Middle Kingdom, and particularly from the beginning of the 18th Dynasty when it became the capital of Egypt), the temples of Karnak were built, enlarged, pulled down, added to, and restored for more than 2000 years (Baines and Malek 1980).

The complex of monuments at Karnak can conveniently be subdivided into three groups, separated by remains of brick walls enclosing the temple precincts. The central enclosure includes the temple of Amun, the largest and the best preserved. This temple was ideologically and economically the most important temple establishment in the whole of Egypt. The central enclosure also includes the temple of Khonsu-Neferhotep (the third member of the Triad of Thebes), the temple of Ramesses III, the temple

Fig. 12.3 Photographs of degraded foundations of some monuments at Luxor

of Akhenaten, a small way station of Seti II, and numerous other structures. The northern enclosure belongs to Montu (the original local god of the Theban area) and includes the temple of Montu, temple of Osiris, and temple of Horpre. The southern enclosure belongs to Mut (the consort of Amun) includes the temple of Mut and the temple of Ramesses III (Baines and Malek 1980).

Perhaps the greatest factor contributing to the preservation of these monuments is that there was no large city in the vicinity, and the ancient buildings were not salvaged or re-worked to build new structures. The level of the River Nile flood plain has risen considerably since the period of occupation and any remains of secular palaces or dwellings are probably covered by the modern town. Therefore, most of the monuments on the east bank of Luxor were buried, only the largest were left exposed. These were neglected or used as habitations (Kamil 1989).

12.3 Geophysical Data Acquisition and Analysis

Forty VES were acquired at Luxor study area (Fig. 12.4). The maximum current electrode spacing of the VES data varied from 600 to 1000 m, providing 26 measurements in each VES. The distances between the potential electrodes increased only few times by relatively small steps starting from 0.8 to 20 m in order to obtain sufficient potential difference to be measured. The distance between the centers of adjacent soundings was generally 1–3 km.

The automatic interpretation computer program developed by Zohdy and Bisdorf (1989) was used to interpret the field resistivity sounding data. The interpretation procedures in this program are based on obtaining true depths and resistivities from shifted current electrode spacings and adjusted apparent resistivities, respectively. The method is fully automated and does not need an initial guess of the number of layers, their thickness or their resistivities. The number of layers in the interpreted model equals the number of digitized points on the sounding curve. The interpreted resistivity-depth model is assumed to exist directly beneath the sounding station.

The results obtained from Zohdy and Bisdorf (1989) program were used to build up a starting model to be further interpreted by the automated inversion technique developed by Schlinv (1977, modified by Sauck 1990). In this interpretation procedure, a linear filter theory has been coupled with the Marquardt method to automatically interpret the resistivity sounding data. The matching between the field and model data (goodness of fit) is measured by the sum of squares of logarithms of the field and model data. The output interpreted results include layers thicknesses and resistivities.

Seismic refraction data were acquired using Geometrics Strata View portable seismograph with 48 Input/Output 14 Hz geophones. The geophones were attached to a geophone-spread cable at two meters intervals and well coupled with the ground to properly detect the ground movement. Forward-and reverse-direction shots were applied at offset of two meters from each end of the array. This was to determine the true velocity of the dipping layer and to provide sufficient data along each profile to map changes in the subsurface parameters with sufficient confidence. A hammer on striker plate was used as a source during data acquisition. The plate was placed on firm ground and the resulting signal is stacked 10–20 times on the seismograph (Fig. 12.5). The acquired data were automatically stored in the equipment memory and then downloaded on a PC for subsequent processing, interpretation and presentation.

In seismic refraction work, only the first arrivals at each geophones are utilized. The first arrivals resulting from the multiple shots (forward and reverse shots) at each seismic array were used to build seismic velocity profile using the SIP family of routines. The first step was to pick the first arrivals using the SIPIK code. First arrivals acquired along profiles close to the eastern part of the study area were sharp and easily to pick. First arrivals acquired along profiles in the western part of the study area (near the Karnak Temples complex and Luxor Temples) were more difficult to pick due to the presence of

Fig. 12.4 A map of Luxor study area shows the soundings stations locations

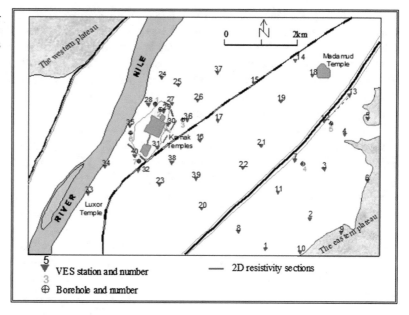

ambient noise. Such noise probably originated from highway traffic or urbanization. The presence of noise can cause errors (on the order of 3 ms) in the picked travel times and consequently the final velocity/depth model may be less precise (on the order of 0.5 m).

The selected first breaks were incorporated into a data file for each profile line, using the

SIPIN and SIPEDT codes. The data file includes precise positions for each geophone, shot point, and all the first arrival picks. Each pick was assigned to a specific subsurface layer in the data file. SIPT-2 processing assumes that the subsurface model consists of laterally continuous discrete layers with constant velocity. Thus the selected layers during data processing have

Fig. 12.5 A map of Luxor study area showing locations of the acquired seismic refraction data

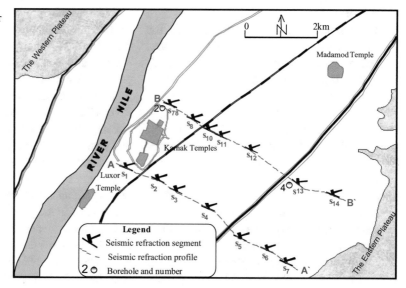

constant velocity. A significant advantage of this interpretation technique over the standard formula interpretations is that the interpretation is based on the data collected at all geophones. This means that the program generates up to 48 depths (one for each geophone location) to each subsurface layer along the same array. Consequently, much more detailed subsurface information can be obtained from the same amount of fieldwork. The program also provides the interpreter with a time-distance curve, the calculated velocity of each layer, and a plot of the computed velocity profile.

12.4 Water Samples Acquisition and Analysis

A total of 39 groundwater and surface water samples were collected in the study area and analyzed for physical and chemical parameters. Locations where samples were collected are shown in Fig. 12.6. The samples coding and sources are shown below.

– Two samples (N1 and N2) were collected from the River Nile.
– Thirteen samples (C1 to C13) were collected from canals.
– Eleven samples (D1 to D11) were collected from open drains.
– One sample (S1) was collected from the Sacred Lake of the Karnak Temples complex.
– Twelve groundwater samples (B1 to B12) were collected from depths ranging from 5 to 35 m.

Surface water sampled from the River Nile, canals, drains, and the Sacred Lake of the Karnak Temples complex were collected by the grab method. All groundwater samples are from the Quaternary Aquifer, and were collected from either hand pumps attached to water wells or from existing piezometers. To ensure the removal of stagnant water in the well bore and hand pumps, the temperature and specific conductance were monitored while pumping. When these parameters stabilized, indicating delivery of

formation water, samples were collected. Analysis was conducted for the physical analysis [pH, specific conductance, and total dissolved salts (TDS)] and chemical analysis (major anions and cations) of groundwater and surface water samples.

12.5 Geophysical and Geochemical Data Interpretation

Acquired data were used to map the different geoelectric (geologic/hydrologic) units in the shallow subsurface. On the basis of the interpretation of the acquired resistivity soundings, the shallow subsurface (<100 m) in the study area was subdivided into units. These units are shown in the schematic diagram of Fig. 12.7 and include: a relatively dry topsoil; a moist silty clay; sand and gravel of Quaternary aquifer; sand and clay of the Plio-Pleistocene Aquifer; fine sand and silt of the newly reclaimed land; sand and gravel of the non-cultivated land; and limestone of the foot of the eastern plateau. Serious ground settlements can occur in areas whereas such deposits lose much of their shear strength when significant vibration or seismic shock occurs. The resistivity trend within the third resistivity layer seems to show a gradual decrease from east to west towards the area of the temples (Fig. 12.8).

The results of seismic refraction survey showed that the depth to the saturated zone ranges from 3 to 9 m, with maximum depth beneath the Karnak and Luxor Temples (Fig. 12.8). The groundwater flow direction was determined to be from the central cultivated areas towards the River Nile, which agrees with conclusions of El Hossary (1994) and RIGW. The River Nile acts in this case as a discharge zone for the shallow groundwater. However, the rise of the River Nile water in the summer time may reverse the primary groundwater flow path resulting in a significant rise in the groundwater level.

Another groundwater flow direction was determined in the study area from the central cultivated part towards the Nile Valley fringes. This flow may have developed as a result of excessive localized groundwater pumping for irrigation of

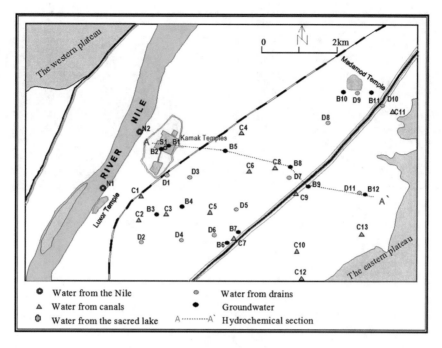

Fig. 12.6 A map of the Luxor study area showing locations of the acquired groundwater and surface water samples for chemical analysis

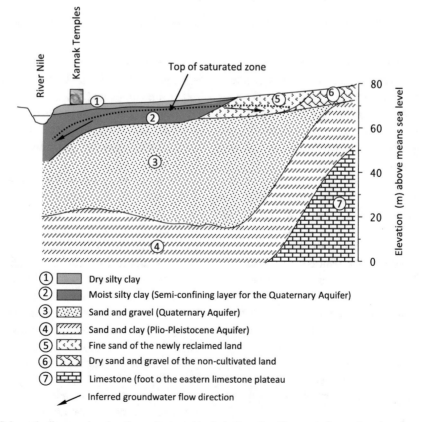

Fig. 12.7 Schematic diagram showing the geologic and hydrologic units at Luxor study area based on the results of the present geophysical study

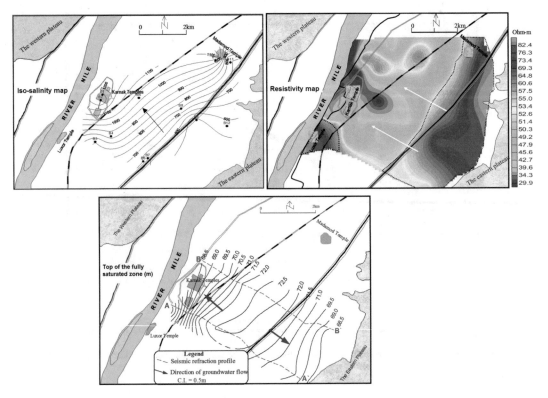

Fig. 12.8 Results of the resistivity, seismic and water analysis data indicating the groundwater flow direction

the newly reclaimed land near the fringes of the Nile Valley. Based on the fact that the groundwater flows from the central cultivated areas towards the Nile Valley fringes, where pumping rates of groundwater is relatively high, the increasing use of pumped groundwater may result in lowering the groundwater level.

The analysis of surface and groundwater analysis showed that the groundwater salinity was found to increase from the Nile Valley fringes towards the River Nile. Salinity increased in the direction of groundwater flow and reached its maximum beneath the Karnak and Madmod Temples. This observation is in contrary to the previous studies (El Hossary 1994), which determined the increase in salinity to be toward the Nile Valley fringes. However, the anomalous increase in groundwater salinity beneath the temples may also be due to the leachate of agricultural chemicals and sewage from urban areas, which increase groundwater salinity along their flow path.

A geochemical relationship was found between the groundwater underneath the Karnak Temples complex and that in the sacred lake of the temples. This relationship indicates that the sacred lake water may represent a local source of charge for the elevated groundwater underneath the temples.

Based on the results of the chemical analysis of groundwater, the salinity within the aquifer ranges from 524 to 1363 ppm with an average value of 870 ppm. The observed geo-chemical relationship between the surface and groundwater within the aquifer implies that the surface water is recharging the aquifer. Note that there is no indication that the Quaternary Aquifer is recharged from the underlying aquifer system through the fault zone as suggested in previous studies in the area (El Hossary 1994).

The increase in the thickness of the silty clay (dry silty clay and moist silty clay) was accompanied by an attendant decrease in the resistivity of the Quaternary Aquifer (which is mostly controlled by water salinity). This indicates that the thicker silty clay might prevent surface water (of lower salinity) from percolating downward into the Quaternary Aquifer and lowering its salinity. This suggested relationship between the silty clay thickness and groundwater salinity is supported by the fact that groundwater salinity increases as silty clay thickness increases.

The determined directions of groundwater flow and increase in salinity are characteristic for the saturated zone. However, in most cases, affected monuments at Luxor are not in the presence of the saturated zone. The groundwater flow in the saturated zone delivers water and salts to the area underneath the monuments, while it is not directly responsible for delivering moisture above the groundwater level and into the monuments' foundations.

Several mechanisms documented in the literatures are responsible for upward mobilization of groundwater and dissolved salts from the saturated soil into the porous stone foundations of monuments. Such mechanisms include capillary action, thermo-migration of water, desiccation, and osmotic flux. Capillarity is an effective mechanism to draw water above the saturated zone. The estimated capillary ability within the silty clay unit at Luxor was 16 m (Sevi 2002), which can draw water from the saturated zone (~9 m depth) to the ground surface. The hot and dry weather conditions of Luxor may also contribute to upward movement of salts and water from the saturated zone through thermo-migration, desiccation, and osmotic flux mechanism.

We also mapped variations in soil moisture, which is mainly caused by capillarity processes within the shallow soil (<7 m depth) in contact with the monuments' foundations. Based on the interpretation of the resistivity soundings, the moisture (upper limit of capillary water) rose from as much as 2 m depth below the ground surface in the study are. The moisture rise was higher in cultivated lands than in urban areas.

The upper limit of capillary water was estimated on the basis of the integrated interpretations of the acquired resistivity soundings, 2D-resistivity and GPR data in the vicinity of the Karnak Temples complex. It has been noticed that the capillary water rises to an average depth of 1.5 m below the surface. At some places, like those to the south of the Karnak Temples complex, the capillary process draws moisture up to the ground surface. This might be attributed to the fact that the area to the south of the Karnak complex is cultivated and is topographically lower (on the order of 1 m) than the surroundings.

On the other hand, surface conditions appear to significantly impact the shallow subsurface moisture. Cultivation increases the soil moisture while urbanization keeps the upper 2 m of the soil relatively dry. Soil moisture variations in the upper 7 m of the soil around the Karnak Temples complex. The mapping of the variations in soil moisture is an important contribution with respect to mitigating the building stone deterioration problem since it is difficult, if not impossible, to obtain such information for a large area using traditional techniques.

12.6 Discussion and Conclusions

The rise in level and increase in salinity of groundwater are causing accelerated deterioration of the stone foundations of visible monuments at Luxor. Groundwater, through capillary action, delivers salts into the porous stone foundations from the underlying soil. Elevated groundwater is thought to cause a loss of cohesion and rigidity of the stone while pressure developed through salt crystallization and hydration within the stone may effectively crumble the stone away. Moreover, monuments that are not yet discovered are also in danger due to the areal expansion of agriculture and urbanization in addition to the rise in level and increase in salinity of groundwater.

Protection of the visible and the buried monuments requires an understanding of the hydrostratigraphy of the Luxor area and the non-destructive identification of sites that may host buried monuments. Therefore, the objectives of this study include: characterizing the subsurface geologic/hydrologic units in the Luxor area, identifying the causes of the rise in level and increase in salinity of groundwater, mapping lateral and vertical variations in soil moisture, evaluating the thickness and water quality of the shallow aquifer system (the source of water causing deterioration of the stone foundations) and mapping yet undiscovered, buried, monuments in the vicinity of the Karnak Temples complex.

In order to accomplish the study objectives, geophysical and hydrological data were acquired from the study area and analyzed. The results of data analysis indicate that the regional increase in groundwater salinity was determined to be towards the River Nile (or area of temples) in the same direction of the groundwater flow with maximum concentration beneath the temples. The elevated groundwater at the area of the temples seems to be caused by the flowing of groundwater from several kilometers apart towards the temples' area. The progressive increase in groundwater salinity along the flow path is believed to cause the increase in salinity beneath the temples (located at the down of the flow path). The sacred lake of the Karnak Temples complex may represent a local source of recharge to the groundwater beneath the temples. Based on these facts, the salt accumulation on the foundations of the monuments is mainly due to salt transport by capillary water from the saline groundwater or saline paleo-water in the thick silty clay unit. On the other hand, the thick silty clay underneath the Karnak and Luxor Temples, can adversely reduce the drainage ability of surface water. Moreover, the thick silty clay may increase the lateral capillary flow from the sewage septic-tanks (mostly used in new urbanizations) toward the relatively lower topographic area of the temples.

References

Baines J, Malek J (1980) Cultural atlas of ancient Egypt. Andromeda Oxford, Oxford

Behlen A, Steiger M, Dannecker W (1997) Quantification of the salt input by wet and dry deposition on a vertical masonry. In: Moropoulou A (ed) Fourth international symposium on the conservation of monuments in the Mediterranean Basin. Technical Chamber of Greece, Rhodes, pp 237–246

Charola AE, Herodotus (2000) Salts in the deterioration of porous materials: an overview. J Am Inst Conserv 39 (3): Article 2, 327–343

El Hosary MM (1994) Hydrogeological and hydrochemical studies on Luxor area, Southern Egypt. MSc thesis. Ain Shams University, Cairo, Egypt

Kamil J (1989) Luxor: a guide to ancient Thebes, 3rd edn. Longman, London

Sauck WA (1990) Modification of the SCHLINV Program. Internal report. Institute for Water Sciences, Western Michigan University, Kalamazoo

Sevi A (2002) Geotechnical investigation of sandstone degradation of antiquities in Luxor, Egypt. MSc thesis. University of Missouri-Rolla, Rolla

Sowers GB, Sowers GF (1970) Introductory soil mechanics and foundations. Macmillan, New York

Strudwick N, Strudwick H (1999) A guide to the ancient tombs and temples of ancient Luxor, Thebes in Egypt. Cornell University Press, Ithaca

Zohdy AAR, Bisdorf RJ (1989) Programs for the automatic processing and interpretation of Schlumberger sounding curves in QuickBASIC 4.0. U.S. Geological Survey, open-file report, 89-137 A and B

Integrated Geophysical Techniques for Archaeological Remains: Real Cases and Full Scale Laboratory Example

13

E. Rizzo and L. Capozzoli

Abstract

The increasing interest in preserving of the archaeological sites requires the integration of a wide spectra of geophysical methodologies for field measurements. In fact, archaeological investigations need multidisciplinary studies to characterize the physical properties of near-surface. In this context, the integration of electromagnetic techniques seems to be one of the most suitable tools. The most suitable geophysical investigation techniques employed for archaeological purposes are the geomagnetic, GPR and resistivity/conductivity (DC and EM) methods. These techniques are not invasive and allow us to obtain high resolution images of subsurface, even if their use is dependent on site and resolution. In general, geomagnetic and EM methods are more adaptive for large survey, in order to obtain fast results with low resolution. On the contrary, GPR shows high resolution information, but for the heavy data process is adapt for small survey areas. The DC methods are not common then the previous ones, but their contribute is important above all in urban area. Anyway, the integration of different geophysical techniques is the best way for field measurements to identify the remains, because

each geophysical technique has the ability to define a variation of the physical parameters (electrical conductivity, magnetic susceptibility, dielectric permittivity) which is able to highlight some pattern of the buried object. This kind of approach was applied in several archaeological site. Moreover, the geophysical contrast between archaeological features and surrounding soils sometimes are difficult to define due to problems of sensitivity and resolution related on the subsoil characteristics and limits of geophysical methods. The results obtained in real and laboratory study cases based on archaeogeophysical approach are here discussed.

13.1 Introduction

The need to have preliminary information to identify archaeological site or manage cultural heritage without invasive excavation activities represents a great challenge for the scientific research. An approach based on the use of geophysical techniques could offer an effective support for the resolution of this challenge. Thanks to their characteristics of non-invasivity and repeatability, geophysical techniques offer benefit to detect contrasts of physical properties of the subsoil associable to the presence of archaeological buried structures allowing the

E. Rizzo (✉) · L. Capozzoli
CNR-IMAA, C.da s.Loja, Tito, Potenza, Italy
e-mail: enzo.rizzo@imaa.cnr.it

© Springer International Publishing AG, part of Springer Nature 2019
G. El-Qady, M. Metwaly (eds.), *Archaeogeophysics*, Natural Science in Archaeology,
https://doi.org/10.1007/978-3-319-78861-6_13

integration of different techniques reducing uncertainties of the interpretation investigating archaeological structures placed in different contexts without interfering with economic and social activities.

An integrated geophysical approach based on the use of electrical and electromagnetic data provides great advantages to overcome resolution and depth investigation limits of each techniques and for this reason it is strongly recommendable (Chianese et al. 2004, 2010; Rizzo et al. 2005, 2010, 2014; Nieto et al. 2005; Bavusi et al. 2009; Lasaponara et al. 2011; and reference therein). Generally the term used to indicate the use of geophysical techniques in the archaeological field is archaeogeophysics that includes all the archaeological prospection techniques that can help archaeologists for the remote sensing of the subsoil in the shallower layers interested by anthropic activities occurred in the past (Piro 2009).

The success of archaeogeophysics is mainly related to the existing physical contrasts between the properties of the hosting subsoil and archaeological buried remains. Furthermore presence of noise caused by anthropic activities as heterogeneity of subsurface material, building artefacts, underground utilities or electromagnetic interference or high attenuation phenomenon due to the characteristics of the soil can reduce the possibility to identify archaeological structures.

In this chapter there is a short discussion of the main geophysical techniques used in archaeology, namely ground penetrating radar, magnetic surveys and resistivity methods. Then, in order to improve the knowledge on the geophysical signature coming from buried remains characterized by low physical parameter contrasts, experimental tests realized on an analogue archaeological full scale laboratory case addressed on the assessment of the capability of geophysical techniques to detect archeological remains placed in humid/saturated subsoil are described. Finally, several interesting real cases where archaeophysical techniques, based on the use of Geomagnetic (M), Ground Penetration Radar (GPR) and electrical resistivity (DC) measurements acquired in different contexts are showed.

The application of GPR for archaeological research had become largely accepted. There are many reasons of this success that made the GPR the most widely used geophysical techniques in the archaeological field: high resolution, an acceptable depth of investigation for archaeological targets: a wide application of the GPR in urban archaeological areas where magnetic method struggles; finally the possibility of integration with the data obtained with other geophysical methods.

The electrical imaging is a geoelectrical method, widely applied in geological research, to analyze electrical behavior of the subsoil and its use is increasing in the archaeological field, in particular in lacustrine and water saturated scenarios as well as in marine and coastal areas. The great advantage of this method is that it is effective in conductive soils and in area characterized by electromagnetic noise supporting for example GPR measurements in clay soils or in marine areas (Passaro 2010; Simyrdanis et al. 2015) or noisier urban context archaeogeophysics (Osella et al. 2015; Papadopoulos et al. 2009).

Magnetic surveys have been often used in the last three decades in the archaeological fields to individuate buried remains and structures providing to have excellent results. Generally in the preliminary phase of archaeological investigation, magnetometric surveys are very able to delimitate archaeological areas of interest with an acceptable resolution reducing time of acquisitions without, nevertheless, give information on anomalies depth (Chianese et al. 2004; Drahor et al. 2011).

13.2 Archaeogeophysical Test in the Full Scale Laboratory

GPR and ERT are two of the most used geophysical techniques in archaeology but each of them suffers of important limitations. In particular this is true for the GPR that, although provide a high resolution, suffers from attenuation of EM energy in conductive scenarios. The EM velocity in soil is influenced by soil water content, at the same time greater is the water content greater is the

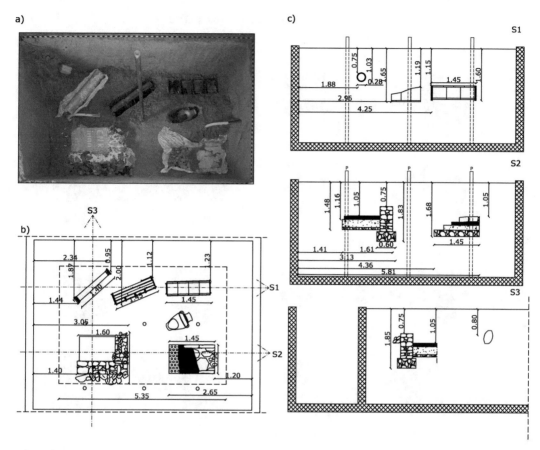

Fig. 13.1 The archaeological framework focused on the Roman age reconstructed in laboratory, orthophoto (**a**), plan of remains (**b**) and sections (**c**). The circles shown in the maps represent the positions of the piezometers. The simulated structures, described clockwise, are a column, capuchin tomb, rectangular tomb and enchytrismos, paved road and stone wall (modified from Capozzoli et al. 2015)

conductive behaviour of the medium. But the most important role in the attenuation phenomenon is imputable to the presence of salts in solution of different nature and the clay content able to produce heavy attenuation losses (Tosti et al. 2013). For this reason GPR is less effective in clay or water saturated soil. The resistivity method could be a good solution to overcome limitations related to GPR. Indeed electrical resistivity tomographies (ERT) are capable to individuate targets placed in conductive soils where the EM signal is absorbed. So geophysical experiences based on the use of GPR and ERT were made to investigate and increase the knowledge about the influence of water content on geophysical response of archaeological remains (Capozzoli

et al. 2015). The archaeogeophysical test was made reconstructing an archaeological framework of Roman times at the Hydrogeosite Laboratory of CNR-IMAA, a pool shape structures of 252 m^3 filled with homogeneous silica sand (95% SiO$_2$) characterized by an average diameter equal to 0.09 mm (very fine sand), a porosity of about $45 \div 50\%$ and a hydraulic conductivity in the order of 10^{-5} ms^{-1}. The archaeological site was composed by different kind of structures as burials, paved road and stone wall partially collapsed located at different depths (see Fig. 13.1). The test was realized in two main phases: in the first one, the top of the shallower structures was at a depth of 0.70 m; in the second phase this depth was limited to 0.20 m.

Depth of the water table from the surface

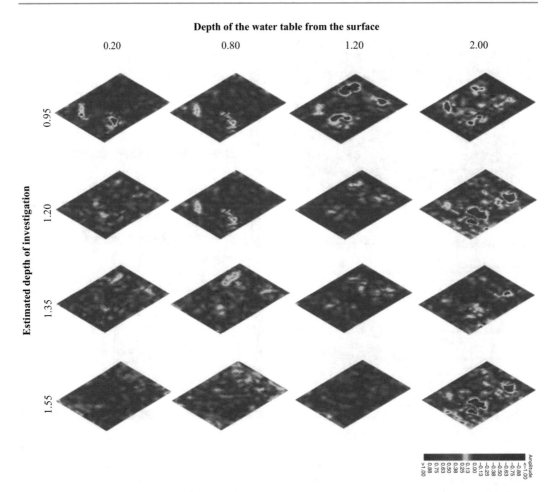

Fig. 13.2 Depth timeslices with water level fixed at the different heights. The depth of the maps was estimated analysing the shape of the hyperbolas present in the 2D radargrams (modified from Capozzoli et al. 2015)

The geophysical surveys were performed using a GPR System SIR-3000 (GSSI-Instruments) coupled to a 400 MHz antenna and survey wheel and a Syscal Pro Switch 96 (Iris Instruments) georesistivimeter with the water table placed at different heights. GPR data were acquired in continuous and reflection mode with a time window of 70 ns, samples per scan set at 512 with a resolution of 16 bits and a transmit rate of 100 KHz. GPR surveys were carried out using a reference grid where the distance from each line was of 0.20 cm and the investigations, covering an area of 4.2 × 6.2 m, were made in both main directions. The geoelectrical acquisitions were performed in both the phases but in the first one, for the presence of concrete walls of the pool, the length of the profile was too little to give information about the test site.

After the adoption of an outright scanning, 2D radargrams of the site allowed us to identify the archaeological structures in dryer conditions as showed in Fig. 13.2 where depth slices with no interpolation are plotted. When the water level increase upon 0.20 m from the surface the attenuation phenomenon of electromagnetic signals didn't allow to identify the deeper remains.

In the second phase, after the removal of the first 50 cm of sand, different geoelectrical acquisitions were made adopting several types of arrays to carry out 2D and 3D acquisitions, also using loop shape grid. Some acquisitions were made directly on the water. 2D ERTs,

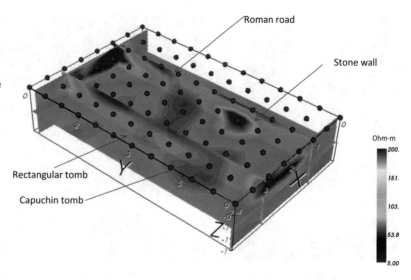

Fig. 13.3 3D electrical resistivity images obtained with a 3D ERT acquisition according to an equatorial dipole–dipole array. The high values of resistivity are due to presence of structures (modified from Capozzoli et al. 2015)

based on use of electrodes placed every 0.20 cm according to Wenner-Schlumberger and Dipole–Dipole arrays, have showed their capability to identify, with an acceptable resolution, the archaeological buried remains. While 3D acquisitions, obtained with a grid of electrodes placed according a square mesh of 0.60 m have allowed to identify only the bigger structures as the stone wall. Strong uncertainties with soil characterized by high water content are related also for ERT acquisitions. The reason is imputable to the decrease of the electrical contrast between the background and the structure, especially in correspondence of the graves where the voids were filled with water (Fig. 13.3).

13.3 Integration of Geophysical Techniques: Real Cases

13.3.1 Archaeogeophysical Investigation at the Site of Masseria Nigro

During the construction of an oil-pipeline, some important evidences of the past came to light and various geophysical prospections were made. In particular an integrated geophysical survey to detect buried remains was realized into the archaeological site of 'Masseria Nigro',

Viggiano, southern Italy, to identify ancient structures developed between the fourth and the third centuries BC, destroyed after the Roman conquest. GPR, ERT and MAG surveys were carried out to study the extent of the archaeological site (Rizzo et al. 2005). Regarding the magnetic measurements, an area of about 800 mq was investigated with a caesium vapour magnetometer G-858 GEOMETRICS (see Fig. 13.4). The mark spacing and profile spacing were set at 0.5 m while the sampling rate was 10 Hz. The acquisitions were made with a gradiometer configuration with two sensors placed at a vertical distance of about 1 m apart. GPR investigations were made with a SIR-2000 (GSSI Instruments) and the whole area was subdivide in two regular sectors. Fifty four parallel profiles equally spaced every meter were acquired and processed with Radan NT software. Finally, on a limited area three short parallel electrical resistivity tomographies (ERT) were carried out with the Syscal R2 multielectrode system (32 channels). A dipole–dipole array layout with an electrode spacing of 0.5 m was adopted.

Thanks to the high electric and electromagnetic contrast between the hosting soil and archaeological remains, it was possible obtain interesting information about shape, dimension and depth of the buried structure before the excavations activities supporting the

☐ **Magnetic investigation area**   ▨ **GPR investigation area**   ▨ **Geoelectrical investigation area**

Fig. 13.4 Localisation of the investigated site with indication of the investigated areas with the different geophysical techniques (modified from Rizzo et al. 2005)

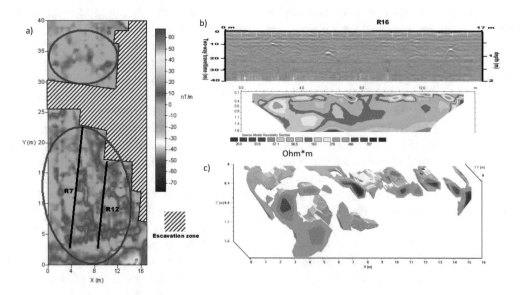

Fig. 13.5 MAG acquisition (**a**) shows several well defined anomalies that provide a good reconstruction of the structure; GPR data (**b**) carried out on a smaller area are characterized by reflections in correspondence of the walls; ERTs have showed an increase of electrical resistivity values where walls are present, the data are then interpolated to define a 3D electrical resistivity volume (modified from Rizzo et al. 2005)

archaeological research. In particular electrical resistivity tomographies (ERT) showed an increase of conductivity values in correspondence of the stone walls. GPR, in addition to recording high amplitude reflections of buried walls, allowed to identify the depth of the structures. Finally, MAG results have been permitted the reconstruction of the ancient house confirming hypothesis of archaeologists (see Fig. 13.5).

13.3.2 Integration of Satellite Investigation and Archaeogeophysics at the Piramide Naranjada Site in Cahuachi (Peru) and Validation with In Situ Excavation

The integration of other remote sensing techniques with archaeogeophysics offers considerable advantages for the detection of archaeological sites as showed at the site of Cahuachi (Nasca, Southern Peru) where Very High Resolution (VHR) satellite imagery, geomagnetic surveys and Ground Penetrating Radar (GPR) were applied. In order to confirm archaeological features as walls detected via satellite, geomagnetic measurements with an optical pumping magnetometer G-858 Geometrics in gradiometric configuration were made to study an area of about 3000 m^2 (Lasaponara et al. 2011).

The focus of MAG acquisitions was to confirm the presence of buried adobe walls individuated by satellite images. The acquired maps have detected linear anomalies associable to shallows and outcropping walls, buried structures, tombs and ceremonial offerings.

In a smaller area as showed at the top of Fig. 13.6, GPR acquisitions were made with a SIR-3000 system (GSSI Instruments) coupled to a 400 MHz antenna without survey wheel. The data were acquired in two main directions mutually perpendicular every 0.30 m in one direction while every 0.50 in the other one. The processed data showed some reflections at a depth lesser than 0.50 m due to remains of huarango (trunks and branches belonging to a wood framework) as subsequent archaeological excavations have demonstrated. Further successive excavations have unearthed a ceremonial offering of the Nasca Culture as depicted in Fig. 13.7.

13.3.3 GPR and MAG Acquisitions to Study a Medieval Monastic Settlement in Basilicata

Archaeogeophysics is often used to support the archaeological reconstruction of disappeared architectural heritage. For this reason GPR and MAG prospections were made in Calvello (Basilicata, Southern Italy), to individuate the medieval monastic plan of San Pietro a Cellaria built between the twelfth and the thirteenth century. Geophysical surveys were supported by UAV based survey that provided to georeference and better visualize the obtained results (Leucci et al. 2015).

A Geometrics G-858 caesium vapour gradiometer was used for magnetometric prospections along parallel lines spaced 1 m apart. Data were acquired in continuous mode with a sampling interval of 0.2 s. The considered processed data were distributed in a range varying between -30 and $+30$ nT/m, which represents values associable to archaeologically significant anomalies. Three areas M1 (11 × 25 m), M2 (15 × 12 m) and M3 (30 × 40 m) distributed around the existing building were analysed and the results, showed in Fig. 13.8, are characterized by weak anomalies (dashed dark line) in area M3, related to the presence of a buried structure. In particular the area M1 is characterized by buried walls as confirmed by GPR surveys.

A georadar Hi Mod (Ingegneria Dei Sistemi-IDS) with 200 and 600 MHz antenna was used to investigate a small are of rectangular shape of size of 12 × 32 m. Parallel radargrams were acquired very 0.50 m according only the greater direction. The comparison between the depth slices extracted by reflection amplitude volume generated with GPR and MAG data confirms the presence of well-visible structures placed in continuity with walls belonging to the existing building. In particular the anomalies indicated with letters a, b, c and d in Fig. 13.9 are present both in MAG map and in GPR results and for the shape suggest a continuation of the actually visible structures.

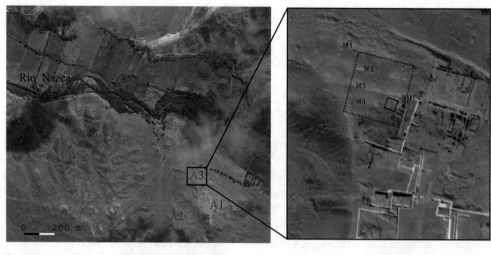

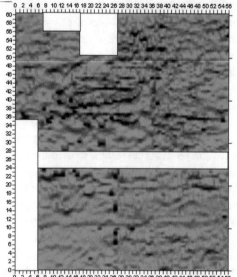

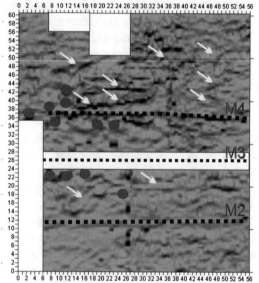

Fig. 13.6 The site was investigated with GPR (continuous line square) and MAG (dashed line square) as depicted at the top of the figure. The MAG results obtained in M2, M3 and M4 show linear anomalies due to shallows and outcropping walls (black dotted line), archaeological feature likely associable to buried structures (white arrows) or potential buried tombs and ceremonial offerings (red circles). At the left lower corner is showed the acquired maps while at the opposite corner there is the interpreted map (modified from Lasaponara et al. 2011)

13.4   Conclusions

Archaeogeophysics based on the integrated use of different geophysical techniques is able to address the problems of archaeological nature overcoming limitations specific to each technique. ERT and MAG surveys can support GPR acquisitions in high attenuating soils, while GPR and ERT represent considerable advantages in urban areas where magnetic surveys are not applicable. Geophysical tests realized in laboratory conditions have showed uncertainties and difficulties of GPR and ERT acquisitions in humid scenarios where physical contrasts are very low; but nevertheless some good results are obtained. The MAG surveys supported by GPR

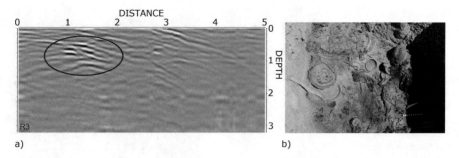

a)   b)

Fig. 13.7 The radargram (**a**) is characterized by strong shallow reflections imputable to the presence of ceremonial objects as confirmed by the archaeological excavations that provided a direct validation of the geophysical data (**b**). (modified from Lasaponara et al. 2011)

Fig. 13.8 Magnetometric results with indication for the three map of the main anomalies (from Leucci et al. 2015)

Fig. 13.9 Comparison between depth-slices (**a–c**) obtained with GPR and magnetic map M1 (**b**). The results show the high integration of the two techniques. In fact the anomalies recorder with GPR are clearly associable to those showed by MAG, in particular for the b that indicates a buried wall (modified from Leucci et al. 2015)

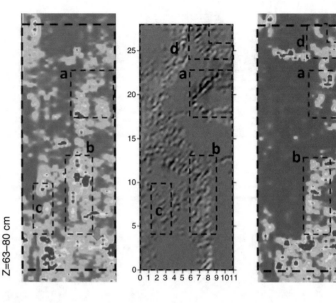

and ERT are applied with success in various archaeological study cases and the effectiveness of their comparison with also other remote sensing techniques (i.e. satellite observation) is showed and strongly recommended for archaeogeophysical issues.

References

Bavusi M, Giocoli A, Rizzo E, Lapenna V (2009) Geophysical characterisation of Carlo's V Castle (Crotone, Italy). J Appl Geophys 67(4):386–401

Capozzoli L, Caputi A, De Martino G, Giampaolo V, Luongo R, Perciante F, Rizzo E (2015) Electrical and electromagnetic techniques applied to an archaeological framework reconstructed in laboratory, Advanced Ground Penetrating Radar (IWAGPR), 2015 8th International Workshop on Advanced Ground Penetrating Radar, IEEE, 7–10 July 2015 Firenze

Chianese D, D'Emilio M, Di Salvia S, Lapenna V, Ragosta M, Rizzo E (2004) Magnetic mapping, ground penetrating radar surveys and magnetic susceptibility measurements for the study of the archaeological site of Serra di Vaglio (Southern Italy). J Archaeol Sci 31:633–643

Chianese D, Lapenna V, Di Salvia S, Perrone A, Rizzo E (2010) Joint geophysical measurements to investigate the Rossano of Vaglio archaeological site (Basilicata Region, Southern Italy). J Archaeol Sci. https://doi.org/10.1016/j.jas.2010.03.021

Drahor MG, Berge MA, Öztürk C (2011) Integrated geophysical surveys for the subsurface mapping of buried structures under and surrounding of the Agios

Voukolos Church in İzmir, Turkey. J Archaeol Sci 38:2231–2242

Lasaponara R, Masini N, Rizzo E, Orefici G (2011) New discoveries in the Piramide Naranjada in Cahuachi (Peru) using satellite, Ground Probing Radar and magnetic investigations. J Archaeol Sci 38(9):2031–2039

Leucci G, Masini N, Rizzo E, Capozzoli L, De Martino G, De Giorgi L, Marzo C, Roubis D, Sogliani F (2015) Integrated archaeogeophysical approach for the study of a medieval monastic settlement in Basilicata. Open Archaeol 1(1):236–246. https://doi.org/10.1515/opar-2015-0014. ISSN (Online) 2300–6560. Nov 2015

Nieto X, Revil A, Morhange C, Vivar G, Rizzo E, Angelo X (2005) La fachada marittima de Ampurias: estudios geoficos Y datos arqueologicos. Empuries 54 (2005):71–100

Osella A, Martinelli P, Grunhut V, de la Vega M, Bonomo N, Weissel M (2015) Electrical imaging for localizing historical tunnels at an urban environment. J Geophys Eng 12(2015):674–685

Papadopoulos N, Sarris A, Yi M-J, Kim J-H (2009) Urban archaeological investigations using surface 3D ground penetrating radar and electrical resistivity tomography methods. Explor Geophys 40:56–68

Passaro S (2010) Marine electrical resistivity tomography for shipwreck detection in very shallow water: a case study from Agropoli (Salerno, Southern Italy). J Archaeol Sci 37(8):1989–1998. https://doi.org/10.1016/j.jas.2010.03.004

Piro S (2009) Introduction to geophysics for archaeology. In: Campana, Piro (eds) Seeing the unseen. Geophysics and landscape archaeology. CRC Press, Taylor & Francis Group, Oxon

Rizzo E, Chianese D, Lapenna V (2005) Integration of magnetometric, GPR and geoelectric measurements applied to the archaeological site of Viggiano

(Southern Italy, Agri Valley-Basilicata). Near Surf Geophys 3:13–19

Rizzo E, Masini N, Lasaponara R, Orefici G (2010) ArchaeoGeophysical methods in the Templo del Escalonado (Cahuachi, Nasca, Perù). Near Surf Geophys 8(5. Oct 2010):433–439. https://doi.org/10.3997/1873-0604.2010030

Rizzo E, Lasaponara R, Capozzoli L, De Martino G, Luongo R, Masini N, Leucci G, Persico R, De Siena A (2014) Non invasive techniques on the detection of buried archaeological structures at Timmari archaeological site (Matera, Italy). Geophys Res Abstracts 16. EGU General Assembly 2014

Simyrdanis K, Papadopoulos N, Kim J-H, Tsourlos P, Moffat I (2015) Archaeological investigations in the shallow seawater environment with electrical resistivity tomography. Near Surf Geophys 13:601–611

Tosti F, Patriarca C, Slob E, Benedetto A, Lambot S (2013) Clay content evaluation in soils through GPR signal processing. J Appl Geophys 97(2013):69–80

What Is Conservation Plan?

14

Mohsen M. Saleh

Abstract

There is no doubt that one of the most difficult challenges facing the monument conservator is how to put a plan for conservation?, What is the criteria that control the success of such plan?.

Without conservation process, archaeological monuments subjected to various distractive factors, caused to loss a part of our cultural heritage. Conservation plan depends on some key stages, such as; registration and documentation process, collecting data (investigation and analysis), analyzing data, interpretation and diagnosis, design conservation plan.

A specialist team must cooperate for putting conservation plan, conservator is a coordinator who collects data from different specializations to be used for conservation plan. In the end we can say that the correct conservation plan help saving monument object, and vice versa.

Keywords

Distractive factors · Monument conservator · Conservation plan · Documentation · Investigation and analysis · Specialist team

14.1 Introduction

A conservation plan is a vital instrument to help conserve of ancient monuments, indoor and outdoor which are considered to be of special interest, and of value to the local community and nation.

A conservation plan sets out why monument is significant and how that significance will be retained in any future use, alteration, development and management. The conservation plan process begins with understanding the monument and moves logically through an assessment of significance, to understanding how that significance might be vulnerable and thus what policies or guidelines are needed to retain that significance.

"The conservation plan can be the first stage of a management plan, but not vice versa". Therefore a heritage plan and a management plan are secondary to making a conservation plan; a conservation plan is produced first, and must include all stakeholders and concerns, as it is public property.

Preparation of a conservation plan encourages those with responsibility for the site to think about it in a structured way, to assess how and why it is significant, and how it should be managed in order to conserve that cultural significance. Conservation plans should meet the needs of the

M. M. Saleh (✉)
Conservation Department, Faculty of Archaeology, Cairo University, Giza, Egypt

© Springer International Publishing AG, part of Springer Nature 2019
G. El-Qady, M. Metwaly (eds.), *Archaeogeophysics*, Natural Science in Archaeology,
https://doi.org/10.1007/978-3-319-78861-6_14

monument. A conservation plan should be as comprehensive as is appropriate for the size and complexity of the monument.

A conservation plan must be a living document, having a clearly defined purpose, which used and updated as required. The preparation of a conservation plan must not be an end in itself, but be considered as a necessary management tool. A conservation plan should pay dividends in the long term by providing a firm foundation for management and expenditure decisions.

14.2   Conservation Plan

There is no doubt that conservation plan of the monument "indoor or outdoor" is an essential matter upon which depends the result of the conservation process.

The plan of conservation based on some basics and rules that should include:

1. Registration and documentation the state of the object using internationally accepted scientific methods.
2. Collecting data of the object that will be conserved such as its kind, ore, properties, previous conservation process, the damaged appearances, and the factors that caused these appearances. In order determine these data, a number of investigation and analysis methods should be used such as: field observation (descriptive study—documentation damage appearances), Optical Microscope (OM) to characterize the morphological features, Polarized Microscope (P.M.) to determine the mineralogical composition and the grains features by using many cross sections and thin sections, Scanning Electron Microscope (SEM) equipped with an EDS used to investigate the morphology of the deteriorated surface, X-ray diffraction method (XRD) to identify the chemical composition.
3. Analyzing data and deducting conclusions.
4. Well interpretation of the results to reach the so-called right diagnosis.
5. Design conservation plan (depended on right diagnosis) in the form of steps that can be implemented on the ground.

6. The provision of the needs of each step of conservation plan separately. Every step considered as a stand-alone task, with timetable, budget, trained staff, equipment and raw materials.

We will deal here with how to prepare the conservation plan of the archaeological excavation sites as an example where such sites contain outdoor and indoor archaeological remains.

14.3   Conservation Plan of Object in Archaeological Excavations Sites

Before starting any restoration work for objects in archaeological excavations site, it should be a clear conservation plan adopted by the conservation team. Where the general outlines of the plan agreed to by the entire team, where success factors is available for restoration process. Therefore, the clarity of some information Sequence is very important as follows, Fig. 14.1.

Restoring a stone wall at Dadan archaeological site, King Saud University Excavations, Al-Ulla, Saudi Arabia, will be an application example of object conservation plan, taking into consideration the conservation plan changed from case to another, depending on the type and status of the detected object.

14.4   Restoring a Stone Wall at Dadan Site

14.4.1   Introduction

Dadan archaeological site is located north of the city of Al-Ulla, 26° 39′ Longitude, and 37° 54′ latitude. Located Al-Ulla in the northwestern part of the Kingdom of Saudi Arabia, between Medina and Tabuk, and away from Hail about 416 km in a westerly direction, and away from Al-Wajh seaport about 240 km the east, away from the Taima about 160 km southwest (Fig. 14.2). Dadan (Khraibeh) has gained great importance due to its existence on the ancient trade road, one of

Fig. 14.1 The conservation plan of archaeological excavations sites

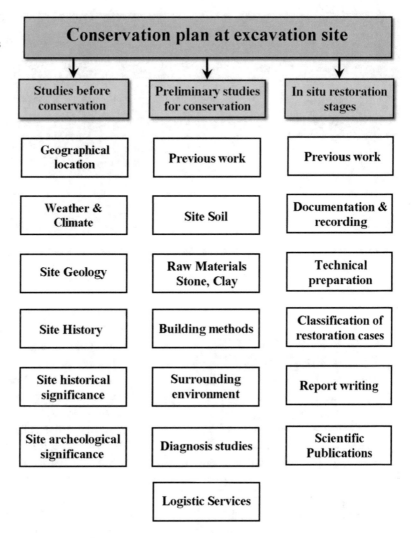

the most important ancient trade routes in the Arabian Peninsula.

Dadan is the capital of the Kingdom of Daydan and Lahyyan after Nabatean control over the north of the Arabian Peninsula, this civilization was moved to Al-Hajr and ended (Al-Anssary and Abo Al-Hassan 2005), and turned the so-called Al-Ulla. Al-Ulla was an agricultural area because of its abundance of waters. In the Islamic era, Al-Ulla was pilgrim's stations for the coming from the Levant, then it is became Hejaz Railway stations, which was established in 1326 AH, (Al-Faqeer 2006).

The excavations of the King Saud University at the Dadan site were discovered many stone ruins that contained various architectural details, such as: lentils, columns, stone pillars, and a lot of archaeological artifacts of stone, pottery, glass, wood and metal. Those discoveries conveyed to Dadan archaeological site its importance in Arabian Peninsula (Al-Zahrani and Saleh 2011).

The Dadan site was exposed to different destructive factors, such as: earthquakes, torrents, rain, wind, storms, etc. (Al-Zahrani 2006), so the walls are in danger of collapsing.

In the sixth season 1430 AH/2009 AD, of archaeology department excavation, college of

Fig. 14.2 (**a**) Al Ulla location in Al Madinah, Saudi Arabia (**b**) Dadan archaeological site, excavated area

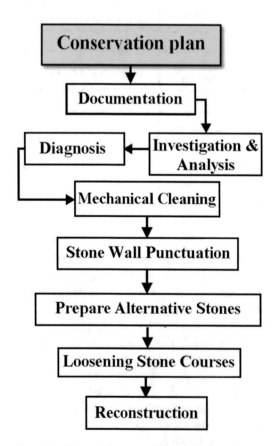

Fig. 14.3 Conservation plan of the leaning wall, Dadan excavations site

tourism and antiquities, king Saud university, a lot of archaeological ruins (walls, floors, and architectural details) had been discovered at Dadan site "Khraibeh". Most of these ruins suffer from deterioration. Leanings of Dadan's walls have been observed toward the west direction, one of these walls was in critical case to the degree of threat of collapse at any moment, square (9H). Therefore, the restoration process should be immediately carried out to save this archaeological wall from collapsing. Documentation, investigation, analysis and diagnosis were carried out in order to determine the wall condition and the main source of damage.

Concerning the case of the leaning wall, archaeologists and conservators specialist has agreed that the case of the leaning wall is a critical case that maybe collapse threatening, so it must intervene to restore the wall to its original state as much as possible, through the disassembly and reconstructive, taking into consideration the documentation for each restoration stages before, during and after restoration.

Referring to similar cases where the conservation team was forced to disassemble and reconstruct some threatened monuments with extinction (Abu Simbel temples, Egypt). The conservation plan of the leaning wall was summarized in Fig. 14.3.

14.5 Materials and Methods

Field observations (a descriptive study—documentation of visible damages) of the stone wall that were discovered by the scientific excavation performed by the college of tourism and, King Saud University. To achieve those objectives, Dadan's stone wall materials samples were collected and analyzed by the following methods:

Optical Microscopy (OM—with a Smart-Eye USB Digital Microscope at various magnifications, from a maximum of 170× fixed magnification), in order to characterize the morphological features (Ghoniem 2011; Ali et al. 2002), superficial shape and the grain size of the stones and mortars samples, Polarized Microscopy (PM—with a MT 9000 Series MEJI Co. LTD Japan) used to determine the mineralogical composition and the grain features, by using many cross section and thin section samples (Liritzis et al. 2008; Ferretti 1993). The Scanning Electron Microscope (SEM) JEOL/EO, JSM-6380 device, equipped with an EDS detector operating up to an accelerating voltage of 25 kV and a working distance of 9 mm was used to investigate the morphology of the deteriorated surfaces (Uda 2004) of the stones and mortars samples and to detect the distribution of the chemical elements on the stones and mortars samples. The X-ray diffraction method (XRD) performed with an Ultima IV, multipurpose X-ray diffraction system, equipped with a copper anticathode. The measuring conditions were set as follows: Cu target, 40 kV accelerating voltage, 40 mA current, the scanning range of 2θ was from 4 to 70° and the scanning speed was 2°/min. It used to identify the chemical compositions (Saleh 2005; Pavlíková et al. 2011) of the stones and mortars samples at Dadan site. Dadan's soil samples analyzed to identify the granular classification and physical properties.

14.6 Results

14.6.1 Field Observations

Through field observations of Dadan site it can be observed the following:

– Most of stone walls ruins are tending towards the West (Fig. 14.4).
– Mortar is friable, weak and mainly consists of clay.
– Partial and total collapses of walls ruins in different parts of the excavation site.
– Double faced wall is the main construction method in Dadan site (Fig. 14.5; Saleh 2011; Barry 1999).
– Difference kinds of stones were founded in Dadan site according to the time periods. Where ashlars sandstones used in the oldest periods but rubble sandstone were used in the newer periods.
– Stone walls ruins have different thickness of 43 cm up to about 278 cm thick.

14.6.2 The Leaning Wall

The field observation notes that most of stone walls ruins are tending to the west, inclination rate have varied from wall to another according to stones quality and walls thickness. Through documentation of 9H square, it was noted great tending of one wall that collapse threatening. Therefore, reinforcement began as fast as possible to save this wall. For the immediate restoration, the stone wall had to be reinforced and collect data as follows:

– Diagonal and horizontal reinforcement were used to support tending wall (Fig. 14.6a and b).

Fig. 14.4 Stone walls ruins tending from the eastern direction to the western direction

Fig. 14.5 Double faced wall (**a**) Three layers thin wall, (**b**) Three layers thick wall, (**c**) Weakly connected, and (**d**) Fairly connected

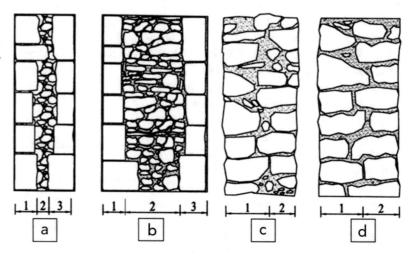

– Eastern and western sides of the wall ruin measurements are 130 cm length, 50 cm width, and 151 cm high.
– Internal fillers layers is heterogeneous consists of clay, different sandstones fragments, and organic remains (Fitzner et al. 2000).
– Determine the damaged stones in the double face of the tending wall.
– Inclination angle was determined 15 degrees towards the west (Figs. 14.7a–c, 14.8a and b).

– Through field observations three kinds of sandstone have been identified in the tending wall, and in the most walls of Dadan excavated site are as follows: (Al-Zahrani 2009).
– Multilayer yellow sandstone, (seen by naked eye).
– Red sandstone.
– White to gray Sandstone. (Fig. 14.9a–c)

The three sandstone samples were prepared in thin and cross-sections to examine under optical microscope (OM) and polarized microscope (PM).

Fig. 14.6 (**a**) Tending wall after discovering, (**b**) Diagonal and horizontal reinforcement

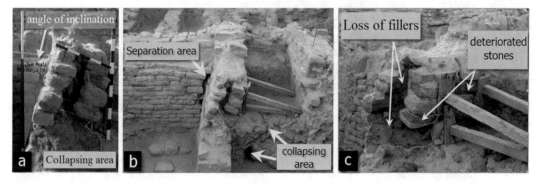

Fig. 14.7 (**a**) Inclination angle of the wall, (**b**) Separation and collapsing areas, (**c**) Loss of internal fillers and stone deterioration

Fig. 14.8 (**a**) Deep crack on stone block, (**b**) Stone flaking

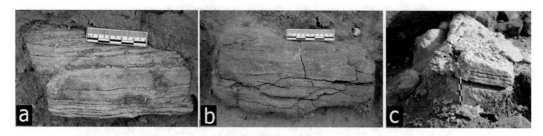

Fig. 14.9 (a) Multilayer yellow sandstone, (b) Red sandstone, (c) White to grey sandstone

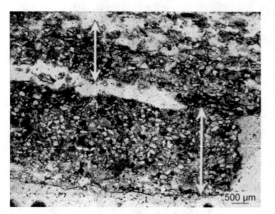

Fig. 14.10 Multilayer yellow sandstone

14.6.3 Optical Microscope Results (OM)

Undoubtedly, the study sample of building material under the microscope helps a lot to know some of their properties also serve as a guide to which analysis that must be followed (Anders and Tronner 1991).

Multilayer yellow sandstone consists of fine quartz grains sharp edges in the ground of bonding material interlaced with quartz grains. Because of the exposure of multi-layered yellow sandstone to the different destructive factors, (the huge difference between the temperature and humidity, and salt water saturation for long time) it was separated significantly into layers (Fig. 14.10). The red sandstone is consists of integration quartz grains, where some grains docked with each other in ground of iron oxides,

quartz grains vary in sizes, shapes and distribution (Fig. 14.11a and b). White to gray sandstone has spherical shape of quartz grains that different in sizes. In addition to the presence of a number of sedimentary and metamorphic rocks crumbs. These were gray and dark brown in color (Fig. 14.12a and b).

14.6.4 Polarized Microscope Results (PM)

Polarizing microscope "PM" used to identify the crystal texture, grain size, grain shape and the type of cementing material of the stone. Polarized microscope is a complementary test and confirmed of the information that is not clear through optical microscopic examination (Adams et al. 1988). The results of polarized microscopic examination as follows:

Polarized microscopy analysis revealed that the weakness point of the yellow sandstone multilayer is the cementing materials. The cementing materials consist of deposits of iron oxides mixed with clay minerals, which is the point of separation of the layers. The polarized microscope describes the crystal texture as microscopic texture where it can be seen only through the microscope (Mackenzie and Adams 1998; Folk 1980; Fig. 14.13a and b). Examination of red sandstone samples shows that the main component is quartz varied in size, shape, that integration with each other in rich ground of iron oxides (Fig. 14.14a and b). The white to gray sandstone consists of quartz grains in rich ground of calcite and small amount of iron oxides (Fig. 14.15a and b).

Fig. 14.11   Red sandstone, (**a**) The integration quartz grains, (**b**) Quartz grains vary in sizes, shapes and distribution

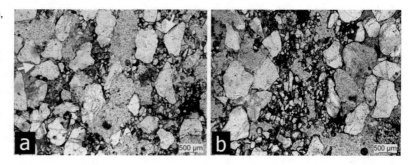

Fig. 14.12   (**a**) and (**b**) White to gray sandstone has spherical shape of quartz grains that different in sizes

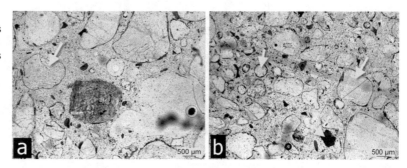

Fig. 14.13   (**a**) Quartz grains integration (C.N.) 100X, (**b**) Separation of sandstone in layers, 50X (C.N.)

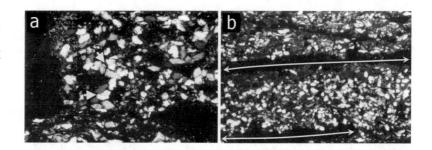

Fig. 14.14   (**a**) and (**b**), Quartz grains varied in size, shape and integration with each other (C.N.) 50X

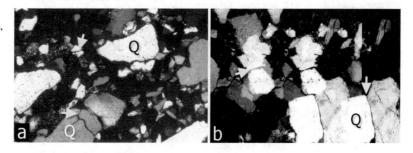

Fig. 14.15 (**a**) and (**b**) Quartz grains varied in size and shape in rich ground of calcite and iron oxides (C.N.) 50X

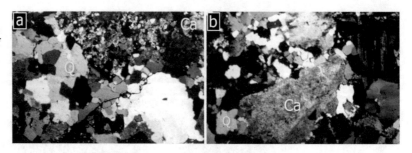

Fig. 14.16 (**a**) and (**b**) Quartz grains in rich ground of clay and iron oxides

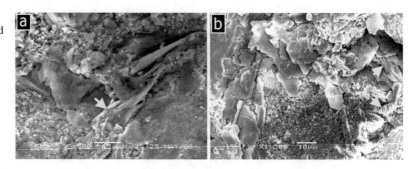

14.6.5   SEM Results

Scanning electron microscope investigations showed that the multilayer yellow sandstone cement materials are clay minerals and iron oxides as a grains binder (Fig. 14.16a and b). The red sandstone cement materials are calcium carbonate (calcite), clay and iron oxides (Fig. 14.17a and b) and white to gray sandstone have calcium carbonate and clay as cementing materials of quartz grains (Fig. 14.18a and b). In the three sandstone kinds it can be seen the deformation of quartz crystals (Ferretti 1993). The results of the EDS analysis carried out on the three sandstone samples shown in Table 14.1.

14.6.6   X-Ray Diffraction Results (XRD)

Two different types of sandstone and mortar were selected and prepared to test with powder method (Saleh 2013). The XRD analysis results were as follows (Table 14.2 and Fig. 14.19a–c).

14.6.7   Dadan's Soil Classification

The results of the soil analysis carried out on Dadan's soil samples are shown in Table 14.3, 14.4, Fig. 14.20a and b (Jain et al. 2005).

14.6.8   Conservation Plan

After studying the previous results a conservation plan of the stone wall summed up in the mechanical cleaning, punctuation, loosening stone courses then sorted and the exclusion of the damaged ones in order to replace with suitable stones then reconstruction with the same old traditional method (Ashurst 2007).

14.6.9   Stone Wall Punctuation

To identify the places of stone it was necessary to punctuation as follows:

– Mechanical cleaning of the stone wall surface to remove dirt and mud so it can be punctuated.

Fig. 14.17 (**a**) The deformation of quartz crystals, (**b**) EDS analysis of red sandstone

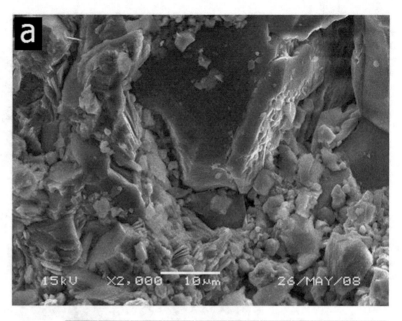

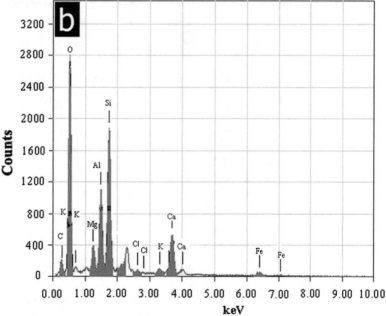

– Use crayons in the punctuation process.
– Punctuation start from top courses to bottom courses (Fig. 14.21).
– Register each stone with number in a detailed diagram of the wall.
– Each stone has number and card, fixed in the stone after its loosening (Saleh 2010).

14.6.10 Reconstruction

Traditional mud mortar was used to rebuild the loosening wall, bearing in mind the principles and rules of ancient construction technique, horizontal and vertical level of the courses, as well as small gravels were fixed on the filling layer surface to distinguish that the wall was restored

Fig. 14.18 (**a**) Quartz
grains in rich ground of
calcite and clay, (**b**) EDS
analysis of white to gray
sandstone

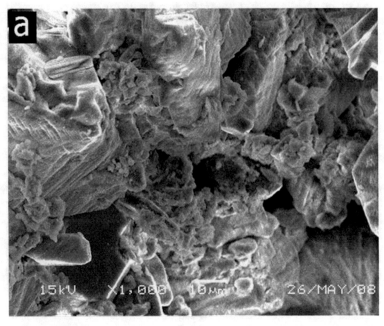

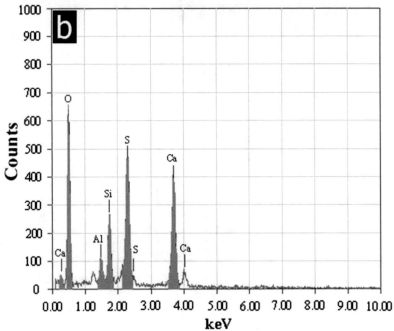

Table 14.1 The elemental composition (wt.%) of the studied stones samples

Sample	C	O	Mg	Al	Si	S	Cl	K	Ca	Fe
Yellow sandstone	05.26	47.94	–	21.55	25.25	–	–	–	–	–
Red sandstone	05.09	34.63	03.55	12.41	23.91	–	01.74	01.51	14.13	03.81
White sandstone	02.95	41.87	–	14.43	38.41	00.08	–	–	02.27	–

Table 14.2 The identify minerals of the stones and mortar samples by XRD analysis

Samples	Minerals	Formula	Index no.
Multilayer yellow sandstone	Chamosite, Iron II	Fe-Mg-Al-Si-Al-O-OH	07-339
	Quartz	SiO_2	05-0490
	Kaolinite	$Al_2Si_2O_5(OH)_4$	01-0527
	Silicon oxide	SiO_2	14-0260
	Iron oxide	Fe_2O_3	16-653
Red sandstone	Quartz	SiO_2	05-0490
	Goethite	α-FeOOH	17-0536
Mortar	Quartz	SiO_2	05-0490
	Goethite	α-FeOOH	17-0536
	Calcite	$CaCO_3$	05-0586
	Orthoclase	$KAlSi_3O_8$	09-0462
	Halite	$NaCl$	05-0628
	Kaolinite	$Al_2Si_2O_5(OH)_4$	01-0527

(Petzet 1999; Lamaee 2005). Figures 14.22, 14.23, 14.24, 14.25 and 14.26 show Dadan's stone wall conservation stages.

14.7   Discussion

Field observations of Dadan site were observed a leaning of the most archaeological stone walls ruins to the western direction. The degrees of inclination varied according to the thickness and build method of each wall. the mountains surrounding the site had many collapse and fall of some rock tombs to the eastern direction (the same direction as Dadan's walls) therefore we think of the possibility that Dadan's site exposed to seismic waves sequential in long term led to such inclination and collapses. Dadan's site abandoned from the population for other reasons than earthquakes where we did not found bodies or daily use during the last ten excavations seasons.

According to the results of the Seismic Studies Center at King Saud University, Al-Ulla has been Subjected to 149 earthquakes of intensity between 1.6 and 5.1 MAG since 641 AD–2003 AD (Fig. 14.27).

From the field observations of Dadan site it can be concluded that ashlars double faced wall masonry with difference thickness from wall to another were the building methods of Dadan's walls. The analyses and investigations detected that Dadan's wall have three kinds of stones. The first kind is multilayer yellow sandstone, which consists of fine quartz grains sharp edges in the ground of bonding material interlaced with quartz grains. This kind of stone separated significantly into layers seen with neck eye. PM analysis was revealed that the weakness point of this stone is the cementing materials that consist of deposits of iron oxides mixed with clay minerals. The polarized microscope describes the crystal texture of this stone as microscopic texture where it can

Fig. 14.19 X-ray patterns of the identified minerals from the studied stones and mortar samples, (**a**) Multilayer yellow sandstone, (**b**) Red sandstone and (**c**) Dadan's mortar

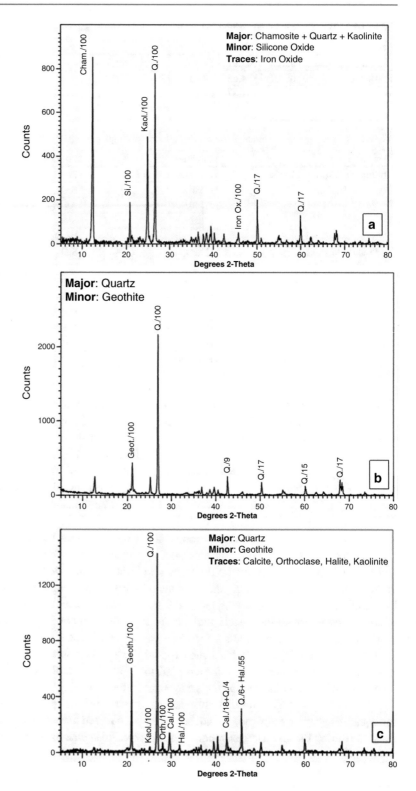

Table 14.3 Grain size analysis test

Sieve No.	Diam. (mm)	Wt. ret.	Retained %	Passing %
	12.5	0	0.0	100.0
	9.500	0	0.0	100.0
# 4	4.760	0.83	0.7	99.3
# 10	2.000	2.22	1.8	98.2
# 16		3.5	2.8	97.2
# 20	0.850	4.88	4.0	96.0
# 40	0.425	16.02	13.0	87.0
# 60	0.250	32.56	26.4	73.6
# 100	0.150	43.83	35.5	64.5
# 200	0.075	50.09	40.6	59.4

Table 14.4 Physical properties of Dadan's soil

Dadan's soil properties	Result
Liquid limit, LL, %	26.400
Plastic limit, PL, %	21.700
Plasticity index, PI, %	04.700
Specific gravity, Gs	02.711
Gravel, %	00.700
Sand, %	39.900
Silt, %	37.400
Clay, %	22.000
% Passing # 200	59.400
Soil classification:	CL-ML

CL-ML = Silty Clay

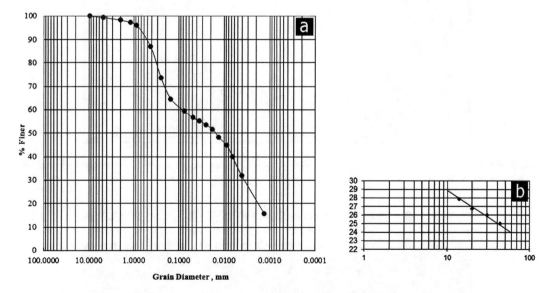

Fig. 14.20 (**a**) Grain size distribution curve for Dadan site, (**b**) Determination of Atterberg limits of soil

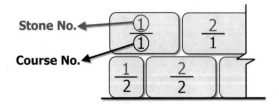

Fig. 14.21 The method of stones punctuation of Dadan's wall

Fig. 14.22 (**a**) Stones numbering, (**b**) The beginning of the manual loosening processes

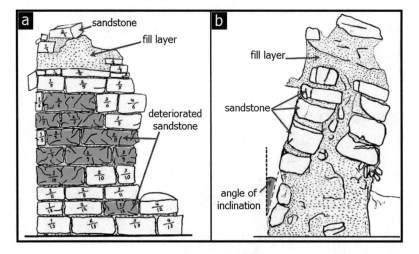

Fig. 14.23 (**a**) Identify deteriorated stones in the eastern direction, (**b**) Detailed drawing shows the stone wall condition from the northern direction

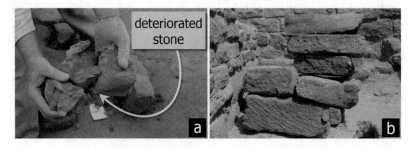

Fig. 14.24 (**a**) Sample of deteriorated stones after loosing, (**b**) Examples of alternative stones

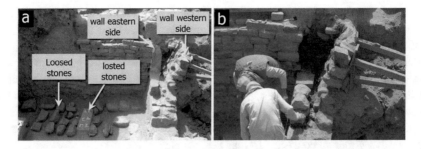

Fig. 14.25 (**a**) Removing the tending part of the western side, (**b**) The beginning of rebuild

Fig. 14.26 (**a**) Removing the tending part of the western side, (**b**) Small gravels were fixed on the filling layer surface to distinguish restored part, (**c**) Dadan's wall after conservation

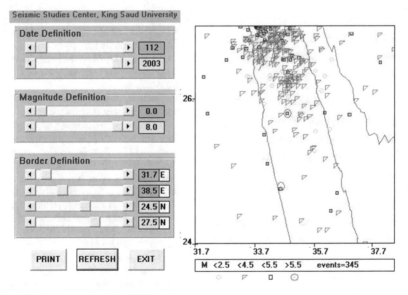

Fig. 14.27 Earthquakes map that struck Al-Ulla site

be seen only through the microscope. The XRD analysis showed that the main components are Chamosite, Iron II (Fe-Mg-Al-Si-Al-O-OH), Quartz (SiO_2), and Kaolinite ($Al_2Si_2O_5(OH)_4$). The second kind is red sandstone; it is the most common stone in Dadan site, which consists of quartz grains vary in sizes, shapes in ground of iron oxides. The XRD analysis showed that the main components are Quartz (SiO_2) and Goethite α-FeOOH. The third stone kind is white to gray sandstone that has spherical shape of quartz grains that different in sizes. In addition to the presence of a number of sedimentary and metamorphic rocks crumbs. These were gray and dark brown in color in rich ground of calcite and small amount of iron oxides.

Mortar samples of Dadan's site were clay mortar mixed with sand and lime. The XRD analysis showed that Dadan's mortar consists of Quartz (SiO_2), Goethite (α-FeOOH), Calcite ($CaCO_3$) and Kaolinite ($Al_2Si_2O_5(OH)_4$). Halite (NaCl) detected in mortar's samples as a salt contamination rather than as a stone component. Its existence is a result of its transferring from underground water in the soil through water capillarity. The question is why the ancient builder used clay mortar instead of using gypsum or lime mortar. Perhaps the answer lies in the fact that the clay mortar is more flexible and absorption of seismic waves and thus makes wall buildings more resistant against earthquakes especially the north west of Saudi Arabia is one of the seismically active places.

For reconstruction traditional clay mortar was used to give the possibility to make any modification may be necessary for the wall where clay mortar is reversible (Maravelaki-Kalaitzaki et al. 2003), also takes into account fixing small stone gravels to be a signal that this stone wall was restored.

The results of the soil analysis of Dadan determined the soil classification as silty clay soil (CL-ML). Inorganic clays, symbol C, group CL. Inorganic silts and very fine sandy soils, silty or clayey fine sands, micaceous and diatomaceous soils, and elastic silts-symbol M, group ML. This type of soil is directly affected by water and moisture, whether increases or decreases affecting the physical and mechanical properties and the consequent impact on the stone walls by the movement and damage.

14.8 Conclusion

Through studying the case of Leaning wall conservation at Dadan excavations, and to be able to

have a conservation plan a number of questions must have answer as follows:

- What is the type of the detected monument?
- What is the virtual case of monument accurately?
- What is the results of the analysis and investigation of the monument?
- What is the correct diagnosis of the monument state?
- Are there similar cases and how they were handled?
- What is exactly required to restore the monument?

Through the answers to these questions and apply the system of SWOT analysis, an appropriate conservation plan is made by specialist team according to the state of monument.

References

Adams AE, Mackenzie WS, Guilford C (1988) Atlas of sedimentary rocks under the microscope. Longman Scientific & Technical, England, pp 19–25

Al-Anssary A, Abo Al-Hassan H (2005) Al-Ulla we Mada'n Salih (Al-Hijr). Dar Al-Qawafil, Riyadh, pp 32–33. (In Arabic language)

Al-Faqeer B (2006) Al-Syaha fe Mohafzet Al-Ulla. King Saud University, University Press, Riyadh, pp 55–58. (In Arabic language)

Ali H, Saleh M, Poksinska M (2002) The use of polarized microscope in the study of the brick, mortar and plaster used in Al-Foustat houses, Old Cairo, Egypt. Conference and workshop on conservation and restoration, Faculty of Fine Arts, Minia Univ, April 2002, pp 1–10

Al-Zahrani A (2006) The destructive factors of archaeological sites: Dadan site – Al-Ulla. Deliberations of Seventh Annual Scientific Meeting of the Society of History and Antiquities, Gulf Council Countries, Manama – Kingdom of Bahrain, 20–23 March 1427/ April 18–21, .2006, pp 515–557. (In Arabic language)

Al-Zahrani A (2009) Studying of building materials used in Dadan site, Al-Ulla, Saudi Arabia. Deliberations tenth annual scientific meeting of the Society for History and Antiquities in the Gulf Council Countries, Abu Dhabi – United Arab Emirates, 24–26 April 2009, pp 243–277. (In Arabic language)

Al-Zahrani A, Saleh M (2011) The preliminary report of restoration and conservation work of Dadan Site (Khraibeh) and Qarh (Al-Mabbiayat) Province of Al-Ulla Seventh Season 2010–2011 AD. King Saud University, College of Tourism and Antiquities, Department of Heritage Resources Management and Tour Guidance, Riyadh. (unpublished). (In Arabic language)

Anders GN, Tronner K (1991) Stone weathering, air pollution effects evidenced by chemical analysis. Central Board of National Antiquities, Stockholm, pp 13–17

Ashurst J (2007) Conservation of ruins. Elsevier Limited, Amsterdam, pp 3–11

Barry R (1999) The construction of buildings, vol 1, 7th edn. Blackwell Science, Paris, pp 119–137

Ferretti M (1993) Scientific investigations of works of art. ICCROM, Rome, pp 39–46

Fitzner B, Heinrichs K, La Bouchardiere D (2000) Damage index for stone monuments. In: Galan E, Zezza F (ed) Protection and conservation of the cultural heritage of the Mediterranean cities. Proceedings of the 5th international symposium on the conservation of monuments in the Mediterranean Basin, Sevilla, Spain, 5–8 April 2000, pp 315–326

Folk LR (1980) Petrology of sedimentary rocks. Hemphill Publishing, Austin, TX, pp 203–272

Ghoniem M (2011) The characterization of a corroded Egyptian bronze statue and a study of the degradation phenomena. Int J Conserv Sci 2(2):95–108

Jain AK, Punmia BC, Jain AK (2005) Soil mechanics and foundations. Firewall Media, pp 111–129

Lamaee S (2005) Principles of monuments restoration, restoration and preservation of architectural heritage – Specialized training course, Handbook of research – Supreme Commission for the development of Riyadh city. pp 6–7

Liritzis I, Sideris C, Vafiadou A, Mitsis J (2008) Mineralogical, petrological and radioactivity aspects of some building material form Egyptian Old Kingdom monuments. J Cult Heritage 9:1–13

Mackenzie WS, Adams AE (1998) A colour atlas of rocks and minerals in thin section. Manson Publishing, London, pp 132–137

Maravelaki-Kalaitzaki P, Bakolas A, Moropoulou A (2003) Physico-chemical study of Cretan ancient mortars. Cem Concr Res 33:651–666

Pavlíková M, Pavlík Z, Keppert M, Černý R (2011) Salt transport and storage parameters of renovation plasters and their possible effects on restored buildings walls. Constr Build Mater 25:1205–1212

Petzet M (1999) Principles of monument conservation. J Ger Natl Commit XXX, ICOMOS, p 33

Saleh M (2005) The problems of soluble salts at The Old Cairo Walls (1176–1193 A.D.), Egypt. Conference and workshop on conservation and restoration, Faculty of Fine Arts, Minia Univ, March 2005, pp 57–69

Saleh M Loosing and preserving the ruins of mud-brick walls: a case study, Faculty of Archaeology Excavation – Cairo University – Saqqara. Adumatu XXII, August 1431–July 2010, pp 29–44. (In Arabic language)

Saleh M Restoration and conservation of Chapel (Mꜣꜥy 𓈖𓏤𓄿) Faculty of Archaeology Excavation – Cairo University at Saqqara'. J King Saud Univ, Tourism and Antiquities XXIII, 1, January 2011–1432, pp 1–16. (In Arabic language)

Saleh M (January–March 2013) Characterization of Qarh's wall plasters, Al-Ulla, Saudi Arabia: a case study. Int J Conserv Sci 4(1):65–80

Uda M (2004) In situ characterization of ancient plaster and pigments on tomb walls in Egypt using energy dispersive X-ray diffraction and fluorescence. Nucl Instr Methods Phys Res B 226:75–82

Index

A
Archaeogeophysics, 244, 249, 250
Archaeological remains, 4, 8, 13, 14, 28, 30, 32, 43, 50, 61, 65, 66, 101, 108, 243–252, 256
Archaeological survey, 19
Archaeology, 1–22, 27, 29, 59, 61, 70, 86–87, 102, 137–166, 169–181, 192, 205–216, 244, 257

B
Buried structures, 9, 17–19, 27, 32, 34, 38, 39, 55, 60, 61, 63, 65, 104, 108, 110, 111, 114–116, 138, 170, 172, 175, 176, 180, 216, 243, 247–250

C
Cemetery, 84, 126, 127, 206
China, 125, 126, 201
Conservation plan, 255–272

D
Deahshour, 5, 7, 154–158, 160–162
Deterioration, 234, 241, 242, 258, 261
Distractive factors, 255
Documentation, 139, 256, 258, 259

E
Egypt, 3, 5, 7, 14, 20, 83–97, 137–166, 233–242, 258
Excavation, 5, 10, 27, 40–43, 45, 49, 51, 55, 58–65, 80, 84, 102–105, 108–110, 115, 126, 128, 133, 138, 139, 142, 143, 149, 154, 158–160, 163–165, 170, 173, 175, 176, 180, 181, 186, 207, 212, 214, 215, 243, 247, 249, 251, 256–259, 267, 272

F
Frequency, 4, 15, 16, 18, 19, 21, 22, 29, 76, 92–94, 128, 130, 171, 184–191, 194, 198–200

G
Geoarchaeology, 174
Geoelectrics, 70, 142, 207, 212–216, 238
Geomagnetics, 132, 143, 145–147, 149, 150, 154, 162, 166, 216, 244, 249

Geophysical prospection, 27–29, 49–51, 55, 58, 115, 170, 176, 247
Geophysics, 3, 8, 10, 29, 34, 61, 62, 66, 101–124, 126, 141–143, 146, 160, 183–203, 213, 234
Ground penetrating radar (GPR), 2, 17–20, 28, 29, 50–52, 54–57, 59, 62, 63, 65, 102, 104–106, 108–112, 114–124, 126–133, 142, 154, 170–173, 175–179, 183–203, 214, 216, 220–222, 225–232, 241, 244–250, 252

H
Han dynasty, 125–127, 133, 134
Hawara, 158–164, 166

I
Investigation and analysis, 256

L
Luxor, 154, 233–242

M
Magnetic, 2–5, 7, 14–16, 18, 28, 30–32, 34–46, 48, 52, 54–56, 58–61, 63, 76, 102, 104, 126–133, 138, 141–162, 164–166, 170–173, 177, 192, 194, 212–214, 216, 220–222, 224–227, 230, 231, 250, 252
Magnetic gradimetry, 138
Mapping, 3, 11, 13, 15, 21, 49, 69–80, 106, 108, 146, 191, 200, 208, 214, 216, 241, 242
Mastaba, 83–97, 156, 158, 161
Mitigation, 233–242
Monument conservator, 255, 256, 258
Mud bricks, 14, 63–65, 84, 86, 138, 146, 147, 149–151, 154, 155, 158–163, 166, 178, 179

N
Neolithic site, 102–104
Nineteenth century crypt, 101
Non-destructive exploration methods, 180
Non-invasive, 8, 10, 84, 220, 222, 234

© Springer International Publishing AG, part of Springer Nature 2019
G. El-Qady, M. Metwaly (eds.), *Archaeogeophysics*, Natural Science in Archaeology,
https://doi.org/10.1007/978-3-319-78861-6

Printed in the United States
By Bookmasters